Current Topics in Microbiology and Immunology

Volume 410

More information about this series at http://www.springer.com/series/82

Akihiko Yoshimura
Editor

Emerging Concepts Targeting Immune Checkpoints in Cancer and Autoimmunity

Responsible series editor: Tasuku Honjo

Springer

Editor
Akihiko Yoshimura
Department of Microbiology and
Immunology
Keio University School of Medicine
Tokyo
Japan

ISSN 0070-217X ISSN 2196-9965 (electronic)
Current Topics in Microbiology and Immunology
ISBN 978-3-319-68928-9 ISBN 978-3-319-68929-6 (eBook)
DOI 10.1007/978-3-319-68929-6

Library of Congress Control Number: 2017959187

Printed on acid-free paper

This Springer imprint is published by Springer Nature
The registered company is Springer International Publishing AG
The registered company address is: Gewerbestrasse 11, 6330 Cham, Switzerland

Preface

The homeostasis of the immune system is maintained by positive and negative factors including effector/regulatory cells as well as positive intracellular signals/negative regulators. Autoimmune diseases and allergic diseases are states in which this balance inclines to the excessive "positive." In contrast, tumor microenvironment provides negative signals for immune systems, which suppress anti-tumor immunity. Those include regulatory T cells (Tregs) and myeloid-derived suppressor cell (MDSC) as well as immune regulatory molecules such as PD-L1, indoleamine 2,3-dioxygenase (IDO), TGF-β. Autoimmunity and anti-tumor immunity are two sides of the same coin. Indeed, knockout mice of various molecules so-called "immune checkpoints" often develop autoimmune diseases. Thus, the understanding of the mechanism of immunological balance is important for the treatment of both autoimmune diseases and cancer.

Typical positive and negative T cells of immune balance are effector T cells and regulatory T cells (Tregs), respectively. Major Treg cells developed in the thymus are called thymus-derived Treg (tTreg) cells. Treg cells are specified by an expression of the transcription factor Forkhead box P3 (Foxp3), which plays crucial roles in the differentiation, maintenance, and function of tTreg cells. Treg cells are believed to be involved in autoimmune diseases and allergy because Treg cells suppress excess immunity against a diverse range of antigens, including self-antigens, commensal bacteria-derived antigens, and environmental allergens. Tregs have been shown to be abundant in tumor tissues and suppress anti-tumor immunity.

In recent years, anti-tumor immunity has attracted attention not only by immunologists but also by cancer researchers. T-cell activation is initiated through antigen recognition by the T-cell receptor (TCR) and co-stimulatory signals such as CD28. On the other hand, the inhibitory signals for T-cell activation (i.e., immune checkpoints) are crucial for the maintenance of self-tolerance and prevention of autoimmunity as well as excess immune responses. The two immune checkpoint receptors, cytotoxic T-lymphocyte-associated antigen 4 (CTLA4, also known as CD152) and programmed cell death protein 1 (PD1, also known as CD279), have been most actively studied in the context of clinical cancer immunotherapy. It has been shown that PD1 recruits the tyrosine phosphatase, which inhibits TCR signaling, while CTLA4 inhibits CD28-mediated co-stimulatory signals. Antibodies against CTLA-4 and PD-1 have been shown to significantly improve survival in patients with metastatic cutaneous melanoma and other cancers. The action of PD-1 and CTLA-4 is now called “immune checkpoints,” since these molecules are involved in T-cell exhaustion and anergy. Clinical efficacies of these antibodies proved that anti-tumor immunity can be enhanced by inhibiting immune checkpoints. However, PD-1 and CTLA4 are not the only molecules that negatively regulate T-cell activation. There are a number of cells and signals that suppress effector T-cell activation.

In addition to TCR and co-stimulatory signals, T-cell activation requires the third signal: signals from the cytokine receptors. For example, IL-2 is necessary for the proliferation of T cells, and IL-12 and IFNγ are important for Th1 differentiation and CTL activation. Various roles of IFNγ in anti-tumor immunity have been established. IL-15 has been shown to be necessary for memory T-cell survival. Thus, negative regulators of the cytokine signaling must be important immune checkpoint molecules that regulate anti-tumor immunity. The suppressors of cytokine signaling (SOCS) family proteins have been shown to negatively regulate cytokine signaling by binding to the receptors and/or JAK tyrosine kinases. Suppression of SOCS1 has been shown to cause autoimmunity and enhance anti-tumor immunity. Therefore, such negative regulators of cytokine signaling can be considered as the third immune checkpoint molecules.

Now we can extend the concept of immune checkpoints to “molecules and cells which negatively regulate T-cell activation.” These molecules and cells must be involved in immune homeostasis and could be new targets of autoimmune diseases and cancer immunotherapy. This book is focusing on molecular and cellular biology of “extracellular” and “intracellular” immune checkpoint regulators. Such factors are regulatory T cells and tolerogenic dendritic cells, as well as signal inhibitors such as SOCS, tyrosine phosphatases, ubiquitin ligases, and miRNAs. I hope this CTMI volume promotes the understanding and application of “extended” immune checkpoint regulators.

Tokyo, Japan Akihiko Yoshimura

Contents

Part I Immune Checkpoint Cells

Regulatory T Cells: Molecular and Cellular Basis for Immunoregulation 3
Yosuke Togashi and Hiroyoshi Nishikawa

Overview of LAG-3-Expressing, IL-10-Producing Regulatory T Cells 29
Keishi Fujio, Kazuhiko Yamamoto and Tomohisa Okamura

Regulatory Dendritic Cells 47
Katsuaki Sato, Tomofumi Uto, Tomohiro Fukaya and Hideaki Takagi

Part II Extracellular Immune Checkpoint Molecules

Role of PD-1 in Immunity and Diseases 75
Kenji Chamoto, Muna Al-Habsi and Tasuku Honjo

CTLA-4, an Essential Immune-Checkpoint for T-Cell Activation 99
Shunsuke Chikuma

Tim-3, Lag-3, and TIGIT 127
Nicole Joller and Vijay K. Kuchroo

Part III Intracelluar Immune Checkpoint Molecules

SOCS1: Regulator of T Cells in Autoimmunity and Cancer 159
Subburaj Ilangumaran, Diwakar Bobbala and Sheela Ramanathan

Mining the Complex Family of Protein Tyrosine Phosphatases for Checkpoint Regulators in Immunity 191
Claudia Penafuerte, Luis Alberto Perez-Quintero, Valerie Vinette, Teri Hatzihristidis and Michel L. Tremblay

Immune Regulation by Ubiquitin Tagging as Checkpoint Code 215
Peng Zeng, Jieyu Ma, Runqing Yang and Yun-Cai Liu

MicroRNA in Immune Regulation 249
Cheng-Jang Wu and Li-Fan Lu

Part I
Immune Checkpoint Cells

Regulatory T Cells: Molecular and Cellular Basis for Immunoregulation

Yosuke Togashi and Hiroyoshi Nishikawa

Abstract $CD4^+$ regulatory T cells (Tregs) are a highly immune-suppressive subset of $CD4^+$ T cells, characterized by expression of the master regulatory transcription factor FOXP3. Tregs are proven to play central roles in the maintenance of self-tolerance in healthy individuals. Tregs are involved in maintaining immune homeostasis: they protect hosts from developing autoimmune diseases and allergy, whereas in malignancies, they promote tumor progression by suppressing anti-tumor immunity. Elucidating factors influencing Treg homeostasis and function have important implications for understanding disease pathogenesis and identifying therapeutic opportunities. Thus, the manipulating Tregs for up- or down-regulation of their suppressive function is a new therapeutic strategy for treating various diseases including autoimmune disorders and cancer. This review will focus on recent advances in how Tregs integrate extracellular and intracellular signals to control their survival and stability. Deeper mechanistic understanding of disease-specific Treg development, maintenance, and function could make disease-specific Treg-targeted therapy more effective, resulting in an increase of efficacy and decrease of side effects related to manipulating Tregs.

Contents

1 Introduction 4
2 Development and Maintenance of Tregs 5

Y. Togashi · H. Nishikawa
Division of Cancer Immunology, Research Institute/EPOC, National Cancer Center, Tokyo, Japan

H. Nishikawa (✉)
Department of Immunology, Nagoya University Graduate School of Medicine, Nagoya, Japan
e-mail: hnishika@east.ncc.go.jp

Current Topics in Microbiology and Immunology (2017) 410:3–28
DOI 10.1007/82_2017_58

Published Online: 07 September 2017

2.1 TCR, CD28, and IL-2 5
2.2 PI3K-Akt-mTOR 6
2.3 Epigenetics 7
3 Functional Classification of Tregs 9
4 Suppressive Mechanism of Tregs 9
5 Tregs and Autoimmune Diseases 12
5.1 Treg Cell Therapy for Autoimmune Diseases 13
5.2 Non-Cell-Based Therapies 13
6 Tregs and Malignant Tumors 15
6.1 Trafficking and Characteristics of Tregs in Cancer 15
6.2 Strategies for Treg-Targeted Therapy 16
6.3 Involvement of Tregs in Immune Checkpoint Inhibitors 17
7 Conclusion 18
References 19

1 Introduction

$CD4^+$ regulatory T cells (Tregs) are a highly immune-suppressive subset of $CD4^+$ T cells, characterized by expression of the master regulatory transcription factor FOXP3 (Sakaguchi et al. 1995; Fontenot et al. 2003; Hori et al. 2003; Khattri et al. 2003). Tregs were originally identified as $CD4^+CD25^+$ T cells by Sakaguchi et al. (1995) and are proven to play central roles in the maintenance of self-tolerance in healthy individuals (Sakaguchi et al. 2010; Wing and Sakaguchi 2010). Mutations of the human FOXP3 result in impaired development or dysfunction of Tregs and, consequently, the occurrence of immunodysregulation polyendocrinopathy enteropathy X-linked syndrome accompanying severe autoimmune diseases, inflammatory bowel disease, and allergy (IPEX syndrome) (Bennett et al. 2001). Likewise, mice that carry a mutation or genetic deletion of FOXP3, called Scurfy mice are deficient in Tregs and develop fatal systemic autoimmunity (Brunkow et al. 2001; Fontenot et al. 2003). In addition, forced expression of FOXP3 is able to confer Treg-like suppressive activity on naive conventional T cells (Tconvs) (Fontenot et al. 2003; Hori et al. 2003). FOXP3 has therefore been considered as a lineage-specifying transcription factor of Tregs or a master regulator of its functions.

Tregs are involved in maintaining immune homeostasis: they protect hosts from developing autoimmune diseases and allergy, whereas in malignancies, they promote tumor progression by suppressing anti-tumor immunity (Onizuka et al. 1999; Shimizu et al. 1999; Sakaguchi et al. 2010; Wing and Sakaguchi 2010). Elucidating factors influencing Treg homeostasis and function have important implications for understanding disease pathogenesis and identifying therapeutic opportunities. Thus, manipulating Tregs for up- or down-regulation of their suppressive function is a new therapeutic strategy for treating various diseases including autoimmune disorders and cancer. This review will focus on recent advances in how Tregs integrate extracellular and intracellular signals to control their survival and stability. We will discuss how these new insights can be utilized for the development of new approaches to promote and stabilize Tregs in many illnesses.

2 Development and Maintenance of Tregs

Tregs are separated into natural/thymic and peripheral/induced Tregs based on the sites where they are generated (Sakaguchi et al. 2010, Adeegbe and Nishikawa 2013). $FOXP3^+$ natural Tregs are generated in the thymus as an antigen-primed and functionally mature T cell subpopulation specialized for immune suppression (natural/thymic Tregs; nTregs). Some of $FOXP3^+$ Tregs also differentiate from Tconvs in the periphery under certain conditions (peripheral/induced Tregs; iTregs). The main task of $FOXP3^+$ nTregs is to migrate to inflammatory sites and suppress various effector lymphocytes, especially helper T (Th) cell subsets and $CD8^+$ cytotoxic T cells (Chaudhry et al. 2009; Koch et al. 2009; Chung et al. 2011; Linterman et al. 2011). nTregs reportedly express high levels of Helios (a member of the Ikaros transcription factor family) and Neuropilin-1 (a type-1 transmembrane protein). In contrast, iTregs that develop in the periphery often lack or have a low-level expression of these molecules. According to data from animal models, these iTregs are readily converted from Tconvs by in vitro stimulation with TGF-β or retinoic acid (Coombes et al. 2007). However, in humans, $FOXP3^+$ T cells induced from Tconvs by in vitro TCR stimulation with TGF-β fail to gain suppressive function and rather produce pro-inflammatory cytokines (Walker et al. 2005; Tran et al. 2007). At present, the function of iTregs such as TGF-β-induced ones in humans is obscure though there are some reports showing that several cytokines or a specific microbiota environment can induce Tregs with a suppressive function from $CD4^+CD25^-$ T cells (Ellis et al. 2012; Atarashi et al. 2013; Hsu et al. 2015). Yet it remains to be determined whether these peripherally induced $FOXP3^+$ Tregs are functionally stable in vivo.

2.1 *TCR, CD28, and IL-2*

nTreg development is initiated by TCR signal followed by a sequential activation of CD25 (IL-2 receptor α chain) expression, IL-2 signal, and FOXP3 expression (Lio and Hsieh 2008; Weissler and Caton 2014). Although not fully clarified in humans, nTregs stem from self-reactive thymocytes present in the thymus (Sakaguchi et al. 2010). A fraction of $CD4^+CD8^-$ thymocytes receive T cell receptor (TCR) stimulation by complexes of major histocompatibility complex (MHC) plus self-peptide and acquire expression of CD25, through which IL-2 signals are delivered via STAT5, resulting in expression FOXP3 and differentiation into Tregs (Jordan et al. 2001; Boyman and Sprent 2012; Malchow et al. 2013). nTreg development can be enhanced through the constitutive activation of STAT5 and directly binds cis elements in the FOXP3 promoter and enhancer to stabilize FOXP3 expression (Burchill et al. 2008). In addition to induction of CD25, TCR and CD28 signal also contribute to establishing and stabilizing the Treg lineage commitment in the thymus by inducing epigenetic and differentiation events in Tregs (Salomon et al.

2000; Tai et al. 2013; Zhang et al. 2013; Franckaert et al. 2015). Thus, antigen and IL-2 signal provided through TCR, CD28, and CD25 are essential for Treg lineage commitment in the thymus.

In the periphery, mature Treg survival for their homeostasis and function depends on TCR, CD28, and CD25, but their roles appear to be distinct from those in the thymus. Tregs proliferate more than Tconvs in steady state in a CD28 dependent fashion, suggesting that Tregs continuously recognize cognate antigens driving their cell cycle progression (Tang et al. 2003; Walker et al. 2003). Indeed, analysis of Treg subsets in the periphery shows that continuous stimulation through the TCR is required to maintain this population (Levine et al. 2014; Vahl et al. 2014). TCR-deficient Tregs proliferated less and expressed fewer effector molecules such as cytotoxic T lymphocyte-associated protein 4 (CTLA-4), IL-10, and Ebi3.

Proliferating Tregs have a tendency to lose their FOXP3 expression and lineage stability in vitro and in vivo in lymphopenic hosts (Hoffmann et al. 2006; Zhou et al. 2009; Rubtsov et al. 2010). The conserved noncoding sequence 2 (CNS2) enhancer element, also known as Treg-specific demethylation region, is crucial for safeguarding lineage stability of proliferating Tregs (Feng et al. 2014; Li et al. 2014). However, stimulation via TCR with limited IL-2 leads to a loss of FOXP3 expression in Tregs with intact CNS2. CNS2 harbors binding sites for both the TCR-triggered transcription factor nuclear factor of activated T cells (NFAT) and IL-2-induced transcription factor STAT5, providing a transcriptional basis for Treg stability by coordinating TCR and IL-2 signal. Interestingly, forced expression of constitutively active STAT5 prevented the loss of FOXP3 in CNS2-deleted Tregs, demonstrating that STAT5 can stabilize FOXP3 expression independent of CNS2 (Feng et al. 2014). This may be explained by the NFAT-mediated looping between CNS2 and the FOXP3 promoter, also having NFAT and STAT5 binding sites (Li et al. 2014). Together, TCR-mediated signals are important for mature Treg function but pose a threat to their stability unless they are balanced by the IL-2 signal.

2.2 PI3K-Akt-mTOR

Phosphatidylinositide 3 kinase (PI3K), protein kinase B (Akt), and mammalian target of rapamycin (mTOR) form an intracellular signal hub common to the TCR, CD28, and IL-2 receptor. PI3K is directly activated when these receptors are engaged, leading to initial activation of Akt by the PH-domain containing protein PDK1 through phosphorylation of threonine 308. Akt is fully activated by additional phosphorylation on serine 473 by the mTOR complex 2 (mTORC2). Akt has many cellular targets; the Forkhead box O (Foxo) transcription factors and mTORC1 are most relevant to Treg biology. Foxo family transcription factors are crucial for Treg lineage commitment (Harada et al. 2010; Ouyang et al. 2012; Samstein et al. 2012) and are inhibited by Akt. mTORC1 coordinates anabolic activities in cells and inactivates mTORC2, limiting further Akt activation. In the thymus, Treg development is enhanced by mutating the p110d catalytic subunit of PI3K (Patton et al. 2006) and is repressed by forced expression of a constitutively active Akt (Haxhinasto et al. 2008), demonstrating a negative role of the PI3K axis on nTreg development. However, deletion of mTOR (thus inactivating both mTORC1 and 2) or individual deletion of mTORC1 or 2 in T cells does not alter thymic development (Delgoffe et al. 2009), suggesting that the negative effect of PI3K and Akt on nTreg development is mTOR independent and mainly due to their role in Foxo1 inactivation.

Activation of PI3K is naturally antagonized by phosphatase and tensin homolog (PTEN). PTEN expression is progressively inhibited by stronger TCR stimulation, permitting efficient T cell activation and effector differentiation, an effect mediated by IL-2-inducible T cell kinase (Itk) (Gomez-Rodriguez et al. 2014). Thus, T cells with Itk deficiency fail to down regulate PTEN after activation and favor FOXP3 induction. In committed Tregs, the PI3K-Akt-mTOR signal axis continues to be repressed by high expression of PTEN. Treg-specific deletion of PTEN disrupted Treg homeostasis, function, and stability (Huynh et al. 2015; Shrestha et al. 2015). These PTEN-deficient Tregs lost both FOXP3 and CD25 expression but had a significant increase of mTORC2, but not mTORC1 activities. Additional deletion of mTORC2 in Tregs largely rescues the phenotype in mice with Treg-specific deletion of PTEN, demonstrating the normal function of PTEN in mature Tregs is to keep mTORC2 in check. In fact, the intact mTORC1 function is required for Treg function because mice with selective deletion of mTORC1 in Tregs die of multi-organ autoimmune diseases similar to FOXP3-deficient mice (Zeng et al. 2013). Mechanistically, mTOR is found to control Treg function in part by regulating metabolic programming. T cells rely on mitochondrial oxidative phosphorylation at steady state and switch to glycolysis after activation, a process essential for effector T cell differentiation (Wang and Green 2012). In contrast, Tregs preferentially use oxidative metabolism even after activation. An emerging concept is that metabolic input can also dictate T cell fate decision (Wang and Green 2012). PTEN-deficient Tregs show exaggerated glycolysis that is thought to contribute to Treg instability (Huynh et al. 2015; Shrestha et al. 2015). Additionally, functional defects in mTORC1-deficient Tregs are associated with disrupted lipid biosynthesis (Zeng et al. 2013). Thus, the impact of PI3K-Akt-mTOR axis on mature Treg function is controversial, while excessive activation of this pathway is clearly detrimental to Treg function as observed in PTEN-deficient Tregs; complete blockade of PI3K impairs Treg function as well (Patton et al. 2006; Patton et al. 2011).

2.3 *Epigenetics*

Epigenetic modifications, which include histone modifications, DNA methylation, microRNAs, nucleosome positioning, chromatin interaction, and chromosome conformational changes, play indispensable roles in cell differentiation, especially for cell-lineage stabilization (Kim et al. 2009). In particular, DNA methylation and histone modifications critically contribute to cell-lineage determination and maintenance because they are heritable through cell divisions. Genomic DNA is mainly methylated by DNA methyltransferases (DNMT family members), whereas it can be demethylated by multiple steps, including methylcytosine hydroxylation mediated by TET family members (Bhutani et al. 2011; Pastor et al. 2011). Similarly, histones are modified for gene activation or repression by acetylation or deacetylation, methylation or demethylation, and phosphorylation or dephosphorylation (Teperino et al. 2010); therefore, epigenetic status is reversible. It is also known, however, that DNA methylation status modified in the early stages of development, such as genomic imprinting, is stably maintained throughout subsequent differentiation processes. Epigenetic changes of some specific loci are also stably sustained in specific cell lineages, including Tregs (Ansel et al. 2003; Schmidl et al. 2009; Ohkura et al. 2012). Recent genome-wide analyses have revealed several regions that show different patterns of DNA methylation or histone modification between Tconvs and Tregs in humans and mice (Floess et al. 2007, Schmidl et al. 2009; Wei et al. 2009; Ohkura et al. 2012). Such genes with Treg-specific DNA hypomethylation include those encoding Treg-function-associated or Treg-specific molecules, such as FOXP3, CTLA-4, and Eos (Ohkura et al. 2012). Furthermore, some Treg-specific changes in DNA methylation are highly stable in Tregs, whereas others are not. For example, *Foxp3* intron 1 (CNS2), *Ctla4* exon 2, and *Ikzf4* (encoding Eos) intron 1, are specifically demethylated in nTreg cells, and the hypomethylation status is stable after TCR stimulation, cell proliferation, or cytokine treatments (e.g., with IL-2 or TGF-β) (Ohkura et al. 2012). In contrast, the DNA methylation status of *Il2ra* intron 1, which is demethylated in nonactivated Tregs, is relatively unstable and demethylated in Tconvs by in vitro culture with or without TCR stimulation. In addition, enhanced H3 K4me3 histone modification of the Treg signature genes detected in nTregs is easily primed in Tconvs under a Th1-, Th2-, or Th17-cell-polarizing or iTreg-inducing condition (Wei et al. 2009). Along with these findings, a high-resolution DNaseI footprint analysis has shown that specific alterations in chromatin accessibility occur in Tregs in the course of their differentiation from their precursors (Samstein et al. 2012), although the DNaseI-hypersensitive regions do not differ mostly between $CD4^{+}FOXP3^{-}$ T cells and $CD4^{+}FOXP3^{+}$ T cells, indicating a small number of genes show increased hypersensitivity in Tregs, specific alterations in local nucleosome positioning and chromatin accessibility. The loci identified as newly accessible in Tregs are enriched in the genes known to be critical for Treg function, such as *Foxp3*, *Ctla4,* and *Ikzf2*. They are also classified as genes possessing Treg-specific DNA hypomethylated regions in Tregs, as discussed above (Schmidl et al. 2009; Ohkura et al. 2012). Together, Tregs acquire and sustain highly specific and stable epigenetic changes as exemplified by DNA hypomethylation at specific loci of a limited number of genes. This Treg-specific DNA hypomethylation is a reliable marker for assessing the epigenetic status of Tregs.

3 Functional Classification of Tregs

While Tregs are originally identified as $CD4^+$ T cells, expressing CD25, as CD25 is an activation marker and its expression is not confined to Tregs, additional markers are needed. Although $CD4^+CD25^+$ T cells with additional low-level expression of CD127 (IL-7 receptor α-chain) were reported to possess FOXP3 expression and suppressive function (Liu et al. 2006; Seddiki et al. 2006), CD127 is also down-regulated following recent activation of naive T cells that also express a low level of FOXP3 (Mazzucchelli and Durum 2007), suggesting a possible contamination of non-Tregs in the $CD127^{low}CD4^+CD25^+$ T cell fraction. FOXP3 is the master regulatory molecule in Tregs, and expression of FOXP3 represents the Treg population in mice. In contrast, to definitely identify Tregs in humans causes difficulty due to the upregulation of FOXP3 upon TCR stimulation of Tconvs (Tran et al. 2007). We have therefore proposed a classification of human Tregs based on the expression levels of a naive marker CD45RA and of FOXP3 (Fig. 1 and Table 1) (Miyara et al. 2009; Sakaguchi et al. 2010; Nishikawa and Sakaguchi 2014; Takeuchi and Nishikawa 2016). $FOXP3^+CD4^+$ T cells can thus be divided into three fractions: naive Tregs ($CD45RA^+FOXP3^{low}CD4^+$); effector Tregs (eTregs: $CD45RA^-FOXP3^{high}CD4^+$); and non-Tregs ($CD45RA^-FOXP3^{low}CD4^+$). The naïve Tregs have recently egressed from the thymus, have not yet been activated in the periphery and possess weak suppressive activity. Upon activation with TCR stimulation, naïve Tregs vigorously proliferate and differentiate into highly suppressive eTregs. In contrast, non-Tregs are not immune suppressive but are rather immune stimulatory T cells, producing inflammatory cytokines including IFN-γ and IL-17 (Miyara et al. 2009). This classification, based on Treg function, reflects the pathophysiology of autoimmune and inflammatory diseases. Both sarcoidosis patients lacking tuberculin reaction due to an immune-suppressive state and systemic lupus erythematosus (SLE) patients with systemic autoimmunity have increased $FOXP3^+CD4^+$ T cells in the peripheral blood (Miyara et al. 2009). In our classification with CD45RA and FOXP3 expression, highly suppressive eTregs ($CD45RA^-FOXP3^{high}CD4^+$) are the dominant component of $FOXP3^+CD4^+$ T cells in sarcoidosis, whereas $FOXP3^+$ non-Tregs ($CD45RA^-FOXP3^{low}CD4^+$) are increased in SLE (Miyara et al. 2009), clearly demonstrating the immune-suppressive state and a dysregulation of self-tolerance in sarcoidosis and SLE, respectively.

4 Suppressive Mechanism of Tregs

Tregs exhibit their suppressive activity by numerous cellular and humoral mechanisms (Fig. 2 and Table 2) such as suppression of antigen-presenting cells (APCs) via CTLA-4, secretion of inhibitory cytokines (IL-10, TGF-β and IL-35), expression of granzyme/perforin, consumption of IL-2, and degradation of ATP

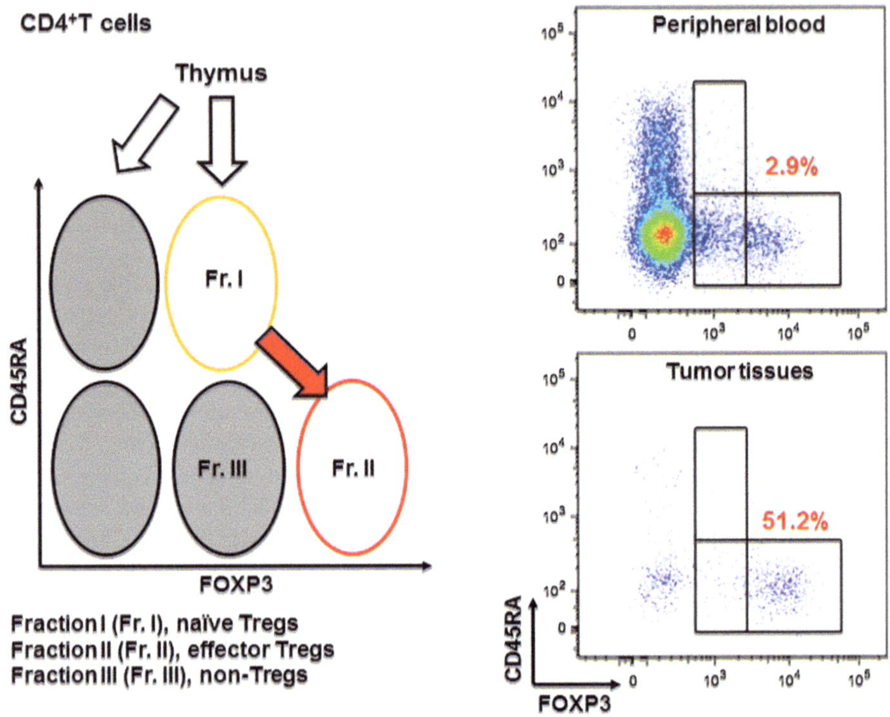

Fig. 1 Functional classification of human FOXP3$^+$CD4$^+$ T cell subpopulations. Human FOXP3$^+$ T cells in the peripheral blood and lymph nodes are composed of heterogeneous subpopulations containing suppressive Tregs (naïve and effector Tregs) and non-Tregs without suppression function. Human Tregs are classified into naive and effector Tregs by the expression levels of a naive marker CD45RA and of FOXP3. These subpopulations are designated as Fraction (Fr.) I, II, and III for naive Tregs, effector Tregs (eTregs), and non-Tregs, respectively. CD25 surface marker can be used in the place of FOXP3 because of their correlative expression in humans. In tumor tissues compared with peripheral blood, naive Tregs (Fr. I) numbers are reduced and highly suppressive eTregs (Fr. II) numbers are increased. The frequency of FOXP3$^+$ non-Tregs (Fr. III) is variable depending on cancer types

(Sakaguchi et al. 2010). Among these mechanisms, suppression via CTLA-4 (a co-inhibitory receptor constitutively expressed by Tregs) and IL-2 consumption via CD25 (IL-2 receptor α-chain, also constitutively expressed by Tregs) appear to play key roles for the following reasons: Treg-specific CTLA-4 deficiency impairs in vitro and in vivo Treg-mediated suppression (Wing et al. 2008); FOXP3 directly suppresses IL-2 gene transcription and upregulates *Ctla4* and *Il2ra* genes transcription (Hori et al. 2003); and high-dose IL-2 neutralizes in vitro Treg-mediated suppression (Takahashi et al. 1998; Thornton and Shevach 1998).

CTLA-4 engages with B7 molecules (i.e., B7-1 and B7-2; CD80 and CD86) on APCs with higher avidity compared with CD28 (Walker and Sansom 2011) and

Table 1 Classification of FOXP3$^+$CD4$^+$ T cells

Subset	Phenotype	Characteristics
Naïve Tregs (Fraction I, CD45RA$^+$FOXP3^{low}CD4$^+$)	CTLA-4^{low}CD25^{high}CD127$^{low/-}$ Ki-67$^-$	• Weak suppressive activity • Differentiate to effector Tregs upon TCR stimulation
Effector Tregs (Fraction II, CD45RA$^-$FOXP3^{high}CD4$^+$)	CTLA-4^{high}CD25^{high} Ki-67$^+$ PD-1$^+$, TIM-3$^+$, GITR$^+$ Fas$^+$, IL-10$^+$, TGF-β$^+$	• Strong suppressive and proliferative activity • Prone to apoptosis • Tend to increase in peripheral blood with aging
Non-Tregs (Fraction III, CD45RA$^-$FOXP3^{low}CD4$^+$)	IL-2$^+$, IFN-γ$^+$, IL-17$^+$	• Heterogeneous population • No suppressive activity

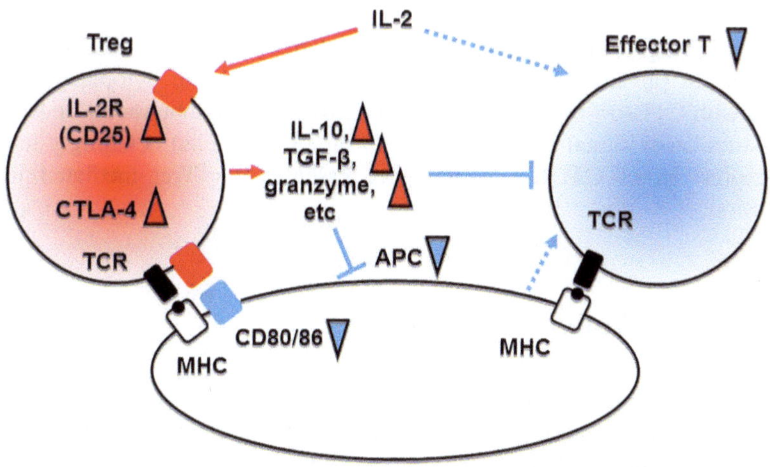

Fig. 2 Treg suppression mechanisms. Treg cells, which scarcely produce IL-2, deprive IL-2 from the surrounding via their high-affinity IL-2 receptor (IL-2R), making it unavailable for effector T cells. They also constitutively express CTLA-4, which down-modulates CD80/CD86 expression by antigen-presenting cells (APCs), thus depriving co-stimulatory signal to effector T cells. Tregs also produce immune-suppressive cytokines such as IL-10 and TGF-β, and secret granzymes, which also down-modulates APC and effector T cell functions. TCR, T cell receptor; MHC, major histocompatibility complex

provides inhibitory signaling to APCs. In mice, Treg-specific deletion of CTLA-4 elicits systemic hyper-proliferation of Tconvs, resulting in fatal autoimmune diseases affecting multiple organs, including severe myocarditis (Wing et al. 2008). Recently, heterozygous CTLA-4 mutations in humans were identified in patients with multiple autoimmune symptoms accompanied by impaired suppressive

Table 2 Key mechanisms of suppression by Tregs

Molecule	Mechanism of suppression
IL-2 receptor/IL-2	Constitutive expression of high-affinity IL-2 receptor α chain (CD25) and dependency on exogenous IL-2 by Treg cells together limit the amount of IL-2 available to Tconvs, thereby hindering the activation and proliferation of the latter
CTLA-4	Constitutively expressed CTLA-4 on Tregs preferentially binds to and downregulates CD80/CD86 co-stimulatory molecules on APCs, depriving Tconvs of the co-stimulatory signal
IL-10, TGF-β and other immune-suppressive cytokines and substances	Tregs produce immune-suppressive cytokines, such as IL-10 and TGF-β, form extracellular adenosine from ATP by CD39 and CD73, and can also mediate direct killing of Tconvs or APCs by secreting granzymes

function of Tregs (Kuehn et al. 2014; Schubert et al. 2014). In addition, B7 molecules are physically transferred to the surface or the cytoplasm of Tregs together with CTLA-4 (Walker and Sansom 2011). Then, maturation of APCs (via the co-stimulatory signal from B7 to CD28 on effector cells) is strongly inhibited. Treg expression of CTLA-4 is therefore essential for Treg-mediated immune suppression.

5 Tregs and Autoimmune Diseases

Reducing the number and function of Tregs compromises self-tolerance, leading to abnormal immune responses to self-antigens, thus resulting in autoimmune diseases (Wing and Sakaguchi 2010). It has been shown that Treg impairment is involved in the pathogenesis of many autoimmune diseases including rheumatoid arthritis (RA), SLE, and ANCA-associated vasculitis, etc. (Scheinecker et al. 2010; Free et al. 2013; Prakken et al. 2013). On the account of heterogeneity and complexity of autoimmune diseases, the function of Tregs needs to be evaluated deliberately. In some autoimmune settings, Treg numbers or frequencies reportedly are reduced compared to healthy controls, while the others not. This may suggest the existence of different Treg phenotypes in disease tissues (Bonelli et al. 2008; Monte et al. 2008). Even though the presence of dysfunctional Tregs in autoimmune diseases is commonly observed and manipulation of these Tregs are an essential issue (Ehrenstein et al. 2004; Venken et al. 2008; Rapetti et al. 2015). Cell therapy and/or reagents manipulating Tregs, therefore, are under intense scrutiny.

5.1 Treg Cell Therapy for Autoimmune Diseases

Mouse models reveal that Treg infusion can prevent/treat autoimmune diseases, and clinical application of Treg administration is now being tested in humans. The first reports of Treg infusion therapy for autoimmune diseases were in the context of type 1 diabetes (Marek-Trzonkowska et al. 2012; Bluestone et al. 2015). In these studies, polyclonal Tregs were administered with no safety concerns. Tracking Tregs with 6,6-$^2H_2$ glucose labeling showed that infused cells are present for at least a year, with no evidence for loss of the expected Treg phenotype (Bluestone et al. 2015).

It is, however, challenging to obtain therapeutic doses of Tregs due to the weak in vitro proliferative capacity. For example, by extrapolating data from mice, the therapeutic dose of polyclonal Tregs is estimated to be 3–5 $\times$ 10^9 Tregs for a 70 kg patient (Tang and Lee 2012). To gain billions of Tregs is laborious and require to develop novel strategies to improve in vitro expansion of Tregs, such as high IL-2 and mTOR inhibition with rapamycin to stimulate Treg division and limit Tconv outgrowth, respectively. In addition to limited cell numbers, polyclonal Tregs carry the risk of non-specific suppression inducing side effects. Indeed a transient increase in viral reactivations was observed in hematopoietic stem cell transplantation patients treated with cord blood-derived Tregs (Brunstein et al. 2013). To overcome these limitations of polyclonal Tregs, methods to generate antigen-specific Tregs are being explored, including antigen-stimulated expansion (Lee et al. 2014), TCR transduction (Kim et al. 2015), and engineering with chimeric antigen receptors (CARs) (MacDonald et al. 2016; Yoon et al. 2017). All of these strategies should maximize disease control with lower numbers of Tregs as in mice, antigen-specific Tregs are 100 fold more potent than polyclonal cells (Hoeppli et al. 2016).

In addition, the suppressive function can be added into Tconvs by overexpressing FOXP3, or by culture with immunosuppressive cytokines such as TGF-β. The stability of cells arising from the latter, however, is unclear, with epigenetic analysis, suggesting that these “induced iTregs” may not be stable in humans (Rossetti et al. 2015). The first application of overexpressing FOXP3 will likely be as gene therapy for IPEX syndrome patients (Bacchetta et al. 2016). For wider application in autoimmunity, a better understanding of Treg function by simple FOXP3 overexpression is necessary to optimize the Treg cell therapy (Bhairavabhotla et al. 2016).

5.2 Non-Cell-Based Therapies

Because of the complexity and highly personalized nature of cell therapy, strategies to augment endogenous Treg numbers and function in vivo may be preferable.

Methods manipulating IL-2 availability are the deeply tested in clinical testing, with other methods promoting FOXP3 expression in the early stage of exploration. The unique requirement of Tregs for exogenous IL-2, constitutive expression of the high-affinity IL-2 receptor, and the association with poor IL-2 response in autoimmunity offers an ideal target for therapeutic manipulation. Whereas high-doses of IL-2 enhance Tconvs in vivo, low doses seem to specifically stimulate Treg survival/expansion. A trial of low dose IL-2 in type 1 diabetes found a dose-dependent increase in numbers of $CD4^{+}FOXP3^{+}$ Tregs and increased CD25, GITR, CTLA-4, and pSTAT5 (Rosenzwajg et al. 2015). Encouragingly, at the highest dose, Tconv responses against beta-cell antigens were suppressed in all patients, leading to the initiation of a larger phase-IIb trial (NCT02411253). This approach has also had success in the treatment of SLE (von Spee-Mayer et al. 2016), with additional trials of low-dose IL-2 planned in RA (NCT02467504), relapsing remitting multiple sclerosis (NCT02424396) and other several autoimmune/autoinflammatory disorders (TRANSREG study, NCT01988506). IL-2 has a short half-life, which can be prolonged through the administration of a cytokine-antibody complex. Careful selection of the anti-IL-2 antibody can allow tailored signal; the JES6-1 anti-mouse IL-2 antibody lowers the affinity of IL-2 for CD25, favoring signal to $CD25^{high}$ Tregs (Spangler et al. 2015). IL-2 itself can also be engineered, creating variants that have more or less affinity for the individual receptor chains, allowing preferential stimulation of Tconvs (Mitra et al. 2015) or, presumably, in the future, of Tregs. It is yet questionable whether these strategies will be feasible in humans due to the high CD25 expression on activated human Tconvs.

Rapamycin (sirolimus) preferentially favors Treg suppression by blocking Tconv proliferation and promoting FOXP3 mRNA expression and is now commonly used as a 'Treg sparing' immuno-suppressant in transplantation. Rapamycin is also being explored in autoimmunity, with a trial in multi-lineage autoimmune cytopenias showing rapid and long-lasting responses in a majority of children with the autoimmune lymphoproliferative syndrome, and encouraging results in those with SLE (Bride et al. 2016). Additionally, clinical trials are ongoing to test the effect of rapamycin in Crohn's disease patients with stenosis (NCT02675153) or in combination with islet transplantation in type 1 diabetes (NCT02505893; NCT00679042). With our knowledge for how peripheral Tregs develop, therapies that harness these processes are also being explored (Hardenberg et al. 2011). For example, Vitamin C can potentiate Tregs by regulating the activity of TET enzymes, which demethylate Treg-specific hypomethylated regions, including the FOXP3 locus (Yue et al. 2016). Similarly, all-trans retinoic acid, the metabolite of vitamin A, prevents human Tregs from becoming unstable by increasing histone acetylation in the FOXP3 promoter and demethylation of the Treg-specific demethylation region (Lu et al. 2014). Overall, there are many complementary strategies to enhance Tregs in vivo and it will be important to compare the effectiveness of these approaches with cell-based therapies.

6 Tregs and Malignant Tumors

The involvement of Tregs in tumor immunity was originally reported in 1999 (Onizuka et al. 1999; Shimizu et al. 1999). Mice treated with the anti-CD25 antibody (which depleted $CD4^+CD25^+$ Tregs) and nude (T cell deficient) mice transferred with splenocytes deprived for $CD25^+$ cells, exhibited tumor rejection and retardation of tumor growth. In the tumor microenvironment (TME) of melanoma, non-small cell lung, gastric and ovarian cancers, eTregs heavily infiltrate and account for 20–50% of $CD4^+$ T cells, as compared with 5–10%in the peripheral blood of healthy individuals (Sakaguchi et al. 2010; Nishikawa and Sakaguchi 2014; Saito et al. 2016; Takeuchi and Nishikawa 2016) (Fig. 1). High infiltration of Tregs in tumors is associated with a poor prognosis in various types of cancers including melanoma, non-small cell lung, gastric, hepatocellular, pancreatic, renal cell, breast and cervical cancers (Fridman et al. 2012; Nishikawa and Sakaguchi 2014). Yet in some cancers such as colorectal, head and neck, and bladder cancers, a higher infiltration of $FOXP3^+$ T cells is reportedly correlated with better prognosis (Fridman et al. 2012). In fact, in colorectal cancer, we have recently shown that $FOXP3^+$ non-Tregs heavily infiltrated a fraction of colorectal cancers containing high levels of inflammatory cytokines such as TGF-β and IL-12 and were associated with a better prognosis (Saito et al. 2016). The difficulty of distinguishing $FOXP3^+$ non-Tregs from $FOXP3^{high}$ eTregs in tumor tissues would have been a major confounding factor in previous studies evaluating the clinical significance of $FOXP3^+CD4^+$ T cells in colorectal cancers using immunohistochemistry.

6.1 Trafficking and Characteristics of Tregs in Cancer

Tregs are chemo-attracted to the TME although the combination of chemokines and their receptors differs in each cancer type (i.e., CCR4 with CCL22 in breast, colorectal, oral and ovarian cancer; CCR10 with CCL28 and CXCR4 with CXCL12 in ovarian cancer; and CCR5 with CCL5 in pancreatic cancer, etc.) (Curiel et al. 2004; Ishida et al. 2006; Wei et al. 2007; Gobert et al. 2009; Tan et al. 2009; Watanabe et al. 2010; Facciabene et al. 2011; Svensson et al. 2012). Blockade of chemotaxis by antibodies or small molecules may result in a reduction in Treg numbers in tumors (Tan et al. 2009; Spranger et al. 2013). These Treg-recruiting chemokines are generated in TMEs by macrophages and/or tumor cells. Additionally, $CD8^+$ T cells in tumors also produced Treg-recruiting chemokines with their exhaustion (Williams et al. 2017). In the TME, highly immune-suppressive eTregs with high-level expression of suppression-related molecules such as CTLA-4 and TIGIT are detected with reduced number of naïve Tregs, indicating a highly activated status of tumor-infiltrating Tregs (Sugiyama et al. 2013; Nishikawa and Sakaguchi 2014, Saito et al. 2016; Takeuchi and Nishikawa 2016). One possible mechanism of Treg activation in tumors is that proliferating and dying tumor cells provide a large amount of self-antigens, which Tregs might recognize, and be activated, as tumors contain a subset of immature dendritic cells that promote the proliferation/stimulation of Tregs in a TGF-β-dependent manner (Ghiringhelli et al. 2005; Nishikawa et al. 2005). In accordance with this, the TCR repertoire of tumor-infiltrating Tregs is skewed and largely distinct from that of tumor-infiltrating Tconvs, suggesting that Tregs recognize certain skewed antigens and clonally expand in the TME (Hindley et al. 2011; Sainz-Perez et al. 2012). Yet whether these antigens are exclusively recognized by Tregs or recognition is shared by Th cells is unclear; however, Tregs usually harbor higher affinity TCRs compared with Tconvs and should be predominantly activated in tumors.

6.2 Strategies for Treg-Targeted Therapy

As discussed above, eTregs are present at a high frequency in tumors and need to be controlled for the generation/activation of anti-tumor immunity. Some clinical studies indicated the potential of depleting CD25-expressing lymphocytes to augment anti-tumor immune responses, yet other similar studies failed. As activated effector T cells also express CD25, CD25-based cell depletion may reduce activated effector T cells as well, canceling the effect of Treg depletion to augment anti-tumor immunity. Treg depletion by the CD25-depleting antibody daclizumab has been evaluated in clinical trials. When daclizumab was administered following dendritic cell vaccination in metastatic melanoma (n = 15), not only Tregs but also activated effector cells were depleted and neither anti-tumor immune responses nor antibody production was observed (Jacobs et al. 2010). In contrast, in breast cancer patients, administration of daclizumab followed by vaccination consisting of multiple tumor-associated peptides succeeded in Treg depletion and demonstrated favorable clinical responses (Rech et al. 2012). Additionally, one plausible concern is increased autoimmunity-related toxicities following Treg depletion. In order to secure the safety of Treg-targeted therapy, selective depletion of eTregs in tumors rather than the entire Treg population can be exploited to augment anti-tumor immunity without eliciting deleterious autoimmunity (Sugiyama et al. 2013). Targeting molecules and signals specific for eTregs is being tested in clinical trials as an effective strategy for eTreg depletion.

We showed that CCR4 was specifically expressed by a subset of suppressive eTregs abundant in melanoma, and treatment using anti-CCR4 antibody depleted the melanoma-infiltrating eTregs with CCR4 expression and efficiently induced/augmented cancer-testis antigen-specific both $CD4^+$ and $CD8^+$ T cells (Sugiyama et al. 2013). Mogamulizumab has been approved in Japan for the treatment of CCR4-expressing adult T cell leukemia/lymphoma (ATLL). Anti-CCR4 antibody markedly reduced eTregs as well as ATLL cells and augmented ATLL antigen (cancer-testis antigen)- specific $CD8^+$ T cell responses in an ATLL patient, possibly in association with the prolonged survival of this patient (Sugiyama et al. 2013). Based on these preclinical data, multiple early phase clinical trials with mogamulizumab as an eTreg depletion reagent are being conducted as monotherapy (trial numbers NCT02281409 and NCT01929486 (Kurose et al. 2015) and in combination with anti-PD-1 antibody (NCT02476123 and NCT02705105), anti-PD-L1 (PD-1 ligand 1) antibody or anti-CTLA-4 antibody (NCT02301130) and anti-4-1BB agonistic antibody (NCT02444793) in advanced solid tumors, and in combination with docetaxel in non-small cell lung cancer (NCT02358473). A recent phase Ia study showed that mogamulizumab administration was safe and well tolerated and that 4 of 10 patients showed stable disease during treatment and were long survivors. The monitoring of eTregs in the peripheral blood mononuclear cells during treatment indicated efficient depletion of those cells, even at the lowest dose (Kurose et al. 2015).

OX-40 and GITR are members of the TNF receptor superfamily and are both co-stimulatory receptors expressed by activated T cells. On Tregs, OX-40 is induced after activation and GITR is constitutively expressed (Shimizu et al. 2002; Griseri et al. 2010). These signals reduce the suppressive activity of Tregs as well as enhancing activation of effector T cells. A phase I trial of an OX-40 agonist demonstrated anti-tumor activity in melanoma and renal cell cancer (Curti et al. 2013). Early phase clinical trials evaluating OX-40 agonists in head and neck, breast and prostate cancer and in B cell lymphoma are also being investigated (NCT01862900, NCT02274155, NCT02318394, and NCT02205333). Additionally, a combination of an OX-40 fusion protein (MEDI6383) and an anti-PD-L1 antibody, durvalumab, is also being investigated (NCT02221960). Similarly, phase-I clinical trials evaluating GITR agonists in solid tumors are under investigation (NCT 02583165 and NCT02628574).

Tregs are highly dependent on PI3K signals for their maintenance and function. Inactivation of PI3K signals in Tregs activates $CD8^+$ T cells and induces tumor regression (Ali et al. 2014). Therefore, not only molecules specifically expressed by Tregs but also signals on which Tregs specifically depend could become targets to control Tregs. CPA is an alkylating agent that reportedly depletes Tregs when used in low doses. In phase II clinical trial, patients with advanced renal cell cancer received a therapeutic vaccination of IMA901 consisting of multiple tumor-associated peptides and GM-CSF with or without preceding CPA administration (Walter et al. 2012). Patients treated with IMA901/GM-CSF/CPA showed Treg reduction with augmented anti-tumor immune responses. The OS tended to be extended in the IMA901/GM-CSF/CPA-treated group (n = 33) compared with the IMA901/GM-CSF-treated group (n = 35) (23.5 months versus 14.8 months). A phase III trial investigating the addition of IMA901/GM-CSF/CPA to the standard care of sunitinib was completed in 2015 and the results are awaited.

6.3 *Involvement of Tregs in Immune Checkpoint Inhibitors*

Immune checkpoint blockade—inhibiting the immunosuppressive signals from co-inhibitory molecules—allows a resurgence in the effector function of tumor-infiltrating T cells and provides clinical success in various types of cancers including malignant melanomas and lung cancers (Hodi et al. 2010; Topalian et al. 2012; Borghaei et al. 2015; Brahmer et al. 2015). As immune checkpoint molecules such as CTLA-4 and PD-1 are expressed by both tumor-infiltrating effector T cells and Tregs, current immune checkpoint blocking agents could target Tregs as well. Analyses of anti-CTLA-4 antibodies in mouse models revealed that the anti-tumor efficacy was dependent on depletion of CTLA-4-expressing Tregs in tumors through the antibody-dependent cellular cytotoxic (ADCC) activity of the anti-CTLA-4 antibody; depletion of Fc function totally abrogated the anti-tumor effect of the anti-CTLA-4 antibody (Simpson et al. 2013; Matheu et al. 2015). Additionally, PD-1-expressing Tregs reportedly possess higher immune-suppressive function than Tregs without PD-1 expression in a mouse model (Park et al. 2015). Therefore, PD-1-blocking antibodies might act on Tregs to augment anti-tumor immunity as well as reversing the effector function of dys-functional effector cells. Yet, more than half of the treated patients did not respond to immune checkpoint blockade therapy, even if combinations. Immuno-monitoring of biomarkers to properly evaluate immune responses in cancer patients is critical for detecting responders.

There are two types of tumor antigens: tumor-specific antigens, which are either oncogenic viral proteins or abnormal proteins from somatic mutations (neoanti-gens); and tumor-associated antigens, which are highly or aberrantly expressed normal proteins. It is not yet determined how $CD8^+$ T cells specific for each antigen contribute to clinical tumor regression and whether activation of these $CD8^+$ T cells specific for self-antigens versus non-self-antigens are controlled differently. In vitro experiments comparing Treg-mediated suppression of self-antigen (Melan-A)-specific $CD8^+$ T cells versus non-self (cytomegalovirus)-specific $CD8^+$ T cells showed that cytomegalovirus-specific $CD8^+$ T cells were resistant to suppression by Tregs (Maeda et al. 2014), indicating that Treg-mediated suppression is more prominent on self-antigen-expressing tumor cells rather than those expressing neoantigens. It is therefore noteworthy that cancers in patients susceptible to immune checkpoint blockade monotherapy contain a large number of neoantigens (Snyder et al. 2014; Rizvi et al. 2015), and that $CD8^+$ T cells specific for the antigens can be resistant to Treg-mediated immune suppression. In contrast, cancers with a lower number of neoantigens did not respond to immune checkpoint blockade (Snyder et al. 2014; Rizvi et al. 2015), and $CD8^+$ T cells can be under the control of Treg-mediated immune suppression. Thus, integration of Treg-targeting therapies that reduce Treg function and/or number may expand the therapeutic spectrum of cancer immunotherapy.

7 Conclusion

Since the discovery of Tregs as a key mediator of immunological self-tolerance, a common immunological basis for Treg-mediated suppression of autoimmunity and tumor immunity has been extensively explored. The manipulating Tregs is under active investigation as a new therapeutic approach for treating a wide variety of diseases including autoimmune diseases and cancer. Deeper mechanistic understanding of disease-specific Treg development, maintenance, and function could make disease-specific Treg-targeted therapy more effective, resulting in an increase of efficacy and decrease of side effects related to manipulating Tregs.

References

Adeegbe DO, Nishikawa H (2013) Natural and induced T regulatory cells in cancer. Front Immunol 4:190

Ali K, Soond DR, Piñeiro R, Hagemann T, Pearce W, Lim EL, Bouabe H, Scudamore CL, Hancox T, Maecker H, Friedman L, Turner M, Okkenhaug K, Vanhaesebroeck B (2014) Inactivation of PI(3)K p110δ breaks regulatory T-cell-mediated immune tolerance to cancer. Nature 510(7505):407–411

Ansel KM, Lee DU, Rao A (2003) An epigenetic view of helper T cell differentiation. Nat Immunol 4(7):616–623

Atarashi K, Tanoue T, Oshima K, Suda W, Nagano Y, Nishikawa H, Fukuda S, Saito T, Narushima S, Hase K, Kim S, Fritz JV, Wilmes P, Ueha S, Matsushima K, Ohno H, Olle B, Sakaguchi S, Taniguchi T, Morita H, Hattori M, Honda K (2013) Treg induction by a rationally selected mixture of Clostridia strains from the human microbiota. Nature 500(7461):232–236

Bacchetta R, Barzaghi F, Roncarolo MG (2016). From IPEX syndrome to FOXP3 mutation: a lesson on immune dysregulation. Ann N Y Acad Sci

Bennett CL, Christie J, Ramsdell F, Brunkow ME, Ferguson PJ, Whitesell L, Kelly TE, Saulsbury FT, Chance PF, Ochs HD (2001) The immune dysregulation, polyendocrinopathy, enteropathy, X-linked syndrome (IPEX) is caused by mutations of FOXP3. Nat Genet 27 (1):20–21

Bhairavabhotla R, Kim YC, Glass DD, Escobar TM, Patel MC, Zahr R, Nguyen CK, Kilaru GK, Muljo SA, Shevach EM (2016) Transcriptome profiling of human FoxP3+ regulatory T cells. Hum Immunol 77(2):201–213

Bhutani N, Burns DM, Blau HM (2011) DNA demethylation dynamics. Cell 146(6):866–872

Bluestone JA, Buckner JH, Fitch M, Gitelman SE, Gupta S, Hellerstein MK, Herold KC, Lares A, Lee MR, Li K, Liu W, Long SA, Masiello LM, Nguyen V, Putnam AL, Rieck M, Sayre PH, Tang Q (2015) Type 1 diabetes immunotherapy using polyclonal regulatory T cells. Sci Transl Med 7(315):315ra189

Bonelli M, Savitskaya A, von Dalwigk K, Steiner CW, Aletaha D, Smolen JS, Scheinecker C (2008) Quantitative and qualitative deficiencies of regulatory T cells in patients with systemic lupus erythematosus (SLE). Int Immunol 20(7):861–868

Borghaei H, Paz-Ares L, Horn L, Spigel DR, Steins M, Ready NE, Chow LQ, Vokes EE, Felip E, Holgado E, Barlesi F, Kohlhäufl M, Arrieta O, Burgio MA, Fayette J, Lena H, Poddubskaya E, Gerber DE, Gettinger SN, Rudin CM, Rizvi N, Crinò L, Blumenschein GR, Antonia SJ, Dorange C, Harbison CT, Graf Finckenstein F, Brahmer JR (2015) Nivolumab versus docetaxel in advanced nonsquamous non-small-cell lung cancer. N Engl J Med 373(17): 1627–1639

Boyman O, Sprent J (2012) The role of interleukin-2 during homeostasis and activation of the immune system. Nat Rev Immunol 12(3):180–190

Brahmer J, Reckamp KL, Baas P, Crinò L, Eberhardt WE, Poddubskaya E, Antonia S, Pluzanski A, Vokes EE, Holgado E, Waterhouse D, Ready N, Gainor J, Arén Frontera O, Havel L, Steins M, Garassino MC, Aerts JG, Domine M, Paz-Ares L, Reck M, Baudelet C, Harbison CT, Lestini B, Spigel DR (2015) Nivolumab versus docetaxel in advanced squamous-cell non-small-cell lung cancer. N Engl J Med 373(2):123–135

Bride KL, Vincent T, Smith-Whitley K, Lambert MP, Bleesing JJ, Seif AE, Manno CS, Casper J, Grupp SA, Teachey DT (2016) Sirolimus is effective in relapsed/refractory autoimmune cytopenias: results of a prospective multi-institutional trial. Blood 127(1):17–28

Brunkow ME, Jeffery EW, Hjerrild KA, Paeper B, Clark LB, Yasayko SA, Wilkinson JE, Galas D, Ziegler SF, Ramsdell F (2001) Disruption of a new forkhead/winged-helix protein, scurfin, results in the fatal lymphoproliferative disorder of the scurfy mouse. Nat Genet 27(1):68–73

Brunstein CG, Blazar BR, Miller JS, Cao Q, Hippen KL, McKenna DH, Curtsinger J, McGlave PB, Wagner JE (2013) Adoptive transfer of umbilical cord blood-derived regulatory T cells and early viral reactivation. Biol Blood Marrow Transplant 19(8):1271–1273

Burchill MA, Yang J, Vang KB, Moon JJ, Chu HH, Lio CW, Vegoe AL, Hsieh CS, Jenkins MK, Farrar MA (2008) Linked T cell receptor and cytokine signaling govern the development of the regulatory T cell repertoire. Immunity 28(1):112–121

Chaudhry A, Rudra D, Treuting P, Samstein RM, Liang Y, Kas A, Rudensky AY (2009) CD4 + regulatory T cells control TH17 responses in a Stat3-dependent manner. Science 326 (5955):986–991

Chung Y, Tanaka S, Chu F, Nurieva RI, Martinez GJ, Rawal S, Wang YH, Lim H, Reynolds JM, Zhou XH, Fan HM, Liu ZM, Neelapu SS, Dong C (2011) Follicular regulatory T cells expressing Foxp3 and Bcl-6 suppress germinal center reactions. Nat Med 17(8):983–988

Coombes JL, Siddiqui KR, Arancibia-Cárcamo CV, Hall J, Sun CM, Belkaid Y, Powrie F (2007) A functionally specialized population of mucosal CD103+ DCs induces Foxp3+ regulatory T cells via a TGF-beta and retinoic acid-dependent mechanism. J Exp Med 204(8):1757–1764

Curiel TJ, Coukos G, Zou L, Alvarez X, Cheng P, Mottram P, Evdemon-Hogan M, Conejo-Garcia JR, Zhang L, Burow M, Zhu Y, Wei S, Kryczek I, Daniel B, Gordon A, Myers L, Lackner A, Disis ML, Knutson KL, Chen L, Zou W (2004) Specific recruitment of regulatory T cells in ovarian carcinoma fosters immune privilege and predicts reduced survival. Nat Med 10 (9):942–949

Curti BD, Kovacsovics-Bankowski M, Morris N, Walker E, Chisholm L, Floyd K, Walker J, Gonzalez I, Meeuwsen T, Fox BA, Moudgil T, Miller W, Haley D, Coffey T, Fisher B, Delanty-Miller L, Rymarchyk N, Kelly T, Crocenzi T, Bernstein E, Sanborn R, Urba WJ, Weinberg AD (2013) OX40 is a potent immune-stimulating target in late-stage cancer patients. Cancer Res 73(24):7189–7198

Delgoffe GM, Kole TP, Zheng Y, Zarek PE, Matthews KL, Xiao B, Worley PF, Kozma SC, Powell JD (2009) The mTOR kinase differentially regulates effector and regulatory T cell lineage commitment. Immunity 30(6):832–844

Ehrenstein MR, Evans JG, Singh A, Moore S, Warnes G, Isenberg DA, Mauri C (2004) Compromised function of regulatory T cells in rheumatoid arthritis and reversal by anti-TNFalpha therapy. J Exp Med 200(3):277–285

Ellis GI, Reneer MC, Vélez-Ortega AC, McCool A, Martí F (2012) Generation of induced regulatory T cells from primary human naïve and memory T cells. J Vis Exp (62)

Facciabene A, Peng X, Hagemann IS, Balint K, Barchetti A, Wang LP, Gimotty PA, Gilks CB, Lal P, Zhang L, Coukos G (2011) Tumour hypoxia promotes tolerance and angiogenesis via CCL28 and T(reg) cells. Nature 475(7355):226–230

Feng Y, Arvey A, Chinen T, van der Veeken J, Gasteiger G, Rudensky AY (2014) Control of the inheritance of regulatory T cell identity by a cis element in the Foxp3 locus. Cell 158(4): 749–763

Floess S, Freyer J, Siewert C, Baron U, Olek S, Polansky J, Schlawe K, Chang HD, Bopp T, Schmitt E, Klein-Hessling S, Serfling E, Hamann A, Huehn J (2007) Epigenetic control of the foxp3 locus in regulatory T cells. PLoS Biol 5(2):e38

Fontenot JD, Gavin MA, Rudensky AY (2003) Foxp3 programs the development and function of CD4+CD25+ regulatory T cells. Nat Immunol 4(4):330–336

Franckaert D, Dooley J, Roos E, Floess S, Huehn J, Luche H, Fehling HJ, Liston A, Linterman MA, Schlenner SM (2015) Promiscuous Foxp3-cre activity reveals a differential requirement for CD28 in Foxp3^{+} and Foxp3^{-}T cells. Immunol Cell Biol 93(4):417–423

Free ME, Bunch DO, McGregor JA, Jones BE, Berg EA, Hogan SL, Hu Y, Preston GA, Jennette JC, Falk RJ, Su MA (2013) Patients with antineutrophil cytoplasmic antibody-associated vasculitis have defective Treg cell function exacerbated by the presence of a suppression-resistant effector cell population. Arthritis Rheum 65(7):1922–1933
Fridman WH, Pagès F, Sautès-Fridman C, Galon J (2012) The immune contexture in human tumours: impact on clinical outcome. Nat Rev Cancer 12(4):298–306
Ghiringhelli F, Puig PE, Roux S, Parcellier A, Schmitt E, Solary E, Kroemer G, Martin F, Chauffert B, Zitvogel L (2005) Tumor cells convert immature myeloid dendritic cells into TGF-beta-secreting cells inducing CD4+CD25+ regulatory T cell proliferation. J Exp Med 202 (7):919–929
Gobert M, Treilleux I, Bendriss-Vermare N, Bachelot T, Goddard-Leon S, Arfi V, Biota C, Doffin AC, Durand I, Olive D, Perez S, Pasqual N, Faure C, Ray-Coquard I, Puisieux A, Caux C, Blay JY, Ménétrier-Caux C (2009) Regulatory T cells recruited through CCL22/CCR4 are selectively activated in lymphoid infiltrates surrounding primary breast tumors and lead to an adverse clinical outcome. Cancer Res 69(5):2000–2009
Gomez-Rodriguez J, Wohlfert EA, Handon R, Meylan F, Wu JZ, Anderson SM, Kirby MR, Belkaid Y, Schwartzberg PL (2014) Itk-mediated integration of T cell receptor and cytokine signaling regulates the balance between Th17 and regulatory T cells. J Exp Med 211(3): 529–543
Griseri T, Asquith M, Thompson C, Powrie F (2010) OX40 is required for regulatory T cell-mediated control of colitis. J Exp Med 207(4):699–709
Harada Y, Elly C, Ying G, Paik JH, DePinho RA, Liu YC (2010) Transcription factors Foxo3a and Foxo1 couple the E3 ligase Cbl-b to the induction of Foxp3 expression in induced regulatory T cells. J Exp Med 207(7):1381–1391
Hardenberg G, Steiner TS, Levings MK (2011) Environmental influences on T regulatory cells in inflammatory bowel disease. Semin Immunol 23(2):130–138
Haxhinasto S, Mathis D, Benoist C (2008) The AKT-mTOR axis regulates de novo differentiation of CD4+Foxp3+ cells. J Exp Med 205(3):565–574
Hindley JP, Ferreira C, Jones E, Lauder SN, Ladell K, Wynn KK, Betts GJ, Singh Y, Price DA, Godkin AJ, Dyson J, Gallimore A (2011) Analysis of the T-cell receptor repertoires of tumor-infiltrating conventional and regulatory T cells reveals no evidence for conversion in carcinogen-induced tumors. Cancer Res 71(3):736–746
Hodi FS, O'Day SJ, McDermott DF, Weber RW, Sosman JA, Haanen JB, Gonzalez R, Robert C, Schadendorf D, Hassel JC, Akerley W, van den Eertwegh AJ, Lutzky J, Lorigan P, Vaubel JM, Linette GP, Hogg D, Ottensmeier CH, Lebbé C, Peschel C, Quirt I, Clark JI, Wolchok JD, Weber JS, Tian J, Yellin MJ, Nichol GM, Hoos A, Urba WJ (2010) Improved survival with ipilimumab in patients with metastatic melanoma. N Engl J Med 363(8):711–723
Hoeppli RE, MacDonald KG, Levings MK, Cook L (2016) How antigen specificity directs regulatory T-cell function: self, foreign and engineered specificity. HLA 88(1–2):3–13
Hoffmann P, Eder R, Boeld TJ, Doser K, Piseshka B, Andreesen R, Edinger M (2006) Only the CD45RA+ subpopulation of CD4+CD25high T cells gives rise to homogeneous regulatory T-cell lines upon in vitro expansion. Blood 108(13):4260–4267
Hori S, Nomura T, Sakaguchi S (2003) Control of regulatory T cell development by the transcription factor Foxp3. Science 299(5609):1057–1061
Hsu P, Santner-Nanan B, Hu M, Skarratt K, Lee CH, Stormon M, Wong M, Fuller SJ, Nanan R (2015) IL-10 potentiates differentiation of human induced regulatory T cells via STAT3 and Foxo1. J Immunol 195(8):3665–3674
Huynh A, DuPage M, Priyadharshini B, Sage PT, Quiros J, Borges CM, Townamchai N, Gerriets VA, Rathmell JC, Sharpe AH, Bluestone JA, Turka LA (2015) Control of PI(3) kinase in Treg cells maintains homeostasis and lineage stability. Nat Immunol 16(2):188–196
Ishida T, Ishii T, Inagaki A, Yano H, Komatsu H, Iida S, Inagaki H, Ueda R (2006) Specific recruitment of CC chemokine receptor 4-positive regulatory T cells in Hodgkin lymphoma fosters immune privilege. Cancer Res 66(11):5716–5722

Jacobs JF, Punt CJ, Lesterhuis WJ, Sutmuller RP, Brouwer HM, Scharenborg NM, Klasen IS, Hilbrands LB, Figdor CG, de Vries IJ, Adema GJ (2010) Dendritic cell vaccination in combination with anti-CD25 monoclonal antibody treatment: a phase I/II study in metastatic melanoma patients. Clin Cancer Res 16(20):5067–5078

Jordan MS, Boesteanu A, Reed AJ, Petrone AL, Holenbeck AE, Lerman MA, Naji A, Caton AJ (2001) Thymic selection of CD4+CD25+ regulatory T cells induced by an agonist self-peptide. Nat Immunol 2(4):301–306

Khattri R, Cox T, Yasayko SA, Ramsdell F (2003) An essential role for Scurfin in CD4+CD25+ T regulatory cells. Nat Immunol 4(4):337–342

Kim JK, Samaranayake M, Pradhan S (2009) Epigenetic mechanisms in mammals. Cell Mol Life Sci 66(4):596–612

Kim YC, Zhang AH, Su Y, Rieder SA, Rossi RJ, Ettinger RA, Pratt KP, Shevach EM, Scott DW (2015) Engineered antigen-specific human regulatory T cells: immunosuppression of FVIII-specific T- and B-cell responses. Blood 125(7):1107–1115

Koch MA, Tucker-Heard G, Perdue NR, Killebrew JR, Urdahl KB, Campbell DJ (2009) The transcription factor T-bet controls regulatory T cell homeostasis and function during type 1 inflammation. Nat Immunol 10(6):595–602

Kuehn HS, Ouyang W, Lo B, Deenick EK, Niemela JE, Avery DT, Schickel JN, Tran DQ, Stoddard J, Zhang Y, Frucht DM, Dumitriu B, Scheinberg P, Folio LR, Frein CA, Price S, Koh C, Heller T, Seroogy CM, Huttenlocher A, Rao VK, Su HC, Kleiner D, Notarangelo LD, Rampertaap Y, Olivier KN, McElwee J, Hughes J, Pittaluga S, Oliveira JB, Meffre E, Fleisher TA, Holland SM, Lenardo MJ, Tangye SG, Uzel G (2014) Immune dysregulation in human subjects with heterozygous germline mutations in CTLA4. Science 345(6204): 1623–1627

Kurose K, Ohue Y, Wada H, Iida S, Ishida T, Kojima T, Doi T, Suzuki S, Isobe M, Funakoshi T, Kakimi K, Nishikawa H, Udono H, Oka M, Ueda R, Nakayama E (2015) Phase Ia study of FoxP3+ CD4 Treg depletion by infusion of a humanized anti-CCR4 antibody, KW-0761, in cancer patients. Clin Cancer Res 21(19):4327–4336

Lee K, Nguyen V, Lee KM, Kang SM, Tang Q (2014) Attenuation of donor-reactive T cells allows effective control of allograft rejection using regulatory T cell therapy. Am J Transplant 14 (1):27–38

Levine AG, Arvey A, Jin W, Rudensky AY (2014) Continuous requirement for the TCR in regulatory T cell function. Nat Immunol 15(11):1070–1078

Li X, Liang Y, LeBlanc M, Benner C, Zheng Y (2014) Function of a Foxp3 cis-element in protecting regulatory T cell identity. Cell 158(4):734–748

Linterman MA, Pierson W, Lee SK, Kallies A, Kawamoto S, Rayner TF, Srivastava M, Divekar DP, Beaton L, Hogan JJ, Fagarasan S, Liston A, Smith KG, Vinuesa CG (2011) Foxp3 + follicular regulatory T cells control the germinal center response. Nat Med 17(8):975–982

Lio CW, Hsieh CS (2008) A two-step process for thymic regulatory T cell development. Immunity 28(1):100–111

Liu W, Putnam AL, Xu-Yu Z, Szot GL, Lee MR, Zhu S, Gottlieb PA, Kapranov P, Gingeras TR, Fazekas de St Groth B, Clayberger C, Soper DM, Ziegler SF, Bluestone JA (2006) CD127 expression inversely correlates with FoxP3 and suppressive function of human CD4+ T reg cells. J Exp Med 203(7):1701–1711

Lu L, Lan Q, Li Z, Zhou X, Gu J, Li Q, Wang J, Chen M, Liu Y, Shen Y, Brand DD, Ryffel B, Horwitz DA, Quismorio FP, Liu Z, Li B, Olsen NJ, Zheng SG (2014) Critical role of all-trans retinoic acid in stabilizing human natural regulatory T cells under inflammatory conditions. Proc Natl Acad Sci USA 111(33):E3432–E3440

MacDonald KG, Hoeppli RE, Huang Q, Gillies J, Luciani DS, Orban PC, Broady R, Levings MK (2016) Alloantigen-specific regulatory T cells generated with a chimeric antigen receptor. J Clin Invest 126(4):1413–1424

Maeda Y, Nishikawa H, Sugiyama D, Ha D, Hamaguchi M, Saito T, Nishioka M, Wing JB, Adeegbe D, Katayama I, Sakaguchi S (2014) Detection of self-reactive $CD8^+$ T cells with an anergic phenotype in healthy individuals. Science 346(6216):1536–1540

Malchow S, Leventhal DS, Nishi S, Fischer BI, Shen L, Paner GP, Amit AS, Kang C, Geddes JE, Allison JP, Socci ND, Savage PA (2013) Aire-dependent thymic development of tumor-associated regulatory T cells. Science 339(6124):1219–1224

Marek-Trzonkowska N, Mysliwiec M, Dobyszuk A, Grabowska M, Techmanska I, Juscinska J, Wujtewicz MA, Witkowski P, Mlynarski W, Balcerska A, Mysliwska J, Trzonkowski P (2012) Administration of CD4+CD25highCD127- regulatory T cells preserves β-cell function in type 1 diabetes in children. Diabetes Care 35(9):1817–1820

Matheu MP, Othy S, Greenberg ML, Dong TX, Schuijs M, Deswarte K, Hammad H, Lambrecht BN, Parker I, Cahalan MD (2015) Imaging regulatory T cell dynamics and CTLA4-mediated suppression of T cell priming. Nat Commun 6:6219

Mazzucchelli R, Durum SK (2007) Interleukin-7 receptor expression: intelligent design. Nat Rev Immunol 7(2):144–154

Mitra S, Ring AM, Amarnath S, Spangler JB, Li P, Ju W, Fischer S, Oh J, Spolski R, Weiskopf K, Kohrt H, Foley JE, Rajagopalan S, Long EO, Fowler DH, Waldmann TA, Garcia KC, Leonard WJ (2015) Interleukin-2 activity can be fine tuned with engineered receptor signaling clamps. Immunity 42(5):826–838

Miyara M, Yoshioka Y, Kitoh A, Shima T, Wing K, Niwa A, Parizot C, Taflin C, Heike T, Valeyre D, Mathian A, Nakahata T, Yamaguchi T, Nomura T, Ono M, Amoura Z, Gorochov G, Sakaguchi S (2009) Functional delineation and differentiation dynamics of human CD4+ T cells expressing the FoxP3 transcription factor. Immunity 30(6):899–911

Monte K, Wilson C, Shih FF (2008) Increased number and function of FoxP3 regulatory T cells during experimental arthritis. Arthritis Rheum 58(12):3730–3741

Nishikawa H, Sakaguchi S (2014) Regulatory T cells in cancer immunotherapy. Curr Opin Immunol 27:1–7

Nishikawa H, Kato T, Tawara I, Saito K, Ikeda H, Kuribayashi K, Allen PM, Schreiber RD, Sakaguchi S, Old LJ, Shiku H (2005) Definition of target antigens for naturally occurring CD4 (+) CD25(+) regulatory T cells. J Exp Med 201(5):681–686

Ohkura N, Hamaguchi M, Morikawa H, Sugimura K, Tanaka A, Ito Y, Osaki M, Tanaka Y, Yamashita R, Nakano N, Huehn J, Fehling HJ, Sparwasser T, Nakai K, Sakaguchi S (2012) T cell receptor stimulation-induced epigenetic changes and Foxp3 expression are independent and complementary events required for Treg cell development. Immunity 37(5):785–799

Onizuka S, Tawara I, Shimizu J, Sakaguchi S, Fujita T, Nakayama E (1999) Tumor rejection by in vivo administration of anti-CD25 (interleukin-2 receptor alpha) monoclonal antibody. Cancer Res 59(13):3128–3133

Ouyang W, Liao W, Luo CT, Yin N, Huse M, Kim MV, Peng M, Chan P, Ma Q, Mo Y, Meijer D, Zhao K, Rudensky AY, Atwal G, Zhang MQ, Li MO (2012) Novel Foxo1-dependent transcriptional programs control T(reg) cell function. Nature 491(7425):554–559

Park HJ, Park JS, Jeong YH, Son J, Ban YH, Lee BH, Chen L, Chang J, Chung DH, Choi I, Ha SJ (2015) PD-1 upregulated on regulatory T cells during chronic virus infection enhances the suppression of CD8+ T cell immune response via the interaction with PD-L1 expressed on CD8+ T cells. J Immunol 194(12):5801–5811

Pastor WA, Pape UJ, Huang Y, Henderson HR, Lister R, Ko M, McLoughlin EM, Brudno Y, Mahapatra S, Kapranov P, Tahiliani M, Daley GQ, Liu XS, Ecker JR, Milos PM, Agarwal S, Rao A (2011) Genome-wide mapping of 5-hydroxymethylcytosine in embryonic stem cells. Nature 473(7347):394–397

Patton DT, Garden OA, Pearce WP, Clough LE, Monk CR, Leung E, Rowan WC, Sancho S, Walker LS, Vanhaesebroeck B, Okkenhaug K (2006) Cutting edge: the phosphoinositide 3-kinase p110 delta is critical for the function of CD4+CD25+Foxp3+ regulatory T cells. J Immunol 177(10):6598–6602

Patton DT, Wilson MD, Rowan WC, Soond DR, Okkenhaug K (2011) The PI3K p110δ regulates expression of CD38 on regulatory T cells. PLoS ONE 6(3):e17359

Prakken B, Wehrens Ellen, van Wijk F (2013) Editorial: Quality or quantity? Unraveling the role of Treg cells in rheumatoid arthritis. Arthritis Rheum 65(3):552–554

Rapetti L, Chavele KM, Evans CM, Ehrenstein MR (2015) B cell resistance to Fas-mediated apoptosis contributes to their ineffective control by regulatory T cells in rheumatoid arthritis. Ann Rheum Dis 74(1):294–302

Rech AJ, Mick R, Martin S, Recio A, Aqui NA, Powell DJ, Colligon TA, Trosko JA, Leinbach LI, Pletcher CH, Tweed CK, DeMichele A, Fox KR, Domchek SM, Riley JL, Vonderheide RH (2012) CD25 blockade depletes and selectively reprograms regulatory T cells in concert with immunotherapy in cancer patients. Sci Transl Med 4(134):134ra162

Rizvi NA, Hellmann MD, Snyder A, Kvistborg P, Makarov V, Havel JJ, Lee W, Yuan J, Wong P, Ho TS, Miller ML, Rekhtman N, Moreira AL, Ibrahim F, Bruggeman C, Gasmi B, Zappasodi R, Maeda Y, Sander C, Garon EB, Merghoub T, Wolchok JD, Schumacher TN, Chan TA (2015) Cancer immunology. Mutational landscape determines sensitivity to PD-1 blockade in non-small cell lung cancer. Science 348(6230):124–128

Rosenzwajg M, Churlaud G, Mallone R, Six A, Dérian N, Chaara W, Lorenzon R, Long SA, Buckner JH, Afonso G, Pham HP, Hartemann A, Yu A, Pugliese A, Malek TR, Klatzmann D (2015) Low-dose interleukin-2 fosters a dose-dependent regulatory T cell tuned milieu in T1D patients. J Autoimmun 58:48–58

Rossetti M, Spreafico R, Saidin S, Chua C, Moshref M, Leong JY, Tan YK, Thumboo J, van Loosdregt J, Albani S (2015) Ex vivo-expanded but not in vitro-induced human regulatory T cells are candidates for cell therapy in autoimmune diseases thanks to stable demethylation of the FOXP3 regulatory T cell-specific demethylated region. J Immunol 194(1):113–124

Rubtsov YP, Niec RE, Josefowicz S, Li L, Darce J, Mathis D, Benoist C, Rudensky AY (2010) Stability of the regulatory T cell lineage in vivo. Science 329(5999):1667–1671

Sainz-Perez A, Lim A, Lemercier B, Leclerc C (2012) The T-cell receptor repertoire of tumor-infiltrating regulatory T lymphocytes is skewed toward public sequences. Cancer Res 72 (14):3557–3569

Saito T, Nishikawa H, Wada H, Nagano Y, Sugiyama D, Atarashi K, Maeda Y, Hamaguchi M, Ohkura N, Sato E, Nagase H, Nishimura J, Yamamoto H, Takiguchi S, Tanoue T, Suda W, Morita H, Hattori M, Honda K, Mori M, Doki Y, Sakaguchi S (2016) Two FOXP3(+)CD4(+) T cell subpopulations distinctly control the prognosis of colorectal cancers. Nat Med 22 (6):679–684

Sakaguchi S, Sakaguchi N, Asano M, Itoh M, Toda M (1995) Immunologic self-tolerance maintained by activated T cells expressing IL-2 receptor alpha-chains (CD25). Breakdown of a single mechanism of self-tolerance causes various autoimmune diseases. J Immunol 155 (3):1151–1164

Sakaguchi S, Miyara M, Costantino CM, Hafler DA (2010) FOXP3+ regulatory T cells in the human immune system. Nat Rev Immunol 10(7):490–500

Salomon B, Lenschow DJ, Rhee L, Ashourian N, Singh B, Sharpe A, Bluestone JA (2000) B7/CD28 costimulation is essential for the homeostasis of the CD4+CD25+ immunoregulatory T cells that control autoimmune diabetes. Immunity 12(4):431–440

Samstein RM, Arvey A, Josefowicz SZ, Peng X, Reynolds A, Sandstrom R, Neph S, Sabo P, Kim JM, Liao W, Li MO, Leslie C, Stamatoyannopoulos JA, Rudensky AY (2012) Foxp3 exploits a pre-existent enhancer landscape for regulatory T cell lineage specification. Cell 151 (1):153–166

Scheinecker C, Bonelli M, Smolen JS (2010) Pathogenetic aspects of systemic lupus erythematosus with an emphasis on regulatory T cells. J Autoimmun 35(3):269–275

Schmidl C, Klug M, Boeld TJ, Andreesen R, Hoffmann P, Edinger M, Rehli M (2009) Lineage-specific DNA methylation in T cells correlates with histone methylation and enhancer activity. Genome Res 19(7):1165–1174

Schubert D, Bode C, Kenefeck R, Hou TZ, Wing JB, Kennedy A, Bulashevska A, Petersen BS, Schäffer AA, Grüning BA, Unger S, Frede N, Baumann U, Witte T, Schmidt RE, Dueckers G, Niehues T, Seneviratne S, Kanariou M, Speckmann C, Ehl S, Rensing-Ehl A, Warnatz K, Rakhmanov M, Thimme R, Hasselblatt P, Emmerich F, Cathomen T, Backofen R, Fisch P, Seidl M, May A, Schmitt-Graeff A, Ikemizu S, Salzer U, Franke A, Sakaguchi S, Walker LS,

Sansom DM, Grimbacher B (2014) Autosomal dominant immune dysregulation syndrome in humans with CTLA4 mutations. Nat Med 20(12):1410–1416

Seddiki, N, Santner-Nanan B, Martinson J, Zaunders J, Sasson S, Landay A, Solomon M, Selby W, Alexander SI, Nanan R, Kelleher A, Fazekas de St Groth B (2006) Expression of interleukin (IL)-2 and IL-7 receptors discriminates between human regulatory and activated T cells. J Exp Med 203(7):1693–1700

Shimizu J, Yamazaki S, Sakaguchi S (1999) Induction of tumor immunity by removing CD25 + CD4+ T cells: a common basis between tumor immunity and autoimmunity. J Immunol 163 (10):5211–5218

Shimizu J, Yamazaki S, Takahashi T, Ishida Y, Sakaguchi S (2002) Stimulation of CD25(+)CD4 (+) regulatory T cells through GITR breaks immunological self-tolerance. Nat Immunol 3 (2):135–142

Shrestha S, Yang K, Guy C, Vogel P, Neale G, Chi H (2015) Treg cells require the phosphatase PTEN to restrain TH1 and TFH cell responses. Nat Immunol 16(2):178–187

Simpson TR, Li F, Montalvo-Ortiz W, Sepulveda MA, Bergerhoff K, Arce F, Roddie C, Henry JY, Yagita H, Wolchok JD, Peggs KS, Ravetch JV, Allison JP, Quezada SA (2013) Fc-dependent depletion of tumor-infiltrating regulatory T cells co-defines the efficacy of anti-CTLA-4 therapy against melanoma. J Exp Med 210(9):1695–1710

Snyder A, Makarov V, Merghoub T, Yuan J, Zaretsky JM, Desrichard A, Walsh LA, Postow MA, Wong P, Ho TS, Hollmann TJ, Bruggeman C, Kannan K, Li Y, Elipenahli C, Liu C, Harbison CT, Wang L, Ribas A, Wolchok JD, Chan TA (2014) Genetic basis for clinical response to CTLA-4 blockade in melanoma. N Engl J Med 371(23):2189–2199

Spranger S, Spaapen RM, Zha Y, Williams J, Meng Y, Ha TT, Gajewski TF (2013) Up-regulation of PD-L1, IDO, and T(regs) in the melanoma tumor microenvironment is driven by CD8(+) T cells. Sci Transl Med 5(200):200ra116

Spangler JB, Tomala J, Luca VC, Jude KM, Dong S, Ring AM, Votavova P, Pepper M, Kovar M, Garcia KC (2015) Antibodies to interleukin-2 elicit selective T cell subset potentiation through distinct conformational mechanisms. Immunity 42(5):815–825

Sugiyama D, Nishikawa H, Maeda Y, Nishioka M, Tanemura A, Katayama I, Ezoe S, Kanakura Y, Sato E, Fukumori Y, Karbach J, Jäger E, Sakaguchi S (2013) Anti-CCR4 mAb selectively depletes effector-type FoxP3+CD4+ regulatory T cells, evoking antitumor immune responses in humans. Proc Natl Acad Sci USA 110(44):17945–17950

Svensson H, Olofsson V, Lundin S, Yakkala C, Björck S, Börjesson L, Gustavsson B, Quiding-Järbrink M (2012) Accumulation of CCR4^{+}CTLA-4 FOXP3^{+}CD25(hi) regulatory T cells in colon adenocarcinomas correlate to reduced activation of conventional T cells. PLoS ONE 7(2):e30695

Tai X, Erman B, Alag A, Mu J, Kimura M, Katz G, Guinter T, McCaughtry T, Etzensperger R, Feigenbaum L, Singer DS, Singer A (2013) Foxp3 transcription factor is proapoptotic and lethal to developing regulatory T cells unless counterbalanced by cytokine survival signals. Immunity 38(6):1116–1128

Takahashi T, Kuniyasu Y, Toda M, Sakaguchi N, Itoh M, Iwata M, Shimizu J, Sakaguchi S (1998) Immunologic self-tolerance maintained by CD25+CD4+ naturally anergic and suppressive T cells: induction of autoimmune disease by breaking their anergic/suppressive state. Int Immunol 10(12):1969–1980

Takeuchi Y, Nishikawa H (2016) Roles of regulatory T cells in cancer immunity. Int Immunol 28 (8):401–409

Tan MC, Goedegebuure PS, Belt BA, Flaherty B, Sankpal N, Gillanders WE, Eberlein TJ, Hsieh CS, Linehan DC (2009) Disruption of CCR5-dependent homing of regulatory T cells inhibits tumor growth in a murine model of pancreatic cancer. J Immunol 182(3):1746–1755

Tang Q, Lee K (2012) Regulatory T-cell therapy for transplantation: how many cells do we need? Curr Opin Organ Transplant 17(4):349–354

Tang Q, Henriksen KJ, Boden EK, Tooley AJ, Ye J, Subudhi SK, Zheng XX, Strom TB, Bluestone JA (2003) Cutting edge: CD28 controls peripheral homeostasis of CD4+CD25+ regulatory T cells. J Immunol 171(7):3348–3352

Teperino R, Schoonjans K, Auwerx J (2010) Histone methyl transferases and demethylases; can they link metabolism and transcription? Cell Metab 12(4):321–327

Thornton AM, Shevach EM (1998) CD4+CD25+ immunoregulatory T cells suppress polyclonal T cell activation in vitro by inhibiting interleukin 2 production. J Exp Med 188(2):287–296

Topalian SL, Hodi FS, Brahmer JR, Gettinger SN, Smith DC, McDermott DF, Powderly JD, Carvajal RD, Sosman JA, Atkins MB, Leming PD, Spigel DR, Antonia SJ, Horn L, Drake CG, Pardoll DM, Chen L, Sharfman WH, Anders RA, Taube JM, McMiller TL, Xu H, Korman AJ, Jure-Kunkel M, Agrawal S, McDonald D, Kollia GD, Gupta A, Wigginton JM, Sznol M (2012) Safety, activity, and immune correlates of anti-PD-1 antibody in cancer. N Engl J Med 366(26):2443–2454

Tran DQ, Ramsey H, Shevach EM (2007) Induction of FOXP3 expression in naive human CD4 +FOXP3 T cells by T-cell receptor stimulation is transforming growth factor-beta dependent but does not confer a regulatory phenotype. Blood 110(8):2983–2990

Vahl JC, Drees C, Heger K, Heink S, Fischer JC, Nedjic J, Ohkura N, Morikawa H, Poeck H, Schallenberg S, Rieß D, Hein MY, Buch T, Polic B, Schönle A, Zeiser R, Schmitt-Gräff A, Kretschmer K, Klein L, Korn T, Sakaguchi S, Schmidt-Supprian M (2014) Continuous T cell receptor signals maintain a functional regulatory T cell pool. Immunity 41(5):722–736

Venken K, Hellings N, Broekmans T, Hensen K, Rummens JL, Stinissen P (2008) Natural naive CD4+CD25+CD127low regulatory T cell (Treg) development and function are disturbed in multiple sclerosis patients: recovery of memory Treg homeostasis during disease progression. J Immunol 180(9):6411–6420

von Spee-Mayer C, Siegert E, Abdirama D, Rose A, Klaus A, Alexander T, Enghard P, Sawitzki B, Hiepe F, Radbruch A, Burmester GR, Riemekasten G, Humrich JY (2016) Low-dose interleukin-2 selectively corrects regulatory T cell defects in patients with systemic lupus erythematosus. Ann Rheum Dis 75(7):1407–1415

Walker LS, Sansom DM (2011) The emerging role of CTLA4 as a cell-extrinsic regulator of T cell responses. Nat Rev Immunol 11(12):852–863

Walker LS, Chodos A, Eggena M, Dooms H, Abbas AK (2003) Antigen-dependent proliferation of CD4+CD25+ regulatory T cells in vivo. J Exp Med 198(2):249–258

Walker MR, Carson BD, Nepom GT, Ziegler SF, Buckner JH (2005) De novo generation of antigen-specific CD4+CD25+ regulatory T cells from human CD4+CD25- cells. Proc Natl Acad Sci USA 102(11):4103–4108

Walter S, Weinschenk T, Stenzl A, Zdrojowy R, Pluzanska A, Szczylik C, Staehler M, Brugger W, Dietrich PY, Mendrzyk R, Hilf N, Schoor O, Fritsche J, Mahr A, Maurer D, Vass V, Trautwein C, Lewandrowski P, Flohr C, Pohla H, Stanczak JJ, Bronte V, Mandruzzato S, Biedermann T, Pawelec G, Derhovanessian E, Yamagishi H, Miki T, Hongo F, Takaha N, Hirakawa K, Tanaka H, Stevanovic S, Frisch J, Mayer-Mokler A, Kirner A, Rammensee HG, Reinhardt C, Singh-Jasuja H (2012) Multipeptide immune response to cancer vaccine IMA901 after single-dose cyclophosphamide associates with longer patient survival. Nat Med 18 (8):1254–1261

Wang R, Green DR (2012) Metabolic checkpoints in activated T cells. Nat Immunol 13(10): 907–915

Watanabe Y, Katou F, Ohtani H, Nakayama T, Yoshie O, Hashimoto K (2010) Tumor-infiltrating lymphocytes, particularly the balance between CD8(+) T cells and CCR4(+) regulatory T cells, affect the survival of patients with oral squamous cell carcinoma. Oral Surg Oral Med Oral Pathol Oral Radiol Endod 109(5):744–752

Wei S, Kryczek I, Edwards RP, Zou L, Szeliga W, Banerjee M, Cost M, Cheng P, Chang A, Redman B, Herberman RB, Zou W (2007) Interleukin-2 administration alters the CD4+FOXP3 + T-cell pool and tumor trafficking in patients with ovarian carcinoma. Cancer Res 67 (15):7487–7494

Wei G, Wei L, Zhu J, Zang C, Hu-Li J, Yao Z, Cui K, Kanno Y, Roh TY, Watford WT, Schones DE, Peng W, Sun HW, Paul WE, O'Shea JJ, Zhao K (2009) Global mapping of H3K4me3 and H3K27me3 reveals specificity and plasticity in lineage fate determination of differentiating CD4+ T cells. Immunity 30(1):155–167

Weissler KA, Caton AJ (2014) The role of T-cell receptor recognition of peptide:MHC complexes in the formation and activity of Foxp3^{+} regulatory T cells. Immunol Rev 259(1):11–22

Williams JB, Horton BL, Zheng Y, Duan Y, Powell JD, Gajewski TF (2017) The EGR2 targets LAG-3 and 4-1BB describe and regulate dysfunctional antigen-specific CD8+ T cells in the tumor microenvironment. J Exp Med 214(2):381–400

Wing K, Sakaguchi S (2010) Regulatory T cells exert checks and balances on self tolerance and autoimmunity. Nat Immunol 11(1):7–13

Wing K, Onishi Y, Prieto-Martin P, Yamaguchi T, Miyara M, Fehervari Z, Nomura T, Sakaguchi S (2008) CTLA-4 control over Foxp3+ regulatory T cell function. Science 322 (5899):271–275

Yoon J, Schmidt A, Zhang AH, Königs C, Kim YC, Scott DW (2017) FVIII-specific human chimeric antigen receptor T-regulatory cells suppress T- and B-cell responses to FVIII. Blood 129(2):238–245

Yue X, Trifari S, Äijö T, Tsagaratou A, Pastor WA, Zepeda-Martínez JA, Lio CW, Li X, Huang Y, Vijayanand P, Lähdesmäki H, Rao A (2016) Control of Foxp3 stability through modulation of TET activity. J Exp Med 213(3):377–397

Zeng H, Yang K, Cloer C, Neale G, Vogel P, Chi H (2013) mTORC1 couples immune signals and metabolic programming to establish T(reg)-cell function. Nature 499(7459):485–490

Zhang R, Huynh A, Whitcher G, Chang J, Maltzman JS, Turka LA (2013) An obligate cell-intrinsic function for CD28 in Tregs. J Clin Invest 123(2):580–593

Zhou X, Bailey-Bucktrout SL, Jeker LT, Penaranda C, Martínez-Llordella M, Ashby M, Nakayama M, Rosenthal W, Bluestone JA (2009) Instability of the transcription factor Foxp3 leads to the generation of pathogenic memory T cells in vivo. Nat Immunol 10(9):1000–1007

Overview of LAG-3-Expressing, IL-10-Producing Regulatory T Cells

Keishi Fujio, Kazuhiko Yamamoto and Tomohisa Okamura

Abstract Regulatory T cells (Treg cells) play crucial roles in the induction of peripheral tolerance to self- and foreign-antigens. IL-10-producing regulatory T cells (IL-10-producing Treg cells) constitute a Treg cell subset characterized by the production of high amounts of IL-10, cytokine-mediated immunosuppressive capabilities, and independence of Foxp3 expression for their suppressive activity. In the past decade, identifying naturally occurring IL-10-producing Treg cells was difficult due to the lack of suitable surface markers. More recently, lymphocyte activation gene 3 (LAG-3) is a CD4 homologue that has been identified as a marker for IL-10-producing Treg cells. $CD4^{+}CD25^{-}LAG3^{+}$ T cells produce large amounts of IL-10 and suppress colitis in a mouse model. These $CD4^{+}CD25^{-}LAG3^{+}$ Treg cells also exhibit suppressive activity in murine models of lupus and humoral immunity in a TGF-β3-dependent manner. Moreover, the combined expression of LAG-3 and CD49b identifies IL-10-producing Treg cells in mice and humans more specifically. Recently, LAG-3 has gained more attention in the context of immune checkpoints because it believed to be related to T cell tolerance and exhausted T cells that infiltrate the tumor microenvironment. Tumors and the tumor microenvironment promote development of IL-10-producing Treg cells and foster tumor growth. This response might interfere with protective immune responses. Understanding LAG-3-expressing IL-10-producing Treg cells may contribute to the development of novel therapeutic strategies in immune-mediated diseases.

Contents

1 Introduction ... 30
2 Features of IL-10-Producing Treg Cells ... 31
3 Molecular Properties of LAG-3 ... 31
4 Identification of LAG-3 as a Marker for IL-10-Producing Treg Cells ... 32

K. Fujio (✉) · K. Yamamoto · T. Okamura
Department of Allergy and Rheumatology, Graduate School of Medicine, The University of Tokyo, 7-3-1 Hongo, Bunkyo-ku, Tokyo, Japan
e-mail: kfujio-tky@umin.ac.jp

Current Topics in Microbiology and Immunology (2017) 410:29–46
DOI 10.1007/82_2017_59

Published Online: 20 September 2017

5 $CD4^+CD25^-LAG3^+$ Treg Cells Control Humoral Immunity Via TGF-β3 33
6 Egr2 and Egr3 Cooperatively Control Systemic Autoimmunity Through TGF-β3 Production by $CD4^+CD25^-LAG3^+$ Treg Cells 34
7 Identification of $CD4^+LAG3^+CD49b^+$ Cells 36
8 Roles of $CD4^+LAG3^+$ Cells or $CD4^+LAG3^+CD49b^+$ Cells 37
9 LAG-3- and/or PD-1-Expressing $CD4^+$ T Cells in Malignancy 39
10 LAG-3 and/or PD-1 Expressing $CD4^+$ T Cells in Infection 40
11 Conclusions 41
References 41

1 Introduction

The concept that T cells are involved in immunological tolerance through active suppression was proposed by Gershon et al. in 1970 and 1971 (Gershon and Kondo 1970, 1971). They showed that thymus-derived lymphocytes were not only required for tolerance induction but that they could adoptively transfer tolerance to naïve recipients. This form of tolerance, termed "infectious tolerance", was antigen-specific and the T cells that inhibited immune responses were called "suppressor T cells" (Gershon et al. 1974). Although enormous effort to identify suppressor T cells was unsuccessful, Sakaguchi et al. provided robust evidence that a distinct subset of naturally occurring $CD4^+CD25^+$ T cells from naïve mice has the capacity to prevent autoimmune disease mediated by endogenous, self-reactive T cells.

$CD4^+CD25^+$ regulatory T cells ($CD25^+$ Treg cells) play crucial roles in preventing autoimmune diseases and maintaining immune homeostasis. $CD25^+$ Treg cells constitute a differentiated cell lineage that develops in the thymus. Their phenotype and function are dependent upon the expression of the transcription factor Forkhead box p3 (Foxp3) (Hori et al. 2003); (Sakaguchi and Powrie 2007). In humans, individuals lacking Foxp3 have an autoimmune disorder, immunodysregulation polyendocrinopathy and enteropathy, X-linked (IPEX) syndrome (Bennett et al. 2001). Scurfy mice, which harbor a frame shift mutation in the *Foxp3* gene, demonstrate extensive lymphoproliferation and severe inflammatory infiltration in some organs, including the lung, skin, and liver (Brunkow et al. 2001). Although these observations indicate that $CD25^+$ Treg cells are an essential population for self-tolerance, one emerging question is whether the $CD25^+$ Treg cell system is the only machinery that is responsible for tolerance to organs. The autoimmune regulator (Aire) gene, which affects the central induction of tolerance by regulating the clonal deletion of self-reactive thymocytes, is responsible for autoimmune polyendocrinopathy-candidiasis-ectodermal dystrophy (APECED) (Liston et al. 2003). Aire regulates the ectopic expression of a battery of peripheral tissue antigens, e.g., insulin, fat acid-binding protein, and salivary protein-1 (Anderson et al. 2005). Surprisingly, an additional defect in central tolerance induction in scurfy mice, which are generated by crossing mice with a null mutation in the Aire gene, did not significantly extend the range of the affected sites, and many organs remained unaffected (Chen et al. 2005). This observation indicates that some additional mechanisms other than central tolerance and the Foxp3 system are required to enforce immunological self-tolerance in peripheral organs. Although Foxp3-independent T cells with regulatory activity may be related to the maintenance of self-tolerance, it is difficult to assess the physiological function of these T cell populations because of the lack of specific markers that can reliably delineate them from the other T cell subsets.

2 Features of IL-10-Producing Treg Cells

In 1998, Groux et al. demonstrated the generation of T cell clones producing IL-10 and transforming growth factor (TGF)-β with ex vivo activation of human or murine $CD4^+$ T cells in the presence of high doses of exogenous IL-10 (Groux et al. 1997; Fujio et al. 2010). These clones suppressed antigen-specific activation of T cells and colitis development via IL-10, and were termed Type-1 T regulatory (Tr1) cells. Upon activation via the T cell receptor (TCR), Tr1 cells characteristically produce large amounts of IL-10, TGF-β, and IL-5, but not interferon (IFN)-γ, IL-2, or IL-4. Tr1 clones were also induced from mice transgenic for a TCR specific for a peptide derived from ovalbumin (OVA). Tr1 cells regulate immune responses through the secretion of the immunosuppressive cytokines IL-10 and TGF-β. Tr1 cell-mediated suppression of $CD4^+$ T cells can be blocked by addition of neutralizing anti-IL-10 or anti-TGF-β antibodies, indicating the importance of cytokines for their suppressive activities (Bacchetta et al. 1994; Groux et al. 1997; Roncarolo et al. 2006). Tr1 cells' dependency on suppressive cytokines contrasts with $CD25^+$ Treg cells' dependency on cell contact (Onishi et al. 2008). It remains unclear whether TGF-β production should be used to classify Tr1 cells because the requirement for TGF-β for the suppressive activity of Tr1 cells depends on the nature of the experimental system. Thus, Battaglia et al. proposed that the suppressive effects mediated by IL-10-secreting Treg cells could be attributed to Tr1 cells regardless of the production of TGF-β, IL-5, and IFN-γ (Battaglia et al. 2006). Thus, IL-10-secreting Treg cells could be a hopeful target for designing antigen-specific therapies to treat a wide array of autoimmune diseases. Although a constitutively expressed surface marker for IL-10-secreting Treg cells would allow us to monitor the emergence and functionality of these cells, surface biomarkers for IL-10-secreting Treg cells remain poorly defined.

3 Molecular Properties of LAG-3

LAG-3 (CD223) is a type I membrane protein belonging to the immunoglobulin (Ig) superfamily that is mainly expressed on activated human natural killer (NK) cells and T cells (Triebel et al. 1990). Subsequently, LAG-3 expression was reported on various immune cell types, including plasmacytoid dendritic cells (pDCs), B cells, NKT cells, γδT cells, tumor-infiltrating lymphocytes, exhausted $CD8^+$ T cells, and Treg cells (Okamura et al. 2012). The position of the gene coding the LAG-3 protein is close to the gene coding for CD4 on human chromosome 12. LAG-3 and CD4 proteins shares approximately 20% homology. Both human and mouse LAG-3 bind to major histocompatibility complex (MHC) class II molecules with higher affinities than their CD4 counterparts (Huard et al. 1996, 1995). Although CD4 molecules are mainly expressed at the cell surface, almost half of LAG-3 molecules are retained intracellularly (Woo et al. 2010). It has been suggested that LAG-3 is continuously recycled and/or rapidly translocated to the plasma membrane in response to cellular stimulation. Moreover, the cell surface expression of LAG-3 is regulated by the cleavage of the extracellular domains by two transmembrane metalloproteases, ADAM10 and ADAM17 (Li et al. 2007). ADAM10 siRNA suppressed T cell proliferation in a LAG-3-dependent manner. Thus, the cell surface expression of LAG-3 is strictly regulated by several mechanisms. LAG-3 regulates both expansion of activated primary T cells and the development of the memory T cell pool (Workman et al. 2004). The crosslinking of the T cell receptor (TCR) and LAG-3 results in less calcium release than TCR stimulation alone (Hannier et al. 1998). CD4 T^+ cells transduced with a non-cleavable LAG-3 mutant vector exhibited a more potent inhibitory effect on their activation than the wild-type LAG-3 vector (Awasthi et al. 2007), confirming an inhibitory role of LAG-3 in $CD4^+$ T cells. The fact that LAG-3-deficient mice displayed an apparently normal phenotype (Miyazaki et al. 1996) highlighted the possibility that the effects of LAG-3 are rather subtle and that the molecule is likely involved in the fine tuning of the immune response.

4 Identification of LAG-3 as a Marker for IL-10-Producing Treg Cells

LAG-3 was reported to be required for the maximal regulatory function of $CD4^+CD25^+Foxp3^+$ Treg cells (Huang et al. 2004). Indeed, ectopic LAG-3 expression confers regulatory activity on naive T cells. LAG-3 on $CD4^+CD25^+$ Treg cells interacts with MHC class II molecules on dendritic cells (DCs), and the binding of LAG-3 to the MHC class II molecules on immature DCs induces Immunoreceptor tyrosine-based activation motif (ITAM)-mediated inhibitory signaling that suppresses DC functions (Liang et al. 2008). However, it was reported that LAG-3 protein is hardly detected on the cell surface of $CD4^+CD25^+Foxp3^+$ Treg cells (Huang et al. 2004). We have found that LAG-3 is specifically expressed on a population of $CD4^+CD25^-CD45RB^{low}$ memory T cells (Okamura et al. 2009), which includes IL-10-secreting $CD4^+Foxp3^-$ T cells (Roncarolo and Battaglia 2007). Approximately, 2% of the $CD4^+CD25^-$ T cell population in the spleen consists of $CD4^+CD25^-LAG3^+$ T cells that express higher levels of IL-10 and Blimp-1 encoded by *Prdm1* mRNA. Blimp-1 is important for IL-10 production by

$CD4^+$ T cells and $CD25^+$ Treg cells (Cretney et al. 2011; Martins et al. 2006). $CD4^+CD25^-LAG3^+$ T cells produce large amounts of IL-10 upon in vitro antigenic stimulation. Moreover, $CD4^+CD25^-LAG3^+$ T cells show an anergic phenotype. These cells suppressed in vivo development of colitis induced in RAG-$1^{-/-}$ recipients by the transfer of naive T cells in an IL-10-dependent manner. Foxp3 is not essential for the function of $CD4^+CD25^-LAG3^+$ T cells because this population does not express Foxp3 protein, and Foxp3-mutated scurfy mice generate functional $CD4^+CD25^-LAG3^+$ T cells. Unlike $CD25^+$ Treg cells, high-affinity interactions with peptide/MHC ligands expressed in the thymus did not induce the development of $CD4^+CD25^-LAG3^+$ T cells. These observations indicate that LAG-3 is a phenotypic marker of IL-10-producing Treg cells. Intriguingly, exposure to viable microbiota affects the development of $CD4^+CD25^-LAG3^+$ T cells, because germ-free mice contained fewer $CD4^+CD25^-LAG3^+$ T cells in spleens and Peyer's patches than specific-pathogen-free mice.

5 $CD4^+CD25^-LAG3^+$ Treg Cells Control Humoral Immunity Via TGF-β3

One notable feature of $CD4^+CD25^-LAG3^+$ Treg cells ($LAG3^+$ Treg cells) is the expression of early growth response gene-2 (Egr2), an anergy-associated transcription factor that confers the phenotype of $LAG3^+$ Treg cells on $CD4^+$ T cells (Okamura et al. 2009). Consistent with previous observations that IL-27 induces a Tr1 phenotype on $CD4^+$ T cells (Awasthi et al. 2007), IL-27 induces LAG-3, IL-10, and Blimp-1 expression by $CD4^+$ T cells in an Egr2-dependent manner (Iwasaki et al. 2013; Heinemann et al. 2014). Moreover, Egr2-mediated control of systemic autoimmunity has been suggested because mice deficient for Egr2 in T cells and B cells develop a systemic autoimmune disease (Zhu et al. 2008). In terms of human genetics, we identified *EGR2* as a genetic risk factor for systemic lupus erythematosus (SLE) and rheumatoid arthritis (RA) in a case control association study (Myouzen et al. 2010). Those results suggested that Egr2-expressing $LAG3^+$ Treg cells are closely associated with the regulation of autoreactive B cells. When we adoptively transferred $LAG3^+$ Treg cells to Fas-mutated MRL/*lpr* lupus-prone mice, these cells strongly suppressed renal disease progression and anti-dsDNA antibody production (Okamura et al. 2015). Transfer of $LAG3^+$ Treg cells efficiently inhibited antibody production and the development of follicular helper T (TFH) cells and germinal center B cells (GCB) in NP-OVA-immunized mice. IL-10 production from $LAG3^+$ Treg cells is not required for $LAG3^+$ Treg cells-mediated suppression of antibody production in vivo.

$LAG3^+$ Treg cells expressed high levels of *Tgfb3* mRNA but not *Tgfb1* or *Tgfb2* mRNA in microarray analysis. In addition, TCR stimulation of $LAG3^+$ Treg cells induced high levels of TGF-β3, but not TGF-β1 or TGF-β2, in the culture supernatants. In contrast, $CD25^+$ Treg cells produce only limited amounts of TGF-β1.

TGF-β3 is responsible for the suppressive activity of LAG3$^+$ Treg cells for humoral immunity under inflammatory and noninflammatory conditions (Okamura et al. 2015).

TGF-β3 was identified in 1988 (Derynck et al. 1988). Its prominent roles in the development of organs such as the heart, lung, and breast have been demonstrated. While TGF-β1-deficient mice showed severe autoimmune inflammation, mice lacking TGF-β3 exhibit cleft palate and die soon after birth (Proetzel et al. 1995). Blockade of TGF-β3, but not TGF-β1 or TGF-β2, abrogates palate fusion. These differences in vivo may be partly explained by the differences in the temporal-spatial expression and activation of latent forms. Otherwise, the difference in the biological activities may play a role. The biological differences between isoforms have been observed in knock-in mice generated by replacing the coding region of the *Tgfb3* gene with the *Tgfb1* cDNA, which resulted in the expression of *Tgfb1* in the *Tgfb3*-expressing site. The knock-in mice display only partial correction of the epithelial fusion defect of *Tgfb3* knockout embryos (Yang and Kaartinen 2007).

Previously, it was thought that the role of TGF-β3 pertained to organ development. In 2008, Shah et al. reported that resting B cells induced the expansion of CD25$^+$ Treg cells by the expression of TGF-β3 (Shah and Qiao 2008). Furthermore, Lee et al. demonstrated that Th17 cells expressed TGF-β3 and that TGF-β3-induced Th17 cells were functionally distinct from TGF-β1-induced Th17 cells in their pathological activity (Lee et al. 2012). TGF-β3-induced Th17 cells possess a gene expression profile similar to that presented by pathogenic effector Th17 cells in autoimmune disease. Moreover, Th17 cells induced by the combination of TGF-β3 and IL-6 exhibited high levels of phosphorylated Smad1 and Smad5 and low levels of phosphorylated Smad2 and Smad3 compared with those induced by TGF-β1 and IL-6. TRIM28 is a component of heterochromatin complexes that regulate IL-2 production. Chikuma et al. showed that TRIM28-deficient mice developed an inflammatory autoimmune disease with enhanced Th17 cell differentiation due to derepression of TGF-β3 (Chikuma et al. 2012). In contrast to these studies indicating a pro-inflammatory role for TGF-β3, we revealed an immunoregulatory role of TGF-β3, especially for humoral immunity (Okamura et al. 2015). Although CD25$^+$ Treg cells are known to express TGF-β1, they produce only limited amounts of the cytokine. Therefore, TGF-β3 could be the major source of TGF-β activity derived from murine CD4$^+$ T cells.

6 Egr2 and Egr3 Cooperatively Control Systemic Autoimmunity Through TGF-β3 Production by CD4$^+$CD25$^-$LAG3$^+$ Treg Cells

Egr2 is a zinc finger transcription factor that was initially identified as a major regulator of myelination and hindbrain development (Schneider-Maunoury et al. 1993; Topilko et al. 1994). Egr2-deficient mice show perinatal or neonatal death due to respiratory or feeding deficits. In immune responses, it has been reported that Egr2 is essential for induction of T cell anergy (Harris et al. 2004; Safford et al. 2005). Egr2 inhibits T cell activation by promoting expression of the E3 ubiquitin ligase Cbl-b, which is related to T cell anergy induction. Among various IL-10-inducing factors, only IL-27 induces Egr2 and LAG-3 expression on $CD4^+$ T cells (Iwasaki et al. 2013). IL-27 suppresses immune responses through inhibition of the development of IL-17-producing Th17 cells and induction of IL-10 producing Tr1-like cells (Awasthi et al. 2007). IL-27 failed to induce IL-10 in Egr-2-deficient T cells, and IL-27-mediated induction of *Prdm1* was also impaired in the absence of Egr2 (Iwasaki et al. 2013). Although IL-27-mediated IL-10 induction was dependent on both STAT1 and STAT3, only STAT3 was required for IL-27-mediated Egr-2 induction. Thus, Egr2 induction could occur in the context of Tr1 cell induction.

The linkage between Egr2 and autoimmunity was revealed by the observation that CD2-Cre-driven lymphocyte-specific Egr2-deficient mice demonstrated systemic autoimmunity (Zhu et al. 2008). However, the mice showed only a mild form of autoimmunity and limited anti-dsDNA antibody production. Therefore, it was possible that an additional regulator contributed to the control of autoimmunity. In the Egr family (Egr1–4), Egr3 is similar to Egr2 in that it induces T cell anergy (Safford et al. 2005), although a systemic deletion of Egr3 induced only gait ataxia in mice due to the lack of muscle spindles (Tourtellotte and Milbrandt 1998). Using a CD2-Cre-driver, Li et al. revealed that Egr2 and Egr3 deletion in both T cells and B cells resulted in a more severe early-onset systemic autoimmune syndrome, compared with deletion of Egr2 alone (Li et al. 2012).These results indicated a compensatory role of Egr3 for Egr2-mediated control of systemic autoimmunity. However, it appears that not only Egr2 and Egr3 expressed in T cells but also those expressed in B cells may modulate systemic autoimmunity, because Egr2 expressed in B cells regulates the development of B cells (Li et al. 2011). In addition, Egr3 is preferentially expressed in follicular and marginal zone B cells, among various B cell populations (Shi et al. 2015). To address these points, we generated T-cell-specific Egr2/Egr3 double-deficient ($Egr2^{fl/fl}CD4Cre^+$: Egr2/3DKO) mice. Egr2/3DKO mice spontaneously developed an early-onset lupus-like disease that was more severe than T-cell-specific Egr2-deficient mice. Egr2/3DKO mice exhibited excessive formation of germinal centers and autoantibody production (Morita et al. 2016). Although $CD25^+$ Treg cells from Egr2/3DKO mice demonstrated no evident functional impairment, $CD4^+CD25^-LAG3^+$ cells from Egr2/3DKO mice completely lost the capacity to suppress B cell functions and failed to produce TGF-β3. The excessive germinal center reaction in Egr2/3DKO mice was suppressed by the adoptive transfer of wild-type $LAG3^+$ Treg cells or treatment with a TGF-β3-expressing vector. The unique attributes of Egr2/Egr3 in T cells may provide an opportunity for developing novel therapeutics for autoantibody-mediated diseases including SLE.

The latent TGF-β binding proteins (LTBP) constitute a family of carrier proteins that control bioavailability of TGF-β. Intriguingly, LTBP3 expression is maintained by Egr2 and Egr3, and they were required for TGF-β3 secretion from

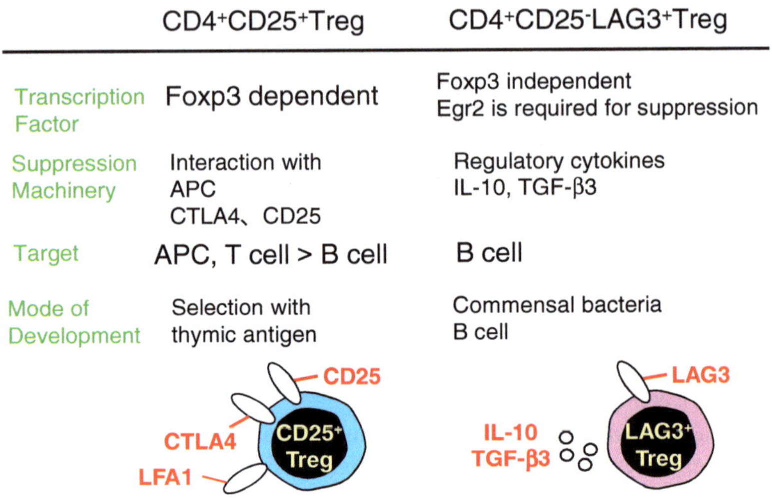

Fig. 1 Comparison between $CD4^+CD25^+$ Treg cells and $CD4^+CD25^-LAG3^+$ Treg cells

$CD4^+CD25^-LAG3^+$ Treg cells. We observed that Egr2 and Egr3 did not demonstrate cell intrinsic suppression of the development of follicular helper T cells. Thus, Egr2- and Egr3-dependent TGF-β3 production by $LAG3^+$ Treg cells is crucial for controlling excessive B cell responses. It can be concluded that Egr2/Egr3 and TGF-β3 are important regulators of systemic autoimmunity.

Collectively, while $CD4^+CD25^+$ Treg cells suppress T cells and antigen presenting cells via CTLA4 and CD25 (Onishi 2008 190), $CD4^+CD25^-LAG3^+$ Treg cells control immune cells including B cells via suppressive cytokines IL-10 and TGF-β3. Moreover, in contrast to $CD4^+CD25^+$ Treg cells that develop through a high-affinity agonistic interaction with self-peptide/MHCs expressed by thymic stromal cells, the number of $CD4^+CD25^-LAG3^+$ Treg cells is influenced by the presence of environmental microbiota (Okamura 2009 29}. Therefore, $LAG3^+$ Treg cells may also be a regulatory mechanism for autoimmunity in addition to $CD25^+$ Treg cells (Fig. 1).

7 Identification of $CD4^+LAG3^+CD49b^+$ Cells

In 2013, Gagliani et al. reported that the surface markers LAG-3 and CD49b were stably and selectively co-expressed on mouse and human Tr1 cells (Gagliani et al. 2013). They identified specific expression of LAG-3 and CD49b on Tr1 cell clones using microarray analysis, and they demonstrated the specificity of these markers in mouse models of inflammation and in the peripheral blood of healthy volunteers. Concomitant expression of LAG-3 and CD49b is specific for Tr1 cells, as Th1, Th2, Th17 and $CD25^+$ Treg cells do not co-express these markers. Co-expression of

LAG-3 and CD49b can be used to purify human Tr1 cells from in vitro Tr1-polarized cell cultures and enables the tracking of Tr1 cells in tolerant subjects after allogeneic hematopoietic stem cell transplantation.

Koch et al. analyzed the frequency of $CD49b^{+}LAG\text{-}3^{+}$ Tr1 cells in the peripheral blood of HIV-infected individuals at different stages of the disease. They found increased levels of $LAG\text{-}3^{+}CD49b^{+}$ Tr1 cells as well as IL-10 in HIV patients (Koch et al. 2015). With disease progression, Tr1 cells negatively correlated with the frequency of pDCs, the main producers of IFN-α. In vitro, IFN-α induces upregulation of IL-10 as well as increases in $LAG\text{-}3^{+}CD49b^{+}$ Tr1 cell counts in healthy controls, recapitulating effects observed in vivo during HIV infection. These results suggest that overexpression of IFN-α during HIV infection drives the generation of $CD49b^{+}LAG\text{-}3^{+}$ Tr1 cells and the immunosuppressive cytokine IL-10. Viral infection may be associated with increased frequency of IL-10-producing Treg cells, similar to the induction of $CD25^{+}$ Treg cells during viral infection (Veiga-Parga et al. 2013).

8 Roles of $CD4^{+}LAG3^{+}$ Cells or $CD4^{+}LAG3^{+}$ $CD49b^{+}$ Cells

Chen et al. reported that B cell-induced subset of Treg (Treg-of-B) cells which expressed LAG-3 but not Foxp3, and they secreted IL-4, IL-10 and TGF-β (Chen et al. 2016). These $LAG3^{+}Foxp3^{-}$ Treg-of-B cells suppressed the proliferation of $CD4^{+}CD25^{-}$ responder T cells through LAG-3 and IL-10 production. Adoptive transfer of $LAG3^{+}$ Treg-of-B cells ameliorated the inflammation and clinical severity of collagen-induced arthritis. Adoptive transfer of Treg-of-B cells protected mice from $CD4^{+}CD45RB^{hi}$ T-cell-induced colitis, including infiltration of leukocytes, depletion of goblet cells, epithelial hyperplasia, and inhibition of Th1 and Th17 cytokines (Shao et al. 2016). Treg-of-B cells protected against experimental colitis through an IL-10-independent mechanism, because IL-10-deficient Treg-of-B cells maintained their suppressive function in vitro as well as in vivo. These $LAG3^{+}$ Treg-of-B cells are similar to $CD4^{+}CD25^{-}LAG3^{+}$ Treg cells because their development is dependent on B cells (Okamura et al. unpublished observation).

Yao et al. reported that LAG-3- and CD49b-expressing Tr1 cells, but not $CD25^{+}$ Treg cells, suppressed the transcription of pro-IL-1β mRNA, inflammasome-mediated activation of caspase-1 and secretion of mature IL-1β. Inhibition of inflammasome activation and IL-1β secretion was abrogated in IL-10R–deficient macrophages, indicating the importance of IL-10 signaling. Moreover, IL-1β production from macrophages derived from $Nlrp3^{A350V}$ knock-in mice, which carry a mutation found in cryopyrin-associated periodic syndrome patients, was suppressed by Tr1 cells but not $CD25^{+}$ Treg cells. Adoptive transfer experiments revealed a direct correlation between Tr1 cell engraftment and protection from weight loss in mice that expressed a gain-of-function NLRP3. This report showed a differential role of Tr1 cells and $CD25^{+}$ Treg cells in regulating innate immune responses (Yao et al. 2015).

Clemente-Casares et al. found that systemic delivery of nanoparticles coated with autoimmune-disease-relevant peptides that bound to MHC class II (pMHC) molecules triggered the generation and expansion of antigen-specific Tr1-like cells in different mouse models, including mice humanized with lymphocytes from patients, leading to resolution of established autoimmune phenomena (Clemente-Casares et al. 2016). Nonobese diabetic (NOD) and NOD Foxp3-eGFP mice (expressing enhanced green fluorescent protein (eGFP) under the control of the mouse *Foxp3* promoter) were treated with uncoated nanoparticles or nanoparticles coated with a pMHC, 2.5mi/IA^{g7}, recognized by the diabetogenic BDC2.5-specific T cell receptor (TCR), or with 2.5mi/IA^{g7} monomers. Notably, expansion was observed in memory-like (CD44^{hi} CD62L^{low}) Foxp3$^-$ Tr1-like cells that expressed inducible T cell co-stimulator (ICOS), latent-associated TGF-β and the Tr1 markers LAG-3 and CD49b. In vivo, splenic CD4$^+$ T cells from donors treated with pMHC-NPs suppressed diabetes development in T-cell-reconstituted NOD *scid* hosts. These nanomedicines promote the differentiation of disease-primed autoreactive T cells into Tr1-like cells, which in turn suppress autoantigen-loaded antigen-presenting cells and drive the differentiation of cognate B cells into disease-suppressing regulatory B cells, without compromising systemic immunity. This study suggests the role of antigen recognition for the induction of IL-10-producing Treg cells.

The LAG-3-expressing IL-10 producing T cells also control allergic responses. Tousa et al. reported that the cytokine activin-A instructs the generation of CD4$^+$ T cells that express IL-10, ICOS, LAG-3, and CD49b, and exert strongly suppressive functions toward allergic responses induced by naive and in vivo primed human Th2 cells. Activin-A signaling induces the activation of the transcription factor interferon regulatory factor (IRF4), which, along with the environmental sensor aryl hydrocarbon receptor (Ahr), forms a multipartite transcriptional complex that binds in IL-10 and ICOS promoter elements and controls gene expression in human CD4$^+$ T cells. Humanized mouse model of allergic asthma indicated that adoptive transfer of human activin-A-induced Tr1 cells confers significant protection against cardinal asthma manifestations, including pulmonary inflammation (Tousa et al. 2017).

Because most reports related to human IL-10-producing Treg cells examined peripheral blood lymphocytes, little is known about the nature of IL-10-producing Treg cells in human lymphoid organs. We identified CD4$^+$CD25$^-$LAG3$^+$ T cells in human tonsils (Sumitomo et al. 2017). This population secreted large amounts of IL-10 and expressed low levels of FOXP3. Surface markers and microarray analysis indicated the uniqueness of this tonsillar CD4$^+$ T cell subset. CD4$^+$CD25$^-$LAG3$^+$ T cells expressed IL-10, PRDM1, and CD274 at high levels and chemokine receptor 5 (CXCR5) at low levels. CD4$^+$CD25$^-$LAG3$^+$ T cells suppressed antibody production more efficiently than did CD25$^+$ Treg cells. In addition, this population inhibited the progression of a graft-versus-host disease (GVHD) in humanized mice. The existence of naturally occurring IL-10-producing Treg cells in human secondary lymphoid tissue suggested an essential role of IL-10-producing Treg cells in immune homeostasis.

9 LAG-3- and/or PD-1-Expressing $CD4^+$ T Cells in Malignancy

Programmed death-1 (PD-1) is an inhibitory molecule that cooperatively suppresses broad immune responses with the LAG-3 system. Okazaki et al. reported that PD-1 and LAG-3 inhibitory co-receptors act synergistically to prevent autoimmunity in mice (Okazaki et al. 2011). They analyzed activation-induced cytidine deaminase-linked autoimmunity (aida) mice that harbor a loss-of-function mutation in the *Lag3* gene. Although *Lag3*-deficiency alone did not induce autoimmunity in non-autoimmune-prone mouse strains, it induced lethal myocarditis in BALB/c mice deficient for the *Pdcdl* gene encoding PD-1. In addition, *Lag3*-deficiency alone accelerated type 1 diabetes mellitus in nonobese diabetic mice. These results demonstrate that *Lag3* acts synergistically with *Pdcdl* and/or other immunoregulatory genes to prevent autoimmunity in mice. Woo et al. found that LAG-3 and PD-1 proteins synergistically regulated T cell function to promote the escape from tumor immunity (Woo et al. 2012). PD-1 and LAG-3 showed extensive co-expression on tumor-infiltrating $CD4^+$ and $CD8^+$ T cells in three distinct transplantable tumors. Dual anti-LAG-3/anti-PD-1 antibody treatment cured most established tumors in mice that were resistant to single antibody treatment. Moreover, $Lag3^{-/-}$ $Pdcd1^{-/-}$ mice showed markedly increased survival and clearance of multiple transplantable tumors. These results define a strong synergy between the PD-1 and LAG-3 inhibitory pathways in immune tolerance.

Recently, the understanding of immune checkpoints has evoked a paradigm shift in cancer treatment (Dempke et al. 2017). These checkpoint mechanisms constitute immunosuppressive mechanisms in tumors. Immune checkpoints exert activity through T-cell-inhibiting and stimulating receptors and their ligands, including cytotoxic T lymphocyte-associated protein 4 (CTLA-4) and PD-1. In addition, other molecules on the surface of T cells could exert inhibitory functions, such as LAG-3 and TIM-3. These 'new' immune inhibitory molecules are gaining more attention since they appear to be related to T cell tolerance and exhausted T cells that infiltrate the tumor microenvironment. Therefore, understanding PD-1- and LAG-3-expressing $CD4^+$ T cells may contribute to the development of novel therapeutic strategies.

Goding et al. reported that melanoma-specific $CD4^+$ T cells in a recurrent tumor model showed traits of chronic exhaustion, as evidenced by high expression of PD-1, LAG-3 and TIM-3 (Goding et al. 2013). A blockade of the PD-1/PD-L1 pathway with either anti-PD-L1 antibodies or by depletion of tumor-specific $CD25^+$ Treg cells failed to reverse tumor recurrence. However, a combination of PD-L1 blockade with tumor-specific Treg depletion effectively mediated disease regression. Notably, blockade with a combination of anti-PD-L1 and anti-LAG-3 antibodies overcame the requirement to deplete tumor-specific $CD25^+$ Treg cells. These data imply a novel paradigm for the use of combinatorial approaches, such as PD-L1 and LAG-3 blockade, in clinical treatment.

Importantly, the tumor microenvironment has the potential to induce $CD4^+LAG3^+$ T cells. Liu et al. found that the frequency of Tr1 cells was significantly increased in diffuse large B cell lymphoma (DLBCL) patients, even during complete remission (Liu et al. 2016). They showed that these Tr1 cells were enriched in the $CD25^{low/-}Foxp3^-CD49b^+LAG3^+$ fraction and could be developed in vitro from naive T cells. When the enriched in vitro induced Tr1 cells were co-cultured with autologous primary DLBCL cells and T cells, these Tr1 cells enhanced the survival of $CD20^+$ DLBCL tumor cells and suppressed the antitumor response of T cells through the production of IL-10. Moreover, the frequency of $CD4^+Foxp3^-IL\text{-}10^+$ Tr1 cells in DLBCL patients during complete remission was directly associated with the risk of relapse. Similar to mouse B cells that induce $CD4^+CD25^-LAG3^+$ Treg cells, DLBCL may expand Tr1 cells to counteract host protective immunity. Rather than the tumor itself, pDCs in the tumor microenvironment also promote immunosuppression by inducing Tr1 cells. Using ex vivo isolated cells from individuals with hepatocellular carcinoma or liver metastases from colorectal cancer, Pedroza-Gonzalez et al. identified a $CD4^+FoxP3^-IL\text{-}13^-IL\text{-}10^+$ T cell population in tumors of individuals with primary or secondary liver cancer that appeared to consist of Tr1 cells based on the expression of LAG-3 and CD49b and strong suppression activity of T cell responses in an IL-10-dependent manner (Pedroza-Gonzalez et al. 2015). pDCs exposed to tumor-derived factors enhance IL-10 production by Tr1 cells through upregulation of the inducible co-stimulatory ligand (ICOS-L). These reports suggest a role for the tumor microenvironment in promoting intra-tumoral immunosuppression by Tr1 cells, which may foster tumor progression and which might interfere with attempted immunotherapeutic intervention.

10 LAG-3 and/or PD-1 Expressing $CD4^+$ T Cells in Infection

With regard to responses to microbiota, we previously identified a requirement for the induction of $CD4^+CD25^-LAG3^+$ Treg cells (Okamura et al. 2009). $CD4^+$ T cells from mice infected with *Plasmodium* parasites expressed PD-1 and LAG-3 as early as 6 days after infection, whereas those from either *Listeria monocytogenes*- or *Leishmania major*-infected mice did not (Doe et al. 2016). Experiments using $Myd88^{-/-}$, $Trif^{-/-}$, and $Irf3^{-/-}$ mice showed that induction of these $CD4^+$ T cells and their ability to produce cytokines were largely independent of Toll-like receptor (TLR) signaling. These studies suggest that the expression of the inhibitory receptors PD-1 and LAG-3 on $CD4^+$ T cells and their reduced IL-2 production are characteristic features of *Plasmodium* infection. Villegas-Mendez et al. reported that $CD4^+IFN\text{-}\gamma^+IL\text{-}10^+$ T cells induced by primary malaria infection expressed high levels of PD-1, LAG-3, and TIGIT, suggestive of cellular exhaustion (Villegas-Mendez et al. 2016). Although $CD4^+IFN\text{-}\gamma^+IL\text{-}10^+$ T cells are unresponsive and fail to proliferate during secondary infection, $CD4^+IFN\text{-}\gamma^+IL\text{-}10^-$ T cells expand rapidly and upregulate IL-10 expression. These observations suggest that IL-10- producing $CD4^+$ T cells contribute to optimized parasite control and prevention of immune-mediated pathology during repeated malaria infections.

With regard to viral infection, Fromentin et al. found that $CD4^+$ T cells expressing PD-1, TIGIT, and LAG-3 contributed to HIV persistence during antiretroviral therapy (Fromentin et al. 2016). Negative binomial regression models revealed that PD-1, TIGIT, and LAG-3 are immune checkpoint molecules positively associated with the frequency of $CD4^+$ T cells harboring integrated HIV DNA. The frequency of $CD4^+$ T cells co-expressing PD-1, TIGIT, and LAG-3 independently predicted the frequency of cells harboring integrated HIV DNA. $CD4^+$ T cells co-expressing the three markers were highly enriched for integrated viral genomes.

11 Conclusions

The concept of LAG-3-expressing IL-10-producing Treg cells has been widely accepted. While $CD25^+$ Treg cells exhibit regulatory activity mainly via cell contact, LAG-3-expressing IL-10-producing Treg cells control immune cells including B cells via suppressive cytokines such as IL-10, TGF-β1, and TGF-β3. The anergy-inducing transcription factor Egr2 is required for the production of IL-10 and TGF-β3. However, the precise molecular basis for the development and functions of IL-10-producing Treg cells remains to be clarified. Elucidation of LAG-3-expressing IL-10-producing Treg cells may contribute to the development of novel therapeutic strategies in immune-mediated diseases.

Acknowledgements This work was supported by Grant-in-Aid for Scientific Research of the Ministry of Education, Culture, Sports, Science and Technology, Japan (MEXT) (Grant Numbers JP16K09918, JP16K15510).

References

Anderson MS, Venanzi ES, Chen Z, Berzins SP, Benoist C, Mathis D (2005) The cellular mechanism of Aire control of T cell tolerance. Immunity 23:227–239

Awasthi A, Carrier Y, Peron JP, Bettelli E, Kamanaka M, Flavell RA, Kuchroo VK, Oukka M, Weiner HL (2007) A dominant function for interleukin 27 in generating interleukin 10-producing anti-inflammatory T cells. Nat Immunol 8:1380–1389

Bacchetta R, Bigler M, Touraine JL, Parkman R, Tovo PA, Abrams J, de Waal Malefyt R, de Vries JE, Roncarolo MG (1994) High levels of interleukin 10 production in vivo are associated with tolerance in SCID patients transplanted with HLA mismatched hematopoietic stem cells. J Exp Med 179:493–502

Battaglia M, Gregori S, Bacchetta R, Roncarolo MG (2006) Tr1 cells: from discovery to their clinical application. Semin Immunol 18:120–127

Bennett CL, Christie J, Ramsdell F, Brunkow ME, Ferguson PJ, Whitesell L, Kelly TE, Saulsbury FT, Chance PF, Ochs HD (2001) The immune dysregulation, polyendocrinopathy, enteropathy, X-linked syndrome (IPEX) is caused by mutations of FOXP3. Nat Genet 27:20–21

Brunkow ME, Jeffery EW, Hjerrild KA, Paeper B, Clark LB, Yasayko SA, Wilkinson JE, Galas D, Ziegler SF, Ramsdell F (2001) Disruption of a new forkhead/winged-helix protein, scurfin, results in the fatal lymphoproliferative disorder of the scurfy mouse. Nat Genet 27:68–73

Chen Z, Benoist C, Mathis D (2005) How defects in central tolerance impinge on a deficiency in regulatory T cells. Proc Natl Acad Sci U S A 102:14735–14740

Chen SY, Hsu WT, Chen YL, Chien CH, Chiang BL (2016) Lymphocyte-activation gene 3(+) (LAG3(+)) forkhead box protein 3(−) (FOXP3(−)) regulatory T cells induced by B cells alleviates joint inflammation in collagen-induced arthritis. J Autoimmun 68:75–85

Chikuma S, Suita N, Okazaki IM, Shibayama S, Honjo T (2012) TRIM28 prevents autoinflammatory T cell development in vivo. Nat Immunol 13:596–603

Clemente-Casares X, Blanco J, Ambalavanan P, Yamanouchi J, Singha S, Fandos C, Tsai S, Wang J, Garabatos N, Izquierdo C et al (2016) Expanding antigen-specific regulatory networks to treat autoimmunity. Nature 530:434–440

Cretney E, Xin A, Shi W, Minnich M, Masson F, Miasari M, Belz GT, Smyth GK, Busslinger M, Nutt SL, Kallies A (2011) The transcription factors Blimp-1 and IRF4 jointly control the differentiation and function of effector regulatory T cells. Nat Immunol 12:304–311

Dempke WC, Fenchel K, Uciechowski P, Dale SP (2017) Second- and third-generation drugs for immuno-oncology treatment-The more the better? Eur J Cancer 74:55–72

Derynck R, Lindquist PB, Lee A, Wen D, Tamm J, Graycar JL, Rhee L, Mason AJ, Miller DA, Coffey RJ et al (1988) A new type of transforming growth factor-beta, TGF-beta 3. EMBO J 7:3737–3743

Doe HT, Kimura D, Miyakoda M, Kimura K, Akbari M, Yui K (2016) Expression of PD-1/LAG-3 and cytokine production by CD4(+) T cells during infection with Plasmodium parasites. Microbiol Immunol 60:121–131

Fromentin R, Bakeman W, Lawani MB, Khoury G, Hartogensis W, DaFonseca S, Killian M, Epling L, Hoh R, Sinclair E et al (2016) CD4 + T Cells Expressing PD-1, TIGIT and LAG-3 Contribute to HIV Persistence during ART. PLoS Pathog 12:e1005761

Fujio K, Okamura T, Yamamoto K (2010) The Family of IL-10-secreting CD4 + T cells. Adv Immunol 105:99–130

Gagliani N, Magnani CF, Huber S, Gianolini ME, Pala M, Licona-Limon P, Guo B, Herbert DR, Bulfone A, Trentini F et al (2013) Coexpression of CD49b and LAG-3 identifies human and mouse T regulatory type 1 cells. Nat Med 19:739–746

Gershon RK, Kondo K (1970) Cell interactions in the induction of tolerance: the role of thymic lymphocytes. Immunology 18:723–737

Gershon RK, Kondo K (1971) Infectious immunological tolerance. Immunology 21:903–914

Gershon RK, Mokyr MB, Mitchell MS (1974) Activation of suppressor T cells by tumour cells and specific antibody. Nature 250:594–596

Goding SR, Wilson KA, Xie Y, Harris KM, Baxi A, Akpinarli A, Fulton A, Tamada K, Strome SE, Antony PA (2013) Restoring immune function of tumor-specific CD4 + T cells during recurrence of melanoma. J Immunol 190:4899–4909

Groux H, O'Garra A, Bigler M, Rouleau M, Antonenko S, de Vries JE, Roncarolo MG (1997) A CD4 + T-cell subset inhibits antigen-specific T-cell responses and prevents colitis. Nature 389:737–742

Hannier S, Tournier M, Bismuth G, Triebel F (1998) CD3/TCR complex-associated lymphocyte activation gene-3 molecules inhibit CD3/TCR signaling. J Immunol 161:4058–4065

Harris JE, Bishop KD, Phillips NE, Mordes JP, Greiner DL, Rossini AA, Czech MP (2004) Early growth response gene-2, a zinc-finger transcription factor, is required for full induction of clonal anergy in CD4+ T cells. J Immunol 173:7331–7338

Heinemann C, Heink S, Petermann F, Vasanthakumar A, Rothhammer V, Doorduijn E, Mitsdoerffer M, Sie C, Prazeres da Costa O, Buch T et al (2014) IL-27 and IL-12 oppose pro-inflammatory IL-23 in CD4+ T cells by inducing Blimp1. Nat Commun 5:3770

Hori S, Nomura T, Sakaguchi S (2003) Control of regulatory T cell development by the transcription factor Foxp3. Science 299:1057–1061

Huang CT, Workman CJ, Flies D, Pan X, Marson AL, Zhou G, Hipkiss EL, Ravi S, Kowalski J, Levitsky HI et al (2004) Role of LAG-3 in regulatory T cells. Immunity 21:503–513

Huard B, Prigent P, Tournier M, Bruniquel D, Triebel F (1995) CD4/major histocompatibility complex class II interaction analyzed with CD4- and lymphocyte activation gene-3 (LAG-3)-Ig fusion proteins. Eur J Immunol 25:2718–2721

Huard B, Prigent P, Pages F, Bruniquel D, Triebel F (1996) T cell major histocompatibility complex class II molecules down-regulate CD4+ T cell clone responses following LAG-3 binding. Eur J Immunol 26:1180–1186

Iwasaki Y, Fujio K, Okamura T, Yanai A, Sumitomo S, Shoda H, Tamura T, Yoshida H, Charnay P, Yamamoto K (2013) Egr-2 transcription factor is required for Blimp-1-mediated IL-10 production in IL-27-stimulated CD4+ T cells. Eur J Immunol 43:1063–1073

Koch K, Koch N, Sandaradura de Silva U, Jung N, Schulze zur Wiesch J, Fatkenheuer G, Hartmann P, Romerio F, and Lehmann C (2015) Increased frequency of CD49b/LAG-3(+) type 1 regulatory T cells in HIV-infected individuals. AIDS Res Hum Retroviruses 31:1238–1246

Lee Y, Awasthi A, Yosef N, Quintana FJ, Xiao S, Peters A, Wu C, Kleinewietfeld M, Kunder S, Hafler DA et al (2012) Induction and molecular signature of pathogenic TH17 cells. Nat Immunol 13:991–999

Li N, Wang Y, Forbes K, Vignali KM, Heale BS, Saftig P, Hartmann D, Black RA, Rossi JJ, Blobel CP et al (2007) Metalloproteases regulate T-cell proliferation and effector function via LAG-3. EMBO J 26:494–504

Li S, Symonds AL, Zhu B, Liu M, Raymond MV, Miao T, Wang P (2011) Early growth response gene-2 (Egr-2) regulates the development of B and T cells. PLoS ONE 6:e18498

Li S, Miao T, Sebastian M, Bhullar P, Ghaffari E, Liu M, Symonds AL, Wang P (2012) The transcription factors Egr2 and Egr3 are essential for the control of inflammation and antigen-induced proliferation of B and T cells. Immunity 37:685–696

Liang B, Workman C, Lee J, Chew C, Dale BM, Colonna L, Flores M, Li N, Schweighoffer E, Greenberg S et al (2008) Regulatory T cells inhibit dendritic cells by lymphocyte activation gene-3 engagement of MHC class II. J Immunol 180:5916–5926

Liston A, Lesage S, Wilson J, Peltonen L, Goodnow CC (2003) Aire regulates negative selection of organ-specific T cells. Nat Immunol 4:350–354

Liu G, Luan J, Li Q (2016) CD4+ Foxp3-IL-10+ Tr1 Cells Promote Relapse of Diffuse Large B Cell Lymphoma by Enhancing the Survival of Malignant B cells and suppressing antitumor T cell immunity. DNA Cell Biol 35:845–852

Martins GA, Cimmino L, Shapiro-Shelef M, Szabolcs M, Herron A, Magnusdottir E, Calame K (2006) Transcriptional repressor Blimp-1 regulates T cell homeostasis and function. Nat Immunol 7:457–465

Miyazaki T, Dierich A, Benoist C, Mathis D (1996) Independent modes of natural killing distinguished in mice lacking Lag3. Science 272:405–408

Morita K, Okamura T, Inoue M, Komai T, Teruya S, Iwasaki Y, Sumitomo S, Shoda H, Yamamoto K, Fujio K (2016) Egr2 and Egr3 in regulatory T cells cooperatively control systemic autoimmunity through Ltbp3-mediated TGF-beta3 production. Proc Natl Acad Sci USA 113:E8131–E8140

Myouzen K, Kochi Y, Shimane K, Fujio K, Okamura T, Okada Y, Suzuki A, Atsumi T, Ito S, Takada K et al (2010) Regulatory polymorphisms in EGR2 are associated with susceptibility to systemic lupus erythematosus. Hum Mol Genet 19:2313–2320

Okamura T, Fujio K, Shibuya M, Sumitomo S, Shoda H, Sakaguchi S, Yamamoto K (2009) CD4 + CD25-LAG3+ regulatory T cells controlled by the transcription factor Egr-2. Proc Natl Acad Sci USA 106:13974–13979

Okamura T, Fujio K, Sumitomo S, Yamamoto K (2012) Roles of LAG3 and EGR2 in regulatory T cells. Ann Rheum Dis 71(Suppl 2):i96–i100

Okamura T, Sumitomo S, Morita K, Iwasaki Y, Inoue M, Nakachi S, Komai T, Shoda H, Miyazaki J, Fujio K, Yamamoto K (2015) TGF-beta3-expressing CD4+CD25(-)LAG3+ regulatory T cells control humoral immune responses. Nat Commun 6:6329

Okazaki T, Okazaki IM, Wang J, Sugiura D, Nakaki F, Yoshida T, Kato Y, Fagarasan S, Muramatsu M, Eto T et al (2011) PD-1 and LAG-3 inhibitory co-receptors act synergistically to prevent autoimmunity in mice. J Exp Med 208:395–407

Onishi Y, Fehervari Z, Yamaguchi T, Sakaguchi S (2008) Foxp3+ natural regulatory T cells preferentially form aggregates on dendritic cells in vitro and actively inhibit their maturation. Proc Natl Acad Sci USA 105:10113–10118

Pedroza-Gonzalez A, Zhou G, Vargas-Mendez E, Boor PP, Mancham S, Verhoef C, Polak WG, Grunhagen D, Pan Q, Janssen H et al (2015) Tumor-infiltrating plasmacytoid dendritic cells promote immunosuppression by Tr1 cells in human liver tumors. Oncoimmunology 4: e1008355

Proetzel G, Pawlowski SA, Wiles MV, Yin M, Boivin GP, Howles PN, Ding J, Ferguson MW, Doetschman T (1995) Transforming growth factor-beta 3 is required for secondary palate fusion. Nat Genet 11:409–414

Roncarolo MG, Battaglia M (2007) Regulatory T-cell immunotherapy for tolerance to self antigens and alloantigens in humans. Nat Rev Immunol 7:585–598

Roncarolo MG, Gregori S, Battaglia M, Bacchetta R, Fleischhauer K, Levings MK (2006) Interleukin-10-secreting type 1 regulatory T cells in rodents and humans. Immunol Rev 212:28–50

Safford M, Collins S, Lutz MA, Allen A, Huang CT, Kowalski J, Blackford A, Horton MR, Drake C, Schwartz RH, Powell JD (2005) Egr-2 and Egr-3 are negative regulators of T cell activation. Nat Immunol 6:472–480

Sakaguchi S, Powrie F (2007) Emerging challenges in regulatory T cell function and biology. Science 317:627–629

Schneider-Maunoury S, Topilko P, Seitandou T, Levi G, Cohen-Tannoudji M, Pournin S, Babinet C, Charnay P (1993) Disruption of Krox-20 results in alteration of rhombomeres 3 and 5 in the developing hindbrain. Cell 75:1199–1214

Shah S, Qiao L (2008) Resting B cells expand a CD4 + CD25 + Foxp3 + Treg population via TGF-beta3. Eur J Immunol 38:2488–2498

Shao TY, Hsu LH, Chien CH, Chiang BL (2016) Novel Foxp3(-) IL-10(-) regulatory T-cells induced by B-Cells alleviate intestinal inflammation in vivo. Sci Rep 6:32415

Shi W, Liao Y, Willis SN, Taubenheim N, Inouye M, Tarlinton DM, Smyth GK, Hodgkin PD, Nutt SL, Corcoran LM (2015) Transcriptional profiling of mouse B cell terminal differentiation defines a signature for antibody-secreting plasma cells. Nat Immunol 16:663–673

Sumitomo S, Nakachi S, Okamura T, Tsuchida Y, Kato R, Shoda H, Furukawa A, Kitahara N, Kondo K, Yamasoba T et al (2017) Identification of tonsillar CD4 + CD25-LAG3 + T cells as naturally occurring IL-10-producing regulatory T cells in human lymphoid tissue. J Autoimmun 76:75–84

Topilko P, Schneider-Maunoury S, Levi G, Baron-Van Evercooren A, Chennoufi AB, Seitanidou T, Babinet C, Charnay P (1994) Krox-20 controls myelination in the peripheral nervous system. Nature 371:796–799

Tourtellotte WG, Milbrandt J (1998) Sensory ataxia and muscle spindle agenesis in mice lacking the transcription factor Egr3. Nat Genet 20:87–91

Tousa S, Semitekolou M, Morianos I, Banos A, Trochoutsou AI, Brodie TM, Poulos N, Samitas K, Kapasa M, Konstantopoulos D et al (2017) Activin-A co-opts IRF4 and AhR signaling to induce human regulatory T cells that restrain asthmatic responses. Proc Natl Acad Sci U S A 114:E2891–E2900

Triebel F, Jitsukawa S, Baixeras E, Roman-Roman S, Genevee C, Viegas-Pequignot E, Hercend T (1990) LAG-3, a novel lymphocyte activation gene closely related to CD4. J Exp Med 171:1393–1405

Veiga-Parga T, Sehrawat S, Rouse BT (2013) Role of regulatory T cells during virus infection. Immunol Rev 255:182–196

Villegas-Mendez A, Inkson CA, Shaw TN, Strangward P, Couper KN (2016) Long-lived CD4 + IFN-gamma + T Cells rather than short-lived CD4 + IFN-gamma + IL-10 + T cells initiate rapid IL-10 production to suppress anamnestic t cell responses during secondary malaria infection. J Immunol 197:3152–3164

Woo SR, Li N, Bruno TC, Forbes K, Brown S, Workman C, Drake CG, Vignali DA (2010) Differential subcellular localization of the regulatory T-cell protein LAG-3 and the coreceptor CD4. Eur J Immunol 40:1768–1777

Woo SR, Turnis ME, Goldberg MV, Bankoti J, Selby M, Nirschl CJ, Bettini ML, Gravano DM, Vogel P, Liu CL et al (2012) Immune inhibitory molecules LAG-3 and PD-1 synergistically regulate T-cell function to promote tumoral immune escape. Cancer Res 72:917–927

Workman CJ, Cauley LS, Kim IJ, Blackman MA, Woodland DL, Vignali DA (2004) Lymphocyte activation gene-3 (CD223) regulates the size of the expanding T cell population following antigen activation in vivo. J Immunol 172:5450–5455

Yang LT, Kaartinen V (2007) Tgfb1 expressed in the Tgfb3 locus partially rescues the cleft palate phenotype of Tgfb3 null mutants. Dev Biol 312:384–395

Yao Y, Vent-Schmidt J, McGeough MD, Wong M, Hoffman HM, Steiner TS, Levings MK (2015) Tr1 cells, but not Foxp3 + regulatory T cells, suppress NLRP3 inflammasome activation via an IL-10-dependent mechanism. J Immunol 195:488–497

Zhu B, Symonds AL, Martin JE, Kioussis D, Wraith DC, Li S, Wang P (2008) Early growth response gene 2 (Egr-2) controls the self-tolerance of T cells and prevents the development of lupuslike autoimmune disease. J Exp Med 205:2295–2307

Regulatory Dendritic Cells

Katsuaki Sato, Tomofumi Uto, Tomohiro Fukaya and Hideaki Takagi

Abstract Dendritic cells (DCs) comprise heterogeneous subsets, functionally classified into conventional DCs (cDCs) and plasmacytoid DCs (pDCs). DCs are considered to be essential antigen (Ag)-presenting cells (APCs) that play crucial roles in activation and fine-tuning of innate and adaptive immunity under inflammatory conditions, as well as induction of immune tolerance to maintain immune homeostasis under steady-state conditions. Furthermore, DC functions can be modified and influenced by stimulation with various extrinsic factors, such as ligands for pattern-recognition receptors (PRRs) and cytokines. On the other hand, treatment of DCs with certain immunosuppressive drugs and molecules leads to the generation of tolerogenic DCs that show downregulation of both the major histocompatibility complex (MHC) and costimulatory molecules, and not only show defective T-cell activation, but also possess tolerogenic properties including the induction of anergic T-cells and regulatory T (T_{reg}) cells. To develop an effective strategy for Ag-specific intervention of T-cell-mediated immune disorders, we have previously established the modified DCs with moderately high levels of MHC molecules that are defective in the expression of costimulatory molecules that had a greater immunoregulatory property than classical tolerogenic DCs, which we therefore designated as regulatory DCs (DC_{reg}). Herein, we integrate the current understanding of the role of DCs in the control of immune responses, and further provide new information of the characteristics of tolerogenic DCs and DC_{reg}, as well as their regulation of immune responses and disorders.

K. Sato (✉) · T. Uto · T. Fukaya · H. Takagi
Division of Immunology, Department of Infectious Diseases, Faculty of Medicine, University of Miyazaki, 5200 Kihara, Kiyotake, Miyazaki 889-1692, Japan
e-mail: katsuaki_sato@med.miyazaki-u.ac.jp

K. Sato · T. Uto · T. Fukaya · H. Takagi
Japan Agency for Medical Research and Development (AMED), 1-7-1 Otemachi, Chiyoda-Ku, Tokyo 100-0004, Japan

Current Topics in Microbiology and Immunology (2017) 410:47–72
DOI 10.1007/82_2017_60

Published Online: 13 September 2017

Contents

1 Introduction 48
2 Overview of DC Biology 50
3 DC Subsets and Functions 53
3.1 Mouse cDC Subsets and Functions 54
3.2 Mouse pDC Subsets and Functions 56
3.3 Mouse Inflammatory DC Subsets and Functions 57
3.4 Human cDC Subsets and Functions 57
3.5 Human pDC Subsets and Functions 58
3.6 Human Inflammatory DC Subsets and Functions 59
4 Tolerogenic DCs and Functions 59
4.1 Tolerogenic DCs and Functions 60
4.2 Clinical Application of Human Tolerogenic DCs 61
5 Regulatory DCs and Functions 62
5.1 Human Regulatory DCs and Functions 63
5.2 Murine Regulatory DCs and Functions 64
6 Conclusions 66
References 66

1 Introduction

It is evident that a critical balance of the innate and adaptive constituents of the immune system is necessary to provide the engagement of the host defense mechanisms against microbial infections, and at the same time to avoid loss of self-tolerance, and much attention has been paid to the bidirectional role of dendritic cells (DCs) in controlling this delicate balance over the past few decades.

The importance of DCs in immunity was originally described by Steinman et al. in the 1970s as being a distinct hematopoietic lineage with unusual morphological features including dendrite-like projections, trafficking, and potent stimulators of primary immune responses (Steinman and Cohn 1973; Steinman and Witmer 1978). It is now recognized that DCs act as essential antigen (Ag)-presenting cells (APCs) that consist of heterogeneous subsets, mainly classified as conventional DCs (cDCs) and plasmacytoid DCs (pDCs) (Sato and Fujita 2007; Shortman and Naik 2007; Villadangos and Schnorrer 2007). DCs serve as sentinels, recognizing the presence of invading pathogens or virus-infected cells through various pattern-recognition receptors (PRRs) (Sato and Fujita 2007; Shortman and Naik 2007; Villadangos and Schnorrer 2007). DCs process such exogenous Ags intracellularly and present them to $CD4^+$ T-cells via major histocompatibility complex class II (MHC II) molecules for induction of various types of $CD4^+$ effector T ($CD4^+$ T_{eff}) cells, depending on intrinsic factors and extrinsic stimulations (Dudziak et al. 2007; Villadangos and Schnorrer 2007; Hildner et al 2008; Merad et al. 2013; Walsh and Mills 2013). DCs also show an unusual specialization in their MHC class I (MHC I)-mediated presentation pathway to prime $CD8^+$ T-cells. Although most cells use their MHC I molecules to present peptides derived from endogenously synthesized proteins, DCs have the capacity to deliver exogenous Ags to the MHC I-dependent pathway, a phenomenon known as cross-presentation, that underlies the generation of cytotoxic T lymphocyte (CTL) immunity (Dudziak et al. 2007; Villadangos and Schnorrer 2007; Hildner et al 2008). DCs thereby play a crucial role in the link between innate and adaptive immunity. Conversely, DCs may be also important for the maintenance of immune homeostasis by promoting immune tolerance via mechanisms including clonal deletion of self-reactive T cells in the thymus (central tolerance), and clonal deletion and anergy, as well as active suppression by the production of regulatory T (T_{reg}) cells (e.g., $CD4^{+}Foxp3^{+}$ T_{reg} cells and interleukin [IL]-10-producing type-1 T_{reg} [Tr1] cells) in the periphery, a function likely of importance in autoimmunity and transplant rejection, as well as self-tolerance (peripheral tolerance) (Roncarolo et al. 2001; Steinman and Nussenzweig 2002; Kretschmer et al. 2005; Manicassamy and Pulendran 2011; Kornete and Piccirillo 2012; Osorio et al. 2015). Therefore, the diverse functions of DCs in immune regulation reflect the heterogeneous subsets with different lineages and maturity, and functional plasticity.

Cumulative studies have shown that the modification of DCs with various immunosuppressive molecules and drugs can generate tolerogenic DCs, which show defective expression of MHC and costimulatory molecules, and lead to the reduction of T-cell stimulatory capacity, as well as generation of Ag-specific anergic T cells and T_{reg} cells (Rutella et al. 2006; Morelli and Thomson 2007; Gordon et al. 2014; Osorio et al. 2015). In addition, treatment with certain tolerogenic DCs reportedly showed protective effects against various immune disorders in animal models (Rutella et al. 2006; Morelli and Thomson 2007; Gordon et al. 2014; Osorio et al. 2015). However, the analyses of these previously known tolerogenic DCs suggested that they comprise relatively heterogeneous subsets with various expression levels of MHC and costimulatory molecules (Rutella et al. 2006; Morelli and Thomson 2007; Gordon et al. 2014; Osorio et al. 2015), which might diminish their potential ability to dampen the pathogenic T-cell responses, leading to inefficient protective effects against immune disorders. Therefore, further advancements in the understanding of DC biology, and techniques for potentiating the ability of these cells to negatively regulate T cells, could facilitate their use for prevention or treatment of immune disorders. To exploit a novel strategy involving the use of DCs for Ag-specific intervention of T-cell-mediated immune disorders, we have previously developed modified human and murine DCs with moderately high levels of MHC molecules and decreased expression of costimulatory molecules, that had a greater capacity to regulate immune responses than classical tolerogenic DCs, resulting from the preferential generation of $CD4^{+}Foxp3^{+}$ T_{reg} cells, and were therefore designated as regulatory DCs (DC_{reg}) (Sato et al. 2003a, b). In this review, we will first describe the current understanding of the role of DC subsets in the control of immune responses. We will also attempt to define the characteristics of DC_{reg} as well as tolerogenic DCs, and discuss their potential applications for the treatment of immune disorders.

2 Overview of DC Biology

Extensive studies on DC biology have provided their comprehensive information on the development and function of DCs, as well as their regulation of immune response (Sato and Fujita 2007; Shortman and Naik 2007; Villadangos and Schnorrer 2007; Merad et al. 2013; Walsh and Mills KH 2013).

Hematopoietic progenitors in bone marrow (BM) give rise to circulating DC precursors that reside as quiescent immature cells in blood as well as non-lymphoid and lymphoid tissues. Immature DCs express high levels of PRRs such as Toll-like receptors (TLRs), retinoic-acid-inducible gene I (RIGI)-like receptors (RLRs), and nucleotide-binding oligomerization domain (NOD)-like receptors (NLRs), Fcγ receptors, complement receptors, mannose-like receptors, and scavenger receptors, which mediate the recognition and endocytosis of potential Ags (van Vliet et al. 2007; Hoshino and Kaisho 2008; Joffre et al. 2009; Uto et al. 2016). In addition, they possess high endocytic and phagocytic capacities, permitting Ag capture through PRRs, but express moderate levels of MHC class II (MHC II) molecules and low levels of costimulatory molecules on their surface (Fig. 1). Following a microbial infection and tissue damage, immature DCs migrate to inflamed regions in response to the production of a large spectrum of inflammatory chemokines (e.g., CCL3, CCL5, and CCL20) upon local inflammation through specific chemokine receptors (e.g., CCR1, CCR5, and CCR6) (Fig. 1). Immature DCs sense a wide range of "danger signals" both from invading microbial components, known as pathogen-associated molecular patterns (PAMPs), and proteins or nuclear products released from host cells in the injured tissues, known as danger-associated molecular patterns (DAMPs), via PRRs (van Vliet et al. 2007; Hoshino and Kaisho 2008; Joffre et al. 2009; Uto et al. 2016) (Fig. 1). Upon recognizing PAMPs and/or DAMPs, PRRs triggers multiple intracellular signaling cascades, leading to activation of pathways for mitogen-activated protein kinase (MAPK), nuclear factor kappa B (NF-κB), and interferon (IFN) regulatory factor (IRF), and subsequently induce the production of large amounts of proinflammatory cytokines, type-I IFN, and chemokines (van Vliet et al. 2007; Hoshino and Kaisho 2008; Uto et al. 2016). These inflammatory events result in the further activation of other innate immune cells, thereby limiting the spread of microbial infection and promoting tissue damage. Additionally, DCs acquire a "mature" phenotype, and the maturation process is associated with several coordinated events such as (*a*) loss of endocytic and phagocytic abilities due to downregulated expression of receptors, (*b*) activation of the Ag-processing machinery for the generation of antigenic peptides, including a shift in lysosomal compartments and an increase in DC-lysosome-associated membrane protein (DC-LAMP), and (*c*) upregulated cell surface expression of Ag-presentation machinery, including processed antigenic peptide-loaded MHC II molecules and costimulatory molecules (e.g., CD40, CD80, and CD86) (Fig. 1). In addition, they reprogram chemokine receptor expression and responsiveness, including (*d*) loss of responsiveness to inflammatory chemokines via either receptor down-regulation or receptor desensitization, and (*e*) acquisition of responsiveness to lymph node-homing chemokines (e.g., CCL19 and CCL21) via upregulation of CCR7 (Fig. 1). Consequently, mature DCs move via afferent lymphatics into the T-cell area of local draining lymph nodes (LNs), where they select rare Ag-specific naïve T cells to present their processed Ags in the context of cell surface MHC II molecules (APC signal 1) together with costimulatory molecules (APC signal 2) and T cell-polarizing cytokine signals (APC signal 3), which are secreted by DCs and other immune cells, as well as non-hematopoietic cells, for differentiation into effector T (T_{eff}) cells (e.g., T_H1, T_H2, and T_H17 cells), thereby initiating primary immune responses (Fig. 1).

Growing evidence suggests that DCs also play a key role in maintaining immune homeostasis via induction of immune tolerance for suppressing T-cell responses, as well as prevention of harmful excessive inflammation (Birnberg et al. 2008; Ohnmacht et al. 2009; Osorio et al. 2015). The regulatory capacity of tissue-resident DCs for the induction of immune tolerance depends on their "immature" state under homeostatic conditions or "semi-mature" phenotype that is influenced by the tissue microenvironment (Roncarolo et al. 2001; Steinman and Nussenzweig 2002; Kretschmer et al. 2005; Joffre et al. 2009; Manicassamy and Pulendran 2011), and they are characterized by a distinct phenotype with moderate expression levels of

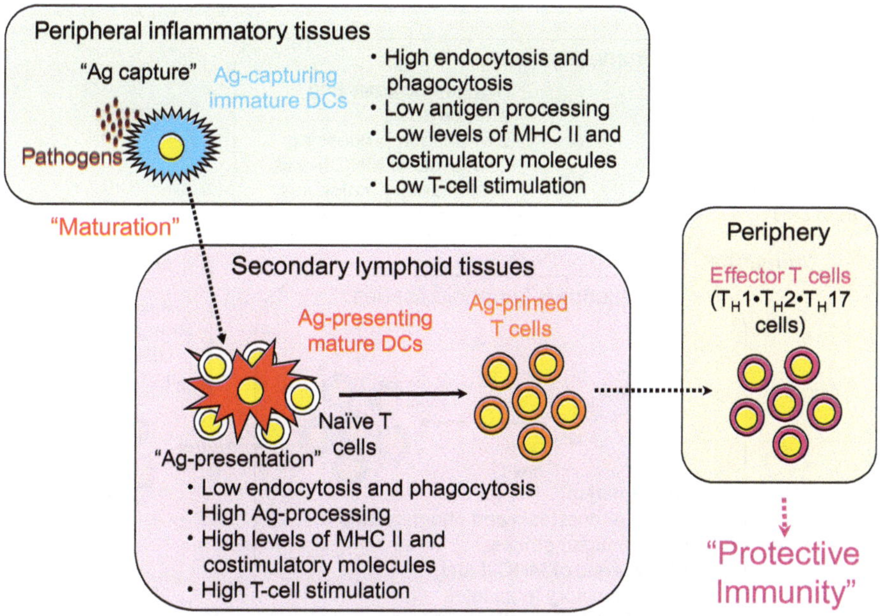

Fig. 1 Induction of adaptive immune responses by DCs under inflammatory conditions. Immature DCs sense the presence of invading pathogens via various PRRs, and process the pathogens intracellularly in inflammatory tissues, and then develop into mature DCs with changes in various functions. Subsequently, mature DCs target secondary lymphoid tissues where they present the processed Ags to naïve T cells to generate effector T cells for protective immunity against microbial infections

antigenic peptide-loaded MHC II molecules and low expression levels of costimulatory molecules and proinflammatory cytokines that could present "APC signal 1", but fail to deliver proper "APC signal 2 plus 3" to naïve T cells. Tissue-resident DCs that acquire innocuous environmental Ags or self-Ags released from host apoptotic cells under steady-state conditions or certain pathophysiological conditions, similarly migrate to the draining LNs as quiescent cells (Fig. 2). Consequently, immature or semi-mature DCs presenting innocuous environmental or self Ags to Ag-specific naïve T cells, participate in the induction of peripheral T-cell tolerance, mediated through the generation of anergic T cells and several types of T_{reg} cells (Fig. 2). Although the induction of unresponsiveness and clonal deletion of T-cells through apoptosis following the cognate interaction between Ag-specific naïve T-cells and quiescent DCs might contribute to the establishment of peripheral T-cell tolerance, quiescent DCs mediated the generation and expansion of $CD4^{+}Foxp3^{+}$ T_{reg} cells for the induction of immune tolerance as a reciprocal crosstalk between DCs and $CD4^{+}Foxp3^{+}$ T_{reg} cells has been reported to be important for maintaining immune homeostasis (Kim et al. 2007; Darrasse-Jèze et al. 2009; Kornete and Piccirillo 2012). In contrast to the requirement of the costimulation through CD80/CD86:CD28 and B7-H2:inducible costimulator (ICOS) for thymic development and peripheral IL-2-mediated expansion of

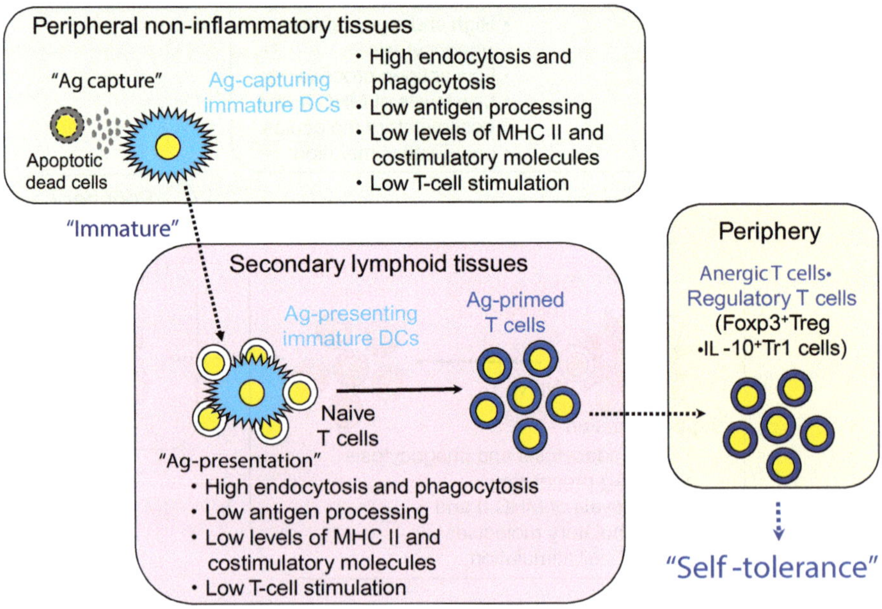

Fig. 2 Induction of immune tolerance by DCs under steady-state conditions. Immature DCs capture apoptotic dead cells via various PRRs, and process them intracellularly in non-inflammatory tissues. Subsequently, immature DCs target secondary lymphoid tissues where they present the processed Ags to naïve T cells to generate anergic T cells and regulatory T cells for induction of self-tolerance

self-reactive naturally occurring $CD4^+Foxp3^+$ T_{reg} ($CD4^+Foxp3^+$ nT_{reg}) cells maintaining peripheral self-tolerance, B7-H1 and B7-DC, which are ligands for programmed death 1 (PD-1), rather than CD80/CD86 expressed on DCs, may be necessary for transforming growth factor (TGF)-β1-mediated conversion of Ag-specific naïve $CD4^+Foxp3^-$T cells into inducible $CD4^+Foxp3^+$ T_{reg} ($CD4^+Foxp3^+$ iT_{reg}) cells in peripheral tissues during the induction of immune tolerance, although interactions mediated by CD80/CD86:CD28 appear to be critical for their expansion (Fontenot et al. 2005; Fukaya et al. 2010; Osorio et al. 2015).

Taken together, these findings indicate that DCs play key roles in dictating the outcome of immune responses by controlling the balance between the initiation of protective immune responses against pathogens and maintenance of immune tolerance via a variety of mechanisms, including the expression of multiple costimulatory and coinhibitory signaling pathways, the secretion of inflammatory and regulatory factors, anatomical localization and migration, and Ag-processing/presentation machineries that are influenced by intrinsic factors (e.g., signaling molecules, transcription factors, and catalytic enzymes) and extrinsic stimulations (e.g., cytokines, and PAMPs/DAMPs), as well as microenvironments.

3 DC Subsets and Functions

DCs are a highly heterogeneous cell population, which contain multiple subsets with distinct origins, markers, and anatomical locations, as well as migratory and functional properties, and are represented by two major lineages, cDCs and pDCs.

cDCs are characterized by their classic dendritic morphology and their outstanding ability to prime naïve T cells to generate various types of T_{eff} cells, owing to the prominent expression of MHC and costimulatory molecules after their activation, as described above. cDCs primarily reside in secondary lymphoid tissues, such as the spleen and LNs, and are called “lymphoid-resident DCs”, while non-lymphoid tissue-resident DCs migrating to the draining LNs from peripheral tissues via the lymphatics are called “migratory DCs”. On the other hand, pDCs represent a distinct class of DCs, which were originally discovered as type-I IFN (IFN-α/β)-producing cells (IPCs) in human LNs in the 1958, as they produce vast amounts of type-I IFN (IFN-α/β) in response to viral infection, and their mouse counterpart was identified in 2001 (Colonna et al. 2004; Gilliet et al. 2008; Swiecki and Colonna 2015). In the steady state, pDCs show plasma cell-like morphology, whereas they show a cDC morphology with dendrites under inflammatory conditions and upon activation. Furthermore, pDCs are less efficient at activating T cells than cDCs, even in an activated state, due to their lower expression of MHC and costimulatory molecules, as well as less efficient Ag-processing machinery. While migratory cDC subsets enter LNs from peripheral tissues through lymphatic vessels via CCR7, pDCs enter directly into LNs from blood by crossing the high endothelial venules (HEVs) via CD62L.

3.1 Mouse cDC Subsets and Functions

Mouse DCs are generally defined by their expression of CD11c and MHC II with a lack of lineage markers. Lymphoid-resident cDCs are further classified into three subsets according to the surface expression of CD8α and CD4, including CD8α^+ cDCs, CD4$^+$ cDCs, and CD4$^-$CD8α^-cDCs (Table 1). CD8α^+ cDCs mainly reside in the T-cell zone of LNs whereas CD4$^+$ cDCs and CD4$^-$CD8α^- cDCs localize in the red pulp and the marginal zone. On the other hand, migratory cDCs are defined by their surface expression of CD24 and CD172a (also known as SIRPα and SHPS1), such as CD24$^+$ cDCs, CD172a$^+$ cDCs, and CD24$^-$CD172a$^-$ cDCs (Table 2). While each cDC subsets exerts distinct functions for the control of immune responses by the impact on the preferential T_{eff} -cell differentiation program (Sato and Fujita 2007; Shortman and Naik 2007; Villadangos and Schnorrer 2007; Fukaya et al. 2012; Merad et al. 2013; Walsh and Mills 2013; Guilliams et al. 2014), recent genetic and functional studies have provide further insights into several nonredundant in vivo functions of individual types of cDCs (Bedoui and Heath 2015; Tussiwand et al. 2015; Luda et al. 2016; Murphy et al. 2016).

CD8α^+ cDCs require several transcription factors, including IRF8, BATF3, ID2, and NFIL3 for their development, and preferentially express CD205, CLEC9A (also known as DNGR1), XCR1, and CD24, as well as TLR3 (Table 1). CD8α^+ cDCs play a role in protective immune responses against intracellular bacteria by the generation of T_H1 cells via production of IL-12 and IFN-λ (Table 1). In addition, they also participate in antiviral immunity by the generation of CTL, mediated through the cross-presentation of viral Ag (Table 1). While CD4$^+$ cDCs or CD4$^-$CD8α^- cDCs differentiate from progenitors depending on transcription factors IRF4 and Notch2 or IRF4, and predominantly express CLEC4A4 (also

Table 1 Mouse lymphoid-resident CD11c$^+$ DC subsets

DC subsets	CD8α^+ cDCs	CD4$^+$ cDCs	CD4$^-$CD8α^-cDCs	pDCs
Transcription factors required for the development	IRF8, BATF3, ID2, NFIL3	IRF4, Notch2	IRF4	IRF8, E2-2
Cell surface molecules	CD24, CD205, CLEC9A, XCR1	CLEC4A4, CD11b, CD172a	CLEC4A4, CD11b, CD172a	BST2, Siglec-H, B220
Cytokine	IL-12, IFN-L	IL-23	IL-23	IFN-α/β
Type of induction of T cells	TH1 cells, CTL	TH2 cells, TH17 cells	TH2 cells, TH17 cells	CTL
Type of immunity	Intracellular bacteria, viruses	Allergen, parasites, extracellular bacteria, fungi	Allergen, parasites, extracellular bacteria, fungi	Viruses

Table 2 Mouse migratory CD11c$^+$ DC subsets

DC subsets	CD24$^+$ cDCs	CD24$^+$ cDCs CD172a$^+$ cDCs	CD24$^-$CD172a$^-$cDCs	pDCs
Transcription factors required for the development	IRF8, BATF3, ID2, NFIL3	IRF4, Notch2	IRF4, KLF4	IRF8, E2-2
Cell surface molecules	CD103, CD205, CLEC9A, XCR1	CD11b	CD11b	BST2, Siglec-H, B200
Cytokine	IL-12, IFN-L	IL-23	?	IFN-α/β
Type of induction of T cells	TH1 cells, CTL	TH17 cells	TH2 cells	CTL
Type of immunity	Intracellular bacteria, viruses	Extracellular bacteria, fungi	Allergen, parasites	Viruses

known as DCIR2), CD11b, and CD172a (Table 1). Both CD4$^+$ cDCs and CD4$^-$CD8α$^-$ cDCs produce IL-23 to generate T_H17 cells for the immune protection against fungi and extracellular bacteria (Table 1). In addition these two lymphoid-resident DC subsets also participate in the generation of T_H2 cells for protection against parasites and the induction of allergic responses (Table 1).

CD24$^+$ cDCs are characterized by their surface expression of CD103, CD205, CLEC9A, XCR1, and TLR3, and they have a similar requirement of transcriptional factors for development to CD8α$^+$ cDCs (Table 2). In addition, CD24$^+$ cDCs are functionally specialized in the cross-presentation of viral Ag to CD8$^+$ T cells for the induction of CTL and polarization into the T_H1 cells, emphasizing their crucial role in acting against intracellular pathogens and viruses (Table 2). Although both CD172a$^+$ cDCs and CD24$^-$CD172a$^-$ cDCs express CD11b, the development of CD172a$^+$ cDCs or CD24$^-$CD172a$^-$ cDCs are distinctly impacted by the transcriptional factors Notch2 or KLF4 although both migratory cDC subsets require IRF4 for their differentiation (Table 2). Furthermore, CD172a$^+$ cDCs are associated with IL-23-dependent generation of T_H17 cells and immune protection from infections by fungi and extracellular bacteria, whereas CD24$^-$CD172a$^-$ cDCs are specialized in the preferential promotion of optimal T_H2-cell responses toward anti-parasite protection and allergies (Table 2).

Whereas each lymphoid-resident and migratory cDC subsets serves characteristic functional properties inducing distinct T_{eff}-cell responses upon activation, in the steady states, lymphoid-resident and migratory mucosal CD103$^+$ cDCs, as well as lymphoid-resident CD8α$^+$ cDCs are superior to other DC subsets in the de novo generation of CD4$^+$Foxp3$^+$ iT_{reg} cells, owing to their prominent expression of TGF-β and retinaldehyde dehydrogenase (RALDH) for the synthesis of a vitamin A metabolite, retinoic acid (RA) (Coombes et al. 2007; Sun et al. 2007; Molenaar et al. 2011; Cong 2011; Bekiaris et al. 2014; Schlitzer and Ginhoux 2014), and thus these cDC subsets may be critical for maintaining systemic and mucosal immune homeostasis.

New terminology classifying DC subsets has recently been proposed, BATF3-dependent "cDC1" for CD8α^+ and CD103$^+$ cDCs, and "IRF4-dependent cDC2" for CD11b$^+$ and CD172a$^+$ cDCs, on the basis of their distinct developmental pathways and requirement of transcriptional factors (Guilliams et al. 2014).

3.2 *Mouse pDC Subsets and Functions*

Mouse pDCs are characterized by surface expression of the cell surface markers CD11c, BST2, Siglec-H, and B220, and depend on the transcription factor E2-2 for their development (Tables 1 and 2). pDCs are specialized in rapid and massive secretion of type-I IFN following endosomal TLR7/9-mediated recognition of nucleic acids (NAs) derived from infectious viruses or host cells in tissue injury and autoimmune conditions, and this unique capacity is crucial for the initiation of innate antiviral immunity (Takagi et al. 2011) (Tables 1 and 2) and certain autoimmune inflammation such as systemic lupus erythematosus (SLE) and psoriasis (Glitzner et al. 2014; Rowland et al. 2014; Sisirak et al. 2014; Takagi et al. 2016). In the steady state, pDCs residue mainly in the bone marrow (BM) and other lymphoid tissues, including the spleen and LNs, while upon activation, they also localized at the inflamed sites in non-lymphoid tissues. Although pDCs are poor APCs for the activation of T cells under steady-state conditions, they undergo phenotypic and functional changes with upregulation of MHC and costimulatory molecules, as well as Ag-processing machinery, to acquire the ability to present Ags to T cells after activation, although they are still not as efficient APCs as cDCs. It has been reported that pDCs meditate oral tolerance through the suppression of dietary Ag-specific T-cell responses (Goubier et al. 2008), while they not only induced tolerance to vascularized cardiac allografts, but also protection against experimental autoimmune encephalomyelitis and atherosclerosis via the generation and expansion of CD4$^+$Foxp3$^+$T$_{reg}$ cells through mechanisms inducing the expression of indoleamine 2,3-dioxygenase (IDO) (Ochando et al. 2006; Irla et al. 2010; Pallotta et al. 2011; Yun et al. 2016). Furthermore, pDCs are also involved in the induction of T$_{eff}$-cell responses in the MHC-dependent presentation of Ag for tumor regression following immunization with tumor Ag and CpG-B, as well as the development of atherosclerosis (Guéry et al. 2014; Sage et al. 2014). These findings suggest that the appropriate stimulations are needed for licensing of pDCs acting as APCs to induce T$_{eff}$-cell responses (Takagi et al. 2016). Thus, pDCs can also participate in the priming of both immunogenic and tolerogenic adaptive immune responses in addition to their contribution to eliciting innate inflammatory immune responses.

3.3 *Mouse Inflammatory DC Subsets and Functions*

Inflammatory DCs (infDCs) arising during infection and inflammation are derived from monocytes, and express CD11b, CD11c, MHC II, and CD64. Upon pathogen recognition, iDCs secrete large amounts of proinflammatory cytokines, including tumor necrosis factor (TNF)-α and inducible nitric oxide synthase (iNOS), and are thereby referred to as TNF/iNOS-producing DCs (Tip-DCs), which are important for the microbial clearance and the regulation of IgA production in mucosal immunity (Serbina et al. 2003; Tezuka et al. 2007). Furthermore, infDCs capture Ag and migrate to the draining LN following activation, where they drive T_H1- and T_H17-mediated immunity through the production of IL-12 and IL-23 (León et al. 2007; Hohl et al. 2009; Nakano et al. 2009; Mochizuki et al. 2013).

While cultures of murine progenitor cells in bone marrow (BM) with granulocyte macrophage colony-stimulating factor (GM-CSF) or Fms-related tyrosine kinase 3 ligand (Flt3-L) give rise to cDCs and macrophages or both cDCs and pDCs (Inaba et al. 1992; Bjorck 2001; Gilliet et al. 2002; Guilliams et al. 2014), Flt3-L-differentiated DCs show relatively similar features to lymphoid tissue-resident DCs found during the steady state, whereas GM-CSF–differentiated DCs have a more immunogenic phenotype and functionality, which resembles Tip-DCs (Watowich and Liu 2010; van de Laar et al. 2012).

3.4 *Human cDC Subsets and Functions*

DCs are present in human blood and lymphoid tissues including the tonsils, thymus, and spleen, as well as non-lymphoid tissues such as the lungs, liver, and skin. Similar to mouse DCs, human DCs also comprise multiple subsets in terms of the expression of a range of cell surface markers, but these might reflect differences in the maturation status rather than separate sublineages. Human DCs are also defined by lineage$^-$MHC II$^+$ cells, and all of them express CD4, but lack the expression of CD8 (Table 3). In addition, human DCs have 2 CD11c$^+$ cDCs and CD11c$^-$ pDCs,

Table 3 Human DC subsets

DC subsets	CD141$^+$ cDCs	CD1c$^+$ cDCs	CD303$^+$ CD304$^+$ pDCs
Cell surface molecules	CD4, CD11clow, CD13, CD24, CD33, CD45RO, CD116, CLEC9A, XCR1	CD2, CD4, CD11b, CD11c, CD13, CD32, CD33, CD64, CD45RO, CD116, CD172a	CD4, CD45RA, CD123
Cytokine	IL-12, IFN-L	IL-12, IL-23	IFN-α/β
Type of induction of T cells	TH1 cells, CTL	TH1 cells, TH17 cells	TH1 cells, TH2 cells, CTL?

and cDCs are further classified into 2 subsets based on the expression of CD1c (also known as BDCA-1) and CD141 (also known as BDCA-3) (Sato and Fujita 2007; Schlitzer and Ginhoux 2014; Murphy et al. 2016) (Table 3).

Human CD141$^+$ cDCs are minor cDC subset that expresses CD13, CD33, CD45RO, and CD116, while they express low levels of CD11c and lack the expression of CD2, CD11b, CD32, and CD64 (Table 3). Several studies indicate that CD141$^+$ cDCs are functional and phenotypic equivalents of murine BATF3-dependent cDC1 including lymphoid- resident CD8α^+ cDCs and migratory CD103$^+$ cDCs. In fact, CD141$^+$ cDCs predominantly express CLEC9A, XCR1, and CD24, and they are specialized in cross-presentation of internalized Ag to CD8$^+$ T cells for CTL-responses (Bachem et al. 2010; Jongbloed et al. 2010; Poulin et al. 2010; Haniffa et al. 2012) and production of high levels of IFN-λ upon activation (Lauterbach et al. 2010) (Table 3). On the other hand, human CD1c$^+$ cDCs are a major cDC subset that are characterized by the expression of CD2, CD11b, CD13, CD32, CD33, CD64, CD45RO, CD116, and CD172a with high expression of CD11c (Table 3), and phenotypically and functionally resemble murine IRF4-dependent CD11b$^+$ cDC2. Indeed, CD1c$^+$ cDCs secrete proinflammatory cytokines, including IL-23, to drive T_H17-responses (Schlitzer et al. 2013), whereas CD141$^+$ cDCs and CD1c$^+$ cDCs have a similar capacity to produce IL-12 for the induction of T_H1-cell responses (Mittag et al. 2011) (Table 3). While the predominant role of CD141$^+$ cDCs and CD1c$^+$ cDCs in skewing into T_H2-cell responses is a matter of debate, thymic stromal lymphopoietin (TSLP)-activated cDCs are suggested to be crucial not only in T_H2-cell priming, but also in the maintenance and further polarization of T_H2 central memory cells in allergic diseases (Wang et al. 2006). Furthermore, TSLP-conditioned cDCs are proposed to be involved in secondary positive selection of medium-to-high affinity self-reactive T cells for the generation of CD4$^+$Foxp3$^+$ nT_{reg} cells within the thymus (Watanabe et al. 2005).

3.5 Human pDC Subsets and Functions

Human pDCs are identified by the expression of CD303 (also known as BDCA-2) and CD304 (also known as BDCA-4), in addition to CD45RA and CD123, with a lack of other myeloid markers observed in cDC subsets (Sato and Fujita 2007; Schlitzer and Ginhoux 2014; Murphy et al. 2016) (Table 3). The development of pDCs has a similar requirement of the transcription factor E2-2 to mouse counterparts, as observed in Pitts–Hopkins syndrome (Cisse et al. 2008), and has a unique capacity to secrete high amounts of type-I IFN upon stimulation through endosomal TLR7/9 (Table 3), that is potentially crucial for innate antiviral immune responses and type-I IFN-associated inflammatory autoimmune diseases (Gilliet et al. 2008). Whereas pDCs show a poor ability to induce of T-cell activation in the steady state, pDCs have the potential to elicit T_H1- and T_H2-cell responses, as well as cross-present viral Ags to induce CTL-response after maturation in a context-dependent manner (Grouard et al. 1997; Rissoan et al. 1999; Cella et al. 2000; Hoeffel et al. 2007), although the physiological relevance of these activated DCs is still unclear (Table 3). On the other hand, activated pDCs acquire the ability to not only suppress T-cell activation, but also to induce T-cell anergy and T_{reg} cells mediated by expression of IDO, granzyme B, and inducible T-cell costimulator ligand (ICOSL) (Gilliet and Liu 2002; Moseman et al. 2004; Chen et al. 2008; Jahrsdörfer et al. 2010; Conrad et al. 2012).

3.6 Human Inflammatory DC Subsets and Functions

Several types of human infDCs have been discovered in certain inflammatory conditions, and have been identified as characteristic myeloid cells that express CD1a, CD11b, CD11c, and MHC II molecules with various expression levels of CD1c and CD14, different from other typical myeloid cells such as monocytes and macrophages (Lowes et al. 2005; Hänsel et al. 2011; Segura et al. 2013; Wilsmann-Theis et al. 2013). Tip-DCs represent human infDCs producing TNF-α and iNOS, and have emerged as putative mouse equivalents in pathogenic inflammatory tissues, where they express CD11c, MHC II molecules, costimulatory molecules (e.g., CD40 and CD86), and to some extent, DC-LAMP and CD83, known as DC maturation markers but lack CD1c, as well as CD207 (also known as Langerin), and CD14 as markers of Langerhans cells and monocytes (Lowes et al. 2005; Segura et al. 2013; Wilsmann-Theis et al. 2013). In addition, Tip-DCs are found in psoriatic skin tissues, and produce various sets of proinflammatory cytokines, including IL-12 and IL-23, to drive strong T_H1/T_H17-cell responses (Lowes et al. 2005; Segura et al. 2013; Wilsmann-Theis et al. 2013).

Similar to the generation of murine DCs in vitro, several different precursors have been used to generate human DCs in culture. Culture of $CD34^+$ fraction isolated from BM and umbilical-cord blood with GM-CSF and TNF-α or Flt3-L lead to the generation of cDCs or pDCs, respectively (Caux et al. 1996; Blom et al. 2000), whereas blood monocytes cultured with GM-CSF and IL-4 differentiate into cDCs that resemble Tip-DCs (Sallusto and Lanzavecchia 1994; Watowich and Liu 2010).

4 Tolerogenic DCs and Functions

Given the importance of immature DCs in maintaining self-tolerance under steady-state conditions, much attention has been paid to the potential clinical use of immature DCs generated in vitro for the treatment of immune disorders (Rutella et al. 2006). However, the clinical application of immature DCs may not be suitable for the treatment of immune disorders, because they are not likely to remain "im-maturure" in vivo during recirculation and will reside in the damaged tissue where chronic inflammation is always present after injection into patients. Additionally, their development of "maturity" might rather exacerbate immune disorders through activation of pathogenic T-cells as opposed to regulation of T-cell responses.

Since the tolerogenic capacity of DCs largely depends on their immature state and can be potentiated by immunosuppressive mediators, certain pathogenic components, signals from immune cells or apoptotic cells, as well as the tissue or tumor microenvironment, considerable effort has been made over the past decades to manipulate immature DCs generated in vitro from BM progenitors or blood monocytes of mice or human as a way to retain "immaturity" for the induction of tolerance (Rutella et al. 2006; Morelli and Thomson 2007; Gordon et al. 2014; Osorio et al. 2015). Although there is currently no definition about the phenotype of tolerogenic DCs, they express lower levels of MHC and costimulatory molecules than immature DCs, and are relatively resistant to activation/maturation-inducing signals. While tolerogenic DCs retain the ability to present Ags to Ag-specific T-cells to a lesser extent than their counterparts, the characteristic functions of tolerogenic DCs include the reduction of T-cell activation and induction of T-cell apoptosis, as well as the generation of anergic T-cells and T_{reg} cells (e.g., $CD4^{+}Foxp3^{+}$ iT_{reg} cells and Tr1 cells), and the expansion of $CD4^{+}Foxp3^{+}$ nT_{reg} cells rather than T_{eff} cells. Furthermore, tolerogenic DCs preferentially produce anti-inflammatory cytokines (e.g., IL-10 and TGF-β) rather than proinflammatory cytokines (e.g., IL-12), and are relatively resistant upon activation/maturation-inducing signals. Although, the precise molecular basis underlying tolerogenic DC control of T-cell functions remains to clearly determined, multiple mechanisms likely involve low numbers of antigenic peptide-MHC complexes (APC signal 1) coupled with limited expression of costimulatory molecules (APC signal 2) and proinflammatory cytokines (APC signal 3), and upregulation of several inhibitory molecules (e.g., B7-H1 and IDO) and anti-inflammatory cytokines (e.g., IL-10 and TGF-β).

4.1 Tolerogenic DCs and Functions

A large array of anti-inflammatory and immunosuppressive mediators reportedly drive the tolerogenic phenotype by interfering with checkpoints of DC differentiation and activation (Rutella et al. 2006; Morelli and Thomson 2007; Gordon et al. 2014; Osorio et al. 2015). These include immunosuppressive drugs, such as corticosteroids (e.g., dexamethasone), calcineurin inhibitors (e.g., cyclosporine, tacrolimus, rapamycin), and aspirin; anti-inflammatory factors, such as IL-10, TGF-β1, vascular endothelial growth factor (VEGF), prostaglandin E2 (PGE2), 1α,25-dihydroxyvitamin D_3 (1α,25(OH)2D3), known as vitamin D_3, retinoids, and HLA-G; and tissue factors, such as hepatocyte growth factor (HGF) and vasoactive intestinal peptide (VIP). Such immunosuppressive and pharmacological agents in conjunction with the GM-CSF alone, or GM-CSF and IL-4, are generally used in the ex vivo differentiation of murine BM progenitors and human monocytes into tolerogenic

DCs. Another strategy to maintain the immature phenotype and enhance suppressive T-cell responses is to genetically manipulate immature DCs with viral vectors that overexpress a variety of immunosuppressive molecules, such as anti-inflammatory cytokines (e.g., IL-4, IL-10, TGF-β), inhibitory molecules (e.g., CTLA-4, B7-H1 and IDO), intracellular signaling molecules (e.g., suppressor of cytokine signaling [SOCS]1), and apoptosis-inducing molecules (e.g., Fas/CD95 ligand [FasL/CD95L] and TNF-related apoptosis-inducing ligand [TRAIL]), to inhibit T-cell proliferation, induce and maintain T-cell anergy, and/or promote T-cell apoptosis, or to interfere with the expression of costimulatory molecules (e.g., CD40, CD80, and CD86) and proinflammatory cytokines (e.g., IL-12) to reduce DC immunogenicity. These "induced tolerogenic DCs" exert suppressive function on T-cell responses through various mechanisms in humans and mice in vitro, as described above, and the in vivo application of murine in vitro-generated tolerogenic DCs before the onset of diseases display a protective effect on the initiation and progression of immune pathogenesis in mouse models of autoimmune diseases (e.g., experimental autoimmune encephalomyelitis [EAE] and collagen-induced arthritis [CIA]), allergies (e.g., asthma), graft-versus-host disease (GVHD), and graft rejection (e.g., heart, skin, and pancreatic islet) (Rutella et al. 2006; Morelli and Thomson 2007; Gordon et al. 2014; Osorio et al. 2015). Interestingly, normal and pathogen-infected tissue stromal microenvironments, as well as tumor microenvironment, can control the program of differentiation in murine natural counterparts of in vitro-generated tolerogenic DCs, including $CD11c^{low}CD45RB^{high}$ DCs and $CD11c^{low}CD11b^{high}Ia^{low}$ DCs, through mechanisms including stroma/tumor cell–derived soluble factors (e.g., IL-10, TGF-β, and macrophage colony-stimulating factor [M-CSF]) and cell–cell contact (Wakkach et al. 2003; Svensson et al. 2004; Zhang et al. 2004; Liu and Cao 2015).

4.2 *Clinical Application of Human Tolerogenic DCs*

To translate the experimental success of rodent models into human applications, the functional stability of human tolerogenic DCs is a major concern in order to avoid the possibility that the infused cells could gain immunogenicity in response to inflammatory signals (e.g. proinflammatory cytokines, endogenous and pathogen-derived PRR ligands, and CD40 ligation) encountered in vivo, leading to the exacerbation of immune disorders in the patient. Furthermore, there are many hurdles to overcome for the clinical applications of human tolerogenic DCs, including standardizing protocols for their preparation by good manufacturing practice (GMP) (e.g., methods for generation, yield, purity, viability, sterility, phenotype, and tolerogenic function), and establishing the correct dosage regimen (e.g., optimal dosage and route of infusion, and the number and period of administrations).

Recent advances in the methods for generation of human DCs with the use of monocytes as sources (e.g., closed cell-culture systems, and current GMP-compliant reagents, and facilities) have allowed new strategies for the preparation of clinically applicable tolerogenic DCs, and their clinical trials as "DC-based negative vaccines" in certain immune disorders (e.g., type-1 diabetes, rheumatoid arthritis [RA], and renal graft rejection) have been undertaken in several institutions (Hu and Wan 2011; Raïch-Regué et al. 2014; Thomson et al. 2016). While these phase I clinical trials have shown that "DC-based negative vaccines" using human tolerogenic DCs are safely tolerated in patients with no notable adverse effects, measurable therapeutic efficacy has not been described in most participants. Thus, further preclinical and clinical studies will be needed to address the feasibility and potential efficacy of "DC-based negative vaccines" using human tolerogenic DCs, as well as the development of modified DCs that can exert more potent immunoregulatory function than the currently known classical tolerogenic DCs, to implement their application for therapy of human immune disorders.

5 Regulatory DCs and Functions

Initial studies have shown that classical tolerogenic DCs generated by the culture of human monocytes with GM-CSF and IL-4 in the presence of IL-10 or vitamin D_3 are characterized as heterogeneous subpopulations with the reduced expression of both MHC and costimulatory molecules, which could provide weak "APC signals 1 and 2" to naïve T cells resulting in the reduction of T-cell activation and the generation of anergic T cells in an Ag-specific manner (Steinbrink et al. 1997; Penna and Adorini 2000). On the other hand, it has been suggested that APCs expressing MHC molecules alone, but lacking costimulatory molecules, sufficiently induce clonal T-cell anergy in both humans and mice (Boussiotis et al. 2000). To strengthen the potential immunoregulatory function of DCs, we have modified the methods for the generation of novel type of immunosuppressive human DCs from the culture of human monocytes with GM-CSF and IL-4 in the presence of IL-10 and TGF-β followed by stimulation with TNF-α, and have shown them to be homogenous populations with high expression of MHC molecules, but low expression of costimulatory molecules (Fig. 3a), that were different from the previously known tolerogenic DCs (Sato et al. 2003a). Concurrently, we also have successfully generated modified murine DCs from the culture of BM cells with GM-CSF in the presence of IL-10 and TGF-β followed by stimulation with lipopolysaccharide (LPS), found them to exhibited similar phenotypes to human counterparts (Sato et al. 2003b) (Fig. 3b). Indeed, these novel types of immunosuppressive human and murine DCs had a greater ability to inhibit the activation of T cells and to induce hypo responsiveness in T cells than the previously known tolerogenic DCs, and retained their potent immunoregulatory properties even under inflammatory conditions (Sato et al. 2003a, b). As a notable function, they could induce $CD4^{+}CD25^{+}CD152^{+}$ T_{reg} cells from naïve $CD4^{+}CD25^{-}CD152^{-}$ T cells, and this is the first description of the contribution of DCs to the generation of $CD4^{+}CD25^{+}Foxp3^{+}$ iT_{reg} cells (Sato et al. 2003a, b). To best our knowledge, we have verified for the first time that immunotherapy with immunosuppressive DCs

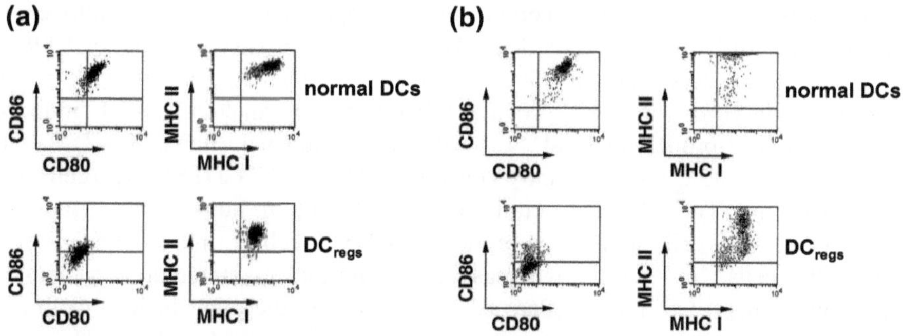

Fig. 3 Phenotype of human and murine DC_{regs}. Expression of MHC (class I and class II) and costimulatory molecules (CD80 and CD86) in normal DCs and DC_{regs} derived from human monocytes (**a**) and murine BM cells (**b**) was analyzed by flow cytometry

exerted a prominent therapeutic effect on murine immune disorders in which the treatment begun even after disease progression (Sato et al. 2003a, b), as information about in vivo efficacy of tolerogenic DCs as therapeutic agents for immune disorders had yet to be addressed in humans and animals before this time. Therefore, we have revealed that immunosuppressive DCs not only retain the expression of MHC molecules with the reduced expression of costimulatory molecules, but also have the ability to induce $CD4^{+}CD25^{+}Foxp3^{+}$ iT_{reg} cells as regulatory DCs (DC_{regs}) (Sato et al. 2003a, b). Since we proposed the concept of “regulatory DCs”, this attractive nomenclature has been also widely used for classical tolerogenic DCs and their derivatives when they possess the potential to regulate T-cell responses for protection against several immune disorders (McCurry et al. 2006; Owens and Kaye 2012; Schmidt et al. 2012; Gordon et al. 2014; Raïch-Regué et al. 2014; Liu and Cao 2015; Thomson et al. 2016). Furthermore, the culture of human monocytes with GM-CSF and IL-4 in the presence of VIP followed by stimulation with LPS or TNF-α, as well as the culture of murine BM cells with GM-CSF in the presence of VIP followed by LPS, have also reportedly generated tolerogenic DCs that resembled the originally described human and murine DC_{regs} in terms of the phenotype, the T-cell regulatory functions, and the protective effects on the progression of murine immune pathogenesis (Gonzalez-Rey et al. 2006; Chorny et al. 2006).

5.1 *Human Regulatory DCs and Functions*

Human DC_{regs} showed moderately high expression levels of MHC molecules (e.g., HLA-A/B/C and HLA-DR), whereas they exhibited extremely low levels of costimulatory molecules (e.g., CD40, CD80 and CD86) as compared with their normal counterparts (Sato et al. 2003a) (Fig. 3a). While human DC_{regs} showed a reduced ability to activate allogeneic naïve $CD4^{+}$ T cells and $CD8^{+}$ T cells than their normal counterparts, they not only exerted direct suppressive effects on allogeneic Ag-specific proliferation of activated $CD4^+$ T cells and CTL activity of $CD8^+T$ cells, but also induced potent allogeneic Ag-specific anergic states upon recall stimulations in vitro (Sato et al. 2003a). Unlike human normal DCs which, induced IFN-γ- and IL-2-producing $CD4^+$ T_{eff} cells from allogeneic naïve $CD4^+$ T cells in vitro, human DC_{reg} predominantly generated $CD4^+CD25^+CD152^+$ iT_{reg} cells and IL-10-producing Tr1 cells (Sato et al. 2003a). In addition, human DC_{regs} induced $CD8^+CD28^-$ T_{reg} cells and IL-10-producing $CD8^+T$ cells from allogeneic naïve $CD8^+$ T cells in vitro, whereas their normal counterparts generated IFN-γ-producing $CD8^+T$ cells (Sato et al. 2003a). On the other hand, preclinical studies have revealed that approximately 1×10^8 of GMP-graded human DC_{regs} or 2.7×10^8 of GMP-graded human DCs could be yielded from nearly 7.5×10^8 of human monocytes obtained from peripheral blood mononuclear cells (PBMCs) derived from the leukapheresis products of 10–12L peripheral blood, respectively (unpublished observations).

5.2 *Murine Regulatory DCs and Functions*

Murine DC_{regs} also had a characteristic phenotype with high levels of MHC molecules and reduced levels of costimulatory molecules (Sato et al. 2003a, b; Fujita et al. 2006, 2007, 2008) (Fig. 3b). Furthermore, murine DC_{regs} preferentially produce IL-10 rather than proinflammatory cytokines (e.g., IL-1, IL-6, IL-12, and TNF-α) following stimulation with TLR ligands, and these events involve the expression of an inhibitor of nuclear factor κB (NF-κB)NS (IκBNS) and B-cell CLL/lymphoma-3 (Bcl-3), as well as cyclic AMP (cAMP)-mediated activation of protein kinase A (PKA) in addition to the suppression of NF-κB- and mitogen-activated protein kinase (MAPK)-mediated signaling cascades (Fujita et al. 2006, 2007). On the other hand, murine DC_{regs} impaired the ability to activate Ag-specific T cells, and instead induced Ag-specific T-cell anergy that was associated with the accumulation of cAMP and subsequent upregulation of $p27^{kip1}$ cyclin-dependent kinase (CDK) inhibitor for arrest at the early G1 phase of the cell cycle (Sato et al. 2003b; Fujita et al. 2007, 2008). The culture of Ag-pulsed murine DC_{regs} with $CD4^+CD25^-Foxp3^-$ T cells allowed the preferential Ag-specific generation of $CD4^+CD25^+Foxp3^+$ T_{reg} cells when *Rag2*$^{-/-}$DO11.10 BALB/c mice harboring $CD4^+$ T cells expressing the transgenic ovalbumin (OVA)-specific T cell receptor (TCR), but lacking $CD4^+CD25^+Foxp3^+$ nT_{reg} cells were used for assay of the differentiation of $CD4^+CD25^+Foxp3^+$ iT_{reg} cells (Fujita et al. 2007).

Given the similar phenotype and T-cell regulatory functions in vitro of murine DC_{regs} to human DC_{regs} (Sato et al. 2003a, b), we have addressed the therapeutic potentials of the immunotherapy with murine DC_{regs} for the treatment of acute and chronic immune disorders in experimental murine models. A single injection of recipient-type, but not donor-type, murine DC_{regs} following MHC-incompatible allogeneic bone marrow transplantation (BMT) completely protected the recipient mice bearing leukemia from the lethality caused by acute GVHD and tumor burden (Sato et al. 2003b), indicating that murine DC_{regs} could protect against acute GVHD and leukemia relapse mediated through the control of the ability of the transplanted T-cells to induce acute GVHD and the graft-versus-leukemia (GVL) effect. Similarly, treatment of the recipient mice with murine recipient-type DC_{regs} led to greater suppression of the incidence and severity of cutaneous inflammatory pathogenesis than rapamycin in the major histocompatibility complex (MHC)-compatible and multiple minor histocompatibility Ag (miHAg)-incompatible model of chronic GVHD in allogeneic BMT (Fujita et al. 2007; Sato et al. 2009). Furthermore, treatment with allergic Ag-pulsed murine DC_{regs} after immunization not only impaired the production of Ag-specific IgE, IgG1, and IgG2a, but also abrogated the allergic airway inflammation in sensitized mice in a murine asthmatic model (Fujita et al. 2008). In vivo-blockade experiments with anti-CD25 mAb have suggested that the protective effects of murine DC_{regs} are largely dependent on the generation of $CD4^+CD25^+Foxp3^+T_{reg}$ cells in vivo (Sato et al. 2003b; Fujita et al. 2007, 2008). Indeed, analysis of adoptive transfer experiments with Ag-pulsed DC_{regs} and $CD4^+CD25^-Foxp3^-T$ cells derived from *Rag2*$^{-/-}$DO11.10 BALB/c mice clearly demonstrated that DC_{regs} could directly generate $CD4^+CD25^+$ $Foxp3^+T_{reg}$ cells from $CD4^+CD25^-Foxp3^-T$ cells in an Ag-specific manner in vivo (Fujita et al. 2007). On the other hand, murine DC_{regs} protected mice against septic lethality induced by experimental endotoxemia and bacterial peritonitis mediated

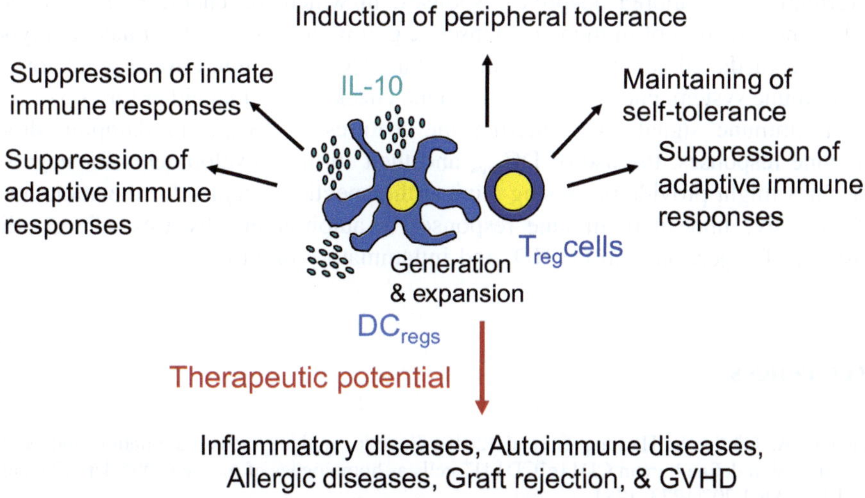

Fig. 4 Regulation of immune responses by DC_{regs}, and their potential clinical application. DC_{regs} directly suppress innate and adaptive immune responses via the production of IL-10, and the induction of anergic T cells, and the generation and expansion of T_{reg} cells, which participate in maintaining of self-tolerance and suppression of adaptive immune responses. Therefore, DC_{reg}-based negative vaccines have the potential to treat autoimmune diseases, allergic diseases, graft rejection, GVHD, and inflammatory diseases

through the IL-10-dependent suppression of systemic inflammatory responses (Fujita et al. 2006). Taken together, DC_{regs} may have preventive and therapeutic potential for the treatment of T-cell-mediated immune disorders, as well as inflammatory diseases (Fig. 4). While the existence of human and murine naturally occurring DC_{regs} remains to be determined, $CD49^{+}CD90^{+}CD200R3^{+}$I-A/I-$E^{+}CD11b^{+}CD11c^{low}CD40^{-}CD80^{-}CD86^{-}$ leukocytes might be a candidate for a natural counterpart of murine DC_{regs} (Sato et al. 2009). Further investigation is needed to clarify their origin, development, and genetic program/transcriptional factor, and specific surface markers that could provide valuable information for the rational design of DC_{reg}-based immunotherapeutic for the control of immune disorders.

6 Conclusions

Extensive recent studies have led to considerable progress in our understanding of various aspects of the biology of DCs, including their development, behavior, and function. Importantly, the diverse functions of DCs are not only controlled by intrinsic factors (i.e., transcription factors, signaling molecules, membrane proteins, and enzymes), but also by stimulation with extrinsic factors (i.e., PRR ligands, cytokines/chemokines, and environmental mediators). Furthermore, DCs appear to play key roles in initiating and orchestrating immune responses, providing innate information for tailored adaptive responses, in which the characteristics of DCs reflect the direction of immune responses (e.g. immunity versus tolerance and types of T cell-mediated adaptive immunity). Thus, DCs constitute critical regulators of the immune system, endowed with "immune checkpoints" that either turn up or turn down immune signals. Considering the features of DC_{regs} in damping down immune responses, the use of DC_{regs} and their further developments for negative vaccines might provide promising clinical therapeutic strategies for the attenuation of excessive undesired immune responses in autoimmune diseases, allergic diseases, graft rejection, and GVHD, and inflammatory diseases.

References

Bachem A, Güttler S, Hartung E et al (2010) Superior antigen cross-presentation and XCR1 expression define human $CD11c^{+}CD141^{+}$ cells as homologues of mouse $CD8^{+}$ dendritic cells. J Exp Med 207:1273–1281

Bedoui S, Heath WR (2015) Krüppel-ling of IRF4-dependent DCs into two functionally distinct DC subsets. Immunity 42:785–787

Bekiaris V, Persson EK, Agace WW (2014) Intestinal dendritic cells in the regulation of mucosal immunity. Immunol Rev 260:86–101

Birnberg T, Bar-On L, Sapoznikov A et al (2008) Lack of conventional dendritic cells is compatible with normal development and T cell homeostasis, but causes myeloid proliferative syndrome. Immunity 29:986–997
Bjorck P (2001) Isolation and characterization of plasmacytoid dendritic cells from Flt3 ligand and granulocyte-macrophage colony-stimulating factor-treated mice. Blood 98:3520–3526
Blom B, Ho S, Antonenko S et al (2000) Generation of interferon α-producing predendritic cell (Pre-DC)2 from human CD34^{+} hematopoietic stem cells. J Exp Med 192:1785–1796
Boussiotis VA, Freeman GJ, Taylor PA et al (2000) p27^{kip1} functions as an anergy factor inhibiting interleukin 2 transcription and clonal expansion of alloreactive human and mouse helper T lymphocytes. Nat Med 6:290–297
Caux C, Vanbervliet B, Massacrier C et al (1996) CD34^{+} hematopoietic progenitors from human cord blood differentiate along two independent dendritic cell pathways in response to GM-CSF + TNFα. J Exp Med 184:695–706
Cella M, Facchetti F, Lanzavecchia A et al (2000) Plasmacytoid dendritic cells activated by influenza virus and CD40L drive a potent TH1 polarization. Nat Immunol 1:305–310
Chen W, Liang X, Peterson AJ et al (2008) The indoleamine 2,3-dioxygenase pathway is essential for human plasmacytoid dendritic cell-induced adaptive T regulatory cell generation. J Immunol 181:5396–5404
Chorny A, Gonzalez-Rey E, Fernandez-Martin A et al (2006) Vasoactive intestinal peptide induces regulatory dendritic cells that prevent acute graft-versus-host disease while maintaining the graft-versus-tumor response. Blood 107:3787–3794
Cisse B, Caton ML, Lehner M et al (2008) Transcription factor E2-2 is an essential and specific regulator of plasmacytoid dendritic cell development. Cell 135:37–48
Colonna M, Trinchieri G, Liu YJ (2004) Plasmacytoid dendritic cells in immunity. Nat Immunol 5:1219–1226
Cong Y (2011) Molecular gastronomy: how to make the critical intestinal Foxp3^{+} Treg cell. Gastroenterology 141:1559–1562
Conrad C, Gregorio J, Wang YH et al (2012) Plasmacytoid dendritic cells promote immunosuppression in ovarian cancer via ICOS costimulation of Foxp3^{+} T-regulatory cells. Cancer Res 72:5240–5249
Coombes JL, Siddiqui KR, Arancibia-Cárcamo CV et al (2007) A functionally specialized population of mucosal CD103^{+} DCs induces Foxp3^{+} regulatory T cells via a TGF-β and retinoic acid-dependent mechanism. J Exp Med 204:1757–1764
Darrasse-Jèze G, Deroubaix S, Mouquet H et al (2009) Feedback control of regulatory T cell homeostasis by dendritic cells in vivo. J Exp Med 206:1853–1862
Dudziak D, Kamphorst AO, Heidkamp GF et al (2007) Differential antigen processing by dendritic cell subsets in vivo. Science 315:107–111
Fontenot JD, Rasmussen JP, Gavin MA et al (2005) A function for interleukin 2 in Foxp3-expressing regulatory T cells. Nat Immunol 6:1142–1151
Fujita S, Seino K, Sato K et al (2006) Regulatory dendritic cells act as regulators of acute lethal systemic inflammatory response. Blood 107:3656–3664
Fujita S, Sato Y, Sato K et al (2007) Regulatory dendritic cells protect against cutaneous chronic graft-versus-host disease mediated through CD4^{+}CD25^{+}Foxp3^{+} regulatory T cells. Blood 110:3793–3803
Fujita S, Yamashita N, Ishii Y et al (2008) Regulatory dendritic cells protect against allergic airway inflammation in a murine asthmatic model. J Allergy Clin Immunol 121:95–104
Fukaya T, Takagi H, Sato Y et al (2010) Crucial roles of B7-H1 and B7-DC expressed on mesenteric lymph node dendritic cells in the generation of antigen-specific CD4^{+} Foxp3^{+} regulatory T cells in the establishment of oral tolerance. Blood 116:2266–2276
Fukaya T, Murakami R, Takagi H et al (2012) Conditional ablation of CD205^{+} conventional dendritic cells impacts the regulation of T-cell immunity and homeostasis in vivo. Proc Natl Acad Sci USA 109:11288–11293
Gilliet M, Liu YJ (2002) Generation of human CD8 T regulatory cells by CD40 ligand-activated plasmacytoid dendritic cells. J Exp Med 195:695–704

Gilliet M, Boonstra A, Paturel C et al (2002) The development of murine plasmacytoid dendritic cell precursors is differentially regulated by FLT3-ligand and granulocyte/ macrophage colony-stimulating factor. J Exp Med 195:953–958

Gilliet M, Cao W, Liu YJ (2008) Plasmacytoid dendritic cells: sensing nucleic acids in viral infection and autoimmune diseases. Nat Rev Immunol 8:594–606

Glitzner E, Korosec A, Brunner PM et al (2014) Specific roles for dendritic cell subsets during initiation and progression of psoriasis. EMBO Mol Med 6:1312–1327

Guilliams M, Ginhoux F, Jakubzick C et al (2014) Dendritic cells, monocytes and macrophages: a unified nomenclature based on ontogeny. Nat Rev Immunol 14:571–578

Gonzalez-Rey E, Chorny A, Fernandez-Martin A et al (2006) Vasoactive intestinal peptide generates human tolerogenic dendritic cells that induce CD4 and CD8 regulatory T cells. Blood 107:3632–3638

Gordon JR, Ma Y, Churchman L et al (2014) Regulatory dendritic cells for immunotherapy in immunologic diseases. Front Immunol 5:7

Goubier A, Dubois B, Gheit H et al (2008) Plasmacytoid dendritic cells mediate oral tolerance. Immunity 29:464–475

Grouard G, Rissoan MC, Filgueira L et al (1997) The enigmatic plasmacytoid T cells develop into dendritic cells with interleukin (IL)-3 and CD40-ligand. J Exp Med 185:1101–1111

Guéry L, Dubrot J, Lippens C et al (2014) Ag-presenting CpG-activated pDCs prime Th17 cells that induce tumor regression. Cancer Res 74:6430–6440

Hänsel A, Günther C, Ingwersen J et al (2011) Human slan (6-sulfo LacNAc) dendritic cells are inflammatory dermal dendritic cells in psoriasis and drive strong T_H17/T_H1 T-cell responses. J Allergy Clin Immunol 127:787–794

Haniffa M, Shin A, Bigley V et al (2012) Human tissues contain CD141^{hi} cross- presenting dendritic cells with functional homology to mouse CD103$^+$ nonlymphoid dendritic cells. Immunity 37:60–73

Hildner K, Edelson BT, Purtha WE et al (2008) Batf3 deficiency reveals a critical role for CD8α^+ dendritic cells in cytotoxic T cell immunity. Science 322:1097–1100

Hoeffel G, Ripoche AC, Matheoud D et al (2007) Antigen cross presentation by human plasmacytoid dendritic cells. Immunity 27:481–492

Hohl TM, Rivera A, Lipuma L et al (2009) Inflammatory monocytes facilitate adaptive CD4 T cell responses during respiratory fungal infection. Cell Host Microbe 6:470–481

Hoshino K, Kaisho T (2008) Nucleic acid sensing Toll-like receptors in dendritic cells. Curr Opin Immunol 20:408–413

Hu J, Wan Y (2011) Tolerogenic dendritic cells and their potential applications. Immunology 132:307–314

Inaba K, Inaba M, Romani N et al (1992) Generation of large numbers of dendritic cells from mouse bone marrow cultures supplemented with granulocyte/macrophage colony-stimulating factor. J Exp Med 176:1693–1702

Irla M, Küpfer N, Suter T et al (2010) MHC class II-restricted antigen presentation by plasmacytoid dendritic cells inhibits T cell-mediated autoimmunity. J Exp Med 207:1891–1905

Jahrsdörfer B, Vollmer A, Blackwell SE et al (2010) Granzyme B produced by human plasmacytoid dendritic cells suppresses T-cell expansion. Blood 115:1156–1165

Joffre O, Nolte MA, Spörri R et al (2009) Inflammatory signals in dendritic cell activation and the induction of adaptive immunity. Immunol Rev 227:234–247

Jongbloed SL, Kassianos AJ, McDonald KJ et al (2010) Human CD141$^+$ (BDCA-3)$^+$ dendritic cells (DCs) represent a unique myeloid DC subset that cross-presents necrotic cell antigens. J Exp Med 207:1247–1260

Kim JM, Rasmussen JP, Rudensky AY (2007) Regulatory T cells prevent catastrophic autoimmunity throughout the lifespan of mice. Nat Immunol 8:191–197

Kornete M, Piccirillo CA (2012) Functional crosstalk between dendritic cells and Foxp3$^+$ regulatory T cells in the maintenance of immune tolerance. Front Immunol 3:165

Kretschmer K, Apostolou I, Hawiger D et al (2005) Inducing and expanding regulatory T cell populations by foreign antigen. Nat Immunol 6:1219–1227

Lauterbach H, Bathke B, Gilles S et al (2010) Mouse CD8α^+ DCs and human BDCA3$^+$ DCs are major producers of IFN-lambda in response to poly IC. J Exp Med 207:2703–2717
León B, López-Bravo M, Ardavín C (2007) Monocyte-derived dendritic cells formed at the infection site control the induction of protective T helper 1 responses against *Leishmania*. Immunity 26:519–531
Liu J, Cao X (2015) Regulatory dendritic cells in autoimmunity: a comprehensive review. J Autoimmun 63:1–12
Lowes MA, Chamian F, Abello MV et al (2005) Increase in TNF-α and inducible nitric oxide synthase-expressing dendritic cells in psoriasis and reduction with efalizumab (anti-CD11a). Proc Natl Acad Sci USA 102:19057–19062
Luda KM, Joeris T, Persson EK et al (2016) IRF8 Transcription-factor-dependent classical dendritic cells are essential for intestinal T cell homeostasis. Immunity 44:860–874
McCurry KR, Colvin BL, Zahorchak AF et al (2006) Regulatory dendritic cell therapy in organ transplantation. Transpl Int 19:525–538
Manicassamy S, Pulendran B (2011) Dendritic cell control of tolerogenic responses. Immunol Rev 241:206–227
Merad M, Sathe P, Helft J et al (2013) The dendritic cell lineage: ontogeny and function of dendritic cells and their subsets in the steady state and the inflamed setting. Annu Rev Immunol 31:563–604
Mittag D, Proietto AI, Loudovaris T et al (2011) Human dendritic cell subsets from spleen and blood are similar in phenotype and function but modified by donor health status. J Immunol 186:6207–6217
Mochizuki K, Xie F, He S et al (2013) Delta-like ligand 4 identifies a previously uncharacterized population of inflammatory dendritic cells that plays important roles in eliciting allogeneic T cell responses in mice. J Immunol 190:3772–3782
Molenaar R, Knippenberg M, Goverse G et al (2011) Expression of retinaldehyde dehydrogenase enzymes in mucosal dendritic cells and gut-draining lymph node stromal cells is controlled by dietary vitamin A. J Immunol 186:1934–1942
Morelli AE, Thomson AW (2007) Tolerogenic dendritic cells and the quest for transplant tolerance. Nat Rev Immunol 7:610–621
Moseman EA, Liang X, Dawson AJ et al (2004) Human plasmacytoid dendritic cells activated by CpG oligodeoxynucleotides induce the generation of CD4$^+$CD25$^+$ regulatory T cells. J Immunol 173:4433–4442
Murphy TL, Grajales-Reyes GE, Wu X et al (2016) Transcriptional control of dendritic cell development. Annu Rev Immunol 34:93–119
Nakano H, Lin KL, Yanagita M et al (2009) Blood-derived inflammatory dendritic cells in lymph nodes stimulate acute T helper type 1immune responses. Nat Immunol 10:394–402
Ochando JC, Homma C, Yang Y et al (2006) Alloantigen-presenting plasmacytoid dendritic cells mediate tolerance to vascularized grafts. Nat Immunol 7:652–662
Ohnmacht C, Pullner A, King SB et al (2009) Constitutive ablation of dendritic cells breaks self-tolerance of CD4 T cells and results in spontaneous fatal autoimmunity. J Exp Med 206:549–559
Osorio F, Fuentes C, López MN et al (2015) Role of dendritic cells in the induction of lymphocyte tolerance. Front Immunol 6:535
Owens BM, Kaye PM (2012) Stromal cell induction of regulatory dendritic cells. Front Immunol 3:262
Pallotta MT, Orabona C, Volpi C et al (2011) Indoleamine 2,3-dioxygenase is a signaling protein in long-term tolerance by dendritic cells. Nat Immunol 12:870–878
Penna G, Adorini L (2000) 1α,25-dihydroxyvitamin D3 inhibits differentiation, maturation, activation, and survival of dendritic cells leading to impaired alloreactive T cell activation. J Immunol 164:2405–2411
Poulin LF, Salio M, Griessinger E et al (2010) Characterization of human DNGR-1$^+$ BDCA3$^+$ leukocytes as putative equivalents of mouseCD8α^+ dendritic cells. J Exp Med 207:1261–1271

Raïch-Regué D, Glancy M, Thomson AW (2014) Regulatory dendritic cell therapy: from rodents to clinical application. Immunol Lett 161:216–221
Rissoan MC, Soumelis V, Kadowaki N et al (1999) Reciprocal control of T helper cell and dendritic cell differentiation. Science 283:1183–1186
Roncarolo MG, Levings MK, Traversari C (2001) Differentiation of T regulatory cells by immature dendritic cells. J Exp Med 193:F5–F9
Rowland SL, Riggs JM, Gilfillan S et al (2014) Early, transient depletion of plasmacytoid dendritic cells ameliorates autoimmunity in a lupus model. J Exp Med 211:1977–1991
Rutella S, Danese S, Leone G (2006) Tolerogenic dendritic cells: cytokine modulation comes of age. Blood 108:1435–1440
Sage AP, Murphy D, Maffia P et al (2014) MHC Class II-restricted antigen presentation by plasmacytoid dendritic cells drives proatherogenic T cell immunity. Circulation 130:1363–1373
Sallusto F, Lanzavecchia A (1994) Efficient presentation of soluble antigen by cultured human dendritic cells is maintained by granulocyte/macrophage colony-stimulating factor plus interleukin 4 and downregulated by tumor necrosis factor α. J Exp Med 179:1109–1118
Sato K, Fujita S (2007) Dendritic cells-nature and classification. Allergol Int 56:183–191
Sato K, Yamashita N, Baba M et al (2003a) Modified myeloid dendritic cells act as regulatory dendritic cells to induce anergic and regulatory T cells. Blood 101:3581–3589
Sato K, Yamashita N, Yamashita N et al (2003b) Regulatory dendritic cells protect mice from murine acute graft-versus-host disease and leukemia relapse. Immunity 18:367–379
Sato K, Eizumi K, Fukaya T et al (2009) Naturally occurring regulatory dendritic cells regulate murine cutaneous chronic graft-versus-host disease. Blood 113:4780–4789
Schlitzer A, McGovern N, Teo P et al (2013) IRF4 transcription factor-dependent $CD11b^{+}$ dendritic cells in human and mouse controlmucosal IL-17 cytokine responses. Immunity 38:970–983
Schlitzer A, Ginhoux F (2014) Organization of the mouse and human DC network. Curr Opin Immunol 26:90–99
Schmidt SV, Nino-Castro AC, Schultze JL (2012) Regulatory dendritic cells: there is more than just immune activation. Front Immunol 3:274
Segura E, Touzot M, Bohineust A et al (2013) Human inflammatory dendritic cells induce Th17 cell differentiation. Immunity 38:336–348
Serbina NV, Salazar-Mather TP, Biron CA et al (2003) TNF/iNOS-producing dendritic cells mediate innate immune defense against bacterial infection. Immunity 19:59–70
Shortman K, Naik SH (2007) Steady-state and inflammatory dendritic-cell development. Nat Rev Immunol 7:19–30
Sisirak V, Ganguly D, Lewis KL et al (2014) Genetic evidence for the role of plasmacytoid dendritic cells in systemic lupus erythematosus. J Exp Med 211:1969–1976
Steinbrink K, Wolfl M, Jonuleit H et al (1997) Induction of tolerance by IL-10-treated dendritic cells. J Immunol 159:4772–4780
Steinman RM, Cohn ZA (1973) Identification of a novel cell type in peripheral lymphoid organs of mice. I. Morphology, quantitation, tissue distribution. J Exp Med 137:1142–1162
Steinman RM, Witmer MD (1978) Lymphoid dendritic cells are potent stimulators of the primary mixed leukocyte reaction in mice. Proc Natl Acad Sci USA 175:5132–5136
Steinman RM, Nussenzweig MC (2002) Avoiding horror autotoxicus: the importance of dendritic cells in peripheral T cell tolerance. Proc Natl Acad Sci USA 99:351–358
Sun CM, Hall JA, Blank RB et al (2007) Small intestine lamina propria dendritic cells promote de novo generation of Foxp3 T reg cells via retinoic acid. J Exp Med 204:1775–1785
Svensson M, Maroof A, Ato M et al (2004) Stromal cells direct local differentiation of regulatory dendritic cells. Immunity 21:805–816
Swiecki M, Colonna M (2015) The multifaceted biology of plasmacytoid dendritic cells. Nat Rev Immunol 15:471–485
Takagi H, Fukaya T, Eizumi K et al (2011) Plasmacytoid dendritic cells are crucial for the initiation of inflammation and T cell immunity in vivo. Immunity 35:958–971

Takagi H, Arimura K, Uto T et al (2016) Plasmacytoid dendritic cells orchestrate TLR7-mediated innate and adaptive immunity for the initiation of autoimmune inflammation. Sci Rep 6:24477
Tezuka H, Abe Y, Iwata M et al (2007) Regulation of IgA production by naturally occurringTNF/iNOS-producing dendritic cells. Nature 448:929–933
Thomson AW, Zahorchak AF, Ezzelarab MB et al (2016) Prospective clinical testing of regulatory dendritic cells in organ transplantation. Front Immunol 7:15
Tussiwand R, Everts B, Grajales-Reyes GE et al (2015) *Klf4* expression in conventional dendritic cells is required for T helper 2 cell responses. Immunity 42:916–928
Uto T, Fukaya T, Takagi H et al (2016) Clec4A4 is a regulatory receptor for dendritic cells that impairs inflammation and T-cell immunity. Nat Commun 7:11273
van de Laar L, Coffer PJ, Woltman AM (2012) Regulation of dendritic cell development by GM-CSF: molecular control and implications for immune homeostasis and therapy. Blood 119:3383–3393
van Vliet SJ, den Dunnen J, Gringhuis SI et al (2007) Innate signaling and regulation of dendritic cell immunity. Curr Opin Immunol 19:435–440
Villadangos JA, Schnorrer P (2007) Intrinsic and cooperative antigen-presenting functions of dendritic-cell subsets in vivo. Nat Rev Immunol 7:543–555
Wakkach A, Fournier N, Brun V et al (2003) Characterization of dendritic cells that induce tolerance and T regulatory 1 cell differentiation in vivo. Immunity 18:605–617
Walsh KP, Mills KH (2013) Dendritic cells and other innate determinants of T helper cell polarisation. Trends Immunol 34:521–530
Wang YH, Ito T, Wang YH et al (2006) Maintenance and polarization of human T_H2 central memory T cells by thymic stromal lymphopoietin-activated dendritic cells. Immunity 24: 827–838
Watanabe N, Wang YH, Lee HK et al (2005) Hassall's corpuscles instruct dendritic cells to induce $CD4^+CD25^+$ regulatory T cells in human thymus. Nature 436:1181–1185
Watowich SS, Liu YJ (2010) Mechanisms regulating dendritic cell specification and development. Immunol Rev 238:76–92
Wilsmann-Theis D, Koch S, Mindnich C et al (2013) Generation and functional analysis of human TNF-α/iNOS-producing dendritic cells (Tip-DC). Allergy 68:890–898
Yun TJ, Lee JS, Machmach K et al (2016) Indoleamine 2,3-dioxygenase-expressing aortic plasmacytoid dendritic cells protect against atherosclerosis by induction of regulatory T cells. Cell Metab 23:852–866
Zhang M, Tang H, Guo Z et al (2004) Splenic stroma drives mature dendritic cells to differentiate into regulatory dendritic cells. Nat Immunol 5:1124–1133

Part II
Extracellular Immune Checkpoint Molecules

Role of PD-1 in Immunity and Diseases

Kenji Chamoto, Muna Al-Habsi and Tasuku Honjo

Abstract Immunity developed to defend our bodies from foreign particles, including bacteria and viruses. Although effector cells responsible for acquired immunity, mainly T cells, and B cells, are able to distinguish self from non-self, they sometimes attack the body's tissues because of imperfect central tolerance. Several immune check points developed to limit overactivation of these cells. One of the most important immune checkpoints is programmed cell death-1 (PD-1), which is expressed mainly on activated lymphocytes. As its ligands (PD-Ls) are expressed widely in the body and affect the responses against self and foreign antigens, controlling PD-1/PD-L interactions enables the management of several immune-related diseases such as autoimmune disease, virus infection, and cancers. Currently, the strategy of PD-1/ PD-L1 blockade has already been applied to clinical cancer therapy, providing evidences that PD-1 signal is one of the main factors of cancer immune escape in humans. The dramatic efficacy of PD-1 blockade in cancer immunotherapy, promises the control of other immune diseases by PD-1 signal modulation. In this review, we summarize the history of PD-1, subsequent basic studies, and their application to the clinic.

Kenji Chamoto and Muna Al-Habsi equally contributed to this work.

K. Chamoto · M. Al-Habsi
Department of Immunology and Genomic Medicine, Graduate School of Medicine, Kyoto University, Kyoto, Japan

T. Honjo (✉)
Department of Immunology and Genomic Medicine, Graduate School of Medicine, Kyoto University, Yoshida Konoe-Cho, Sakyo-Ku, 606-8501 Kyoto, Japan
e-mail: honjo@mfour.med.kyoto-u.ac.jp

Current Topics in Microbiology and Immunology (2017) 410:75–98
DOI 10.1007/82_2017_67

Published Online: 17 September 2017

Contents

1 Introduction 76
2 History of PD-1 and Its Ligands 77
2.1 Discovery 77
2.2 Structure 78
3 Molecular Functions 81
4 Expression and Biological Roles 82
4.1 Expression Profile of PD-1 and PD-Ls 82
4.2 Role in Autoimmunity 83
4.3 Role in Viral Immunology 85
4.4 Role in Tumor Immunology 87
5 Issues and Future Directions in PD-1 Blockade Cancer Immunotherapy 90
References 92

1 Introduction

The immune system functions to defend against foreign particles while avoiding self-recognition. Maintaining this balance requires critically organized interactions between the different components of the immune system, cellular and humoral immunity (Bardhan et al. 2016). The immune system employs regulatory immune cells, discussed in previous chapters, to control the activity of T cells and antibody-producing B cells (Francisco et al. 2009). The activity of T-lymphocytes is regulated by a complex system of signals generated by both stimulatory and inhibitory receptors (Probst et al. 2005). These receptors are expressed on the surface of lymphocytes and mediate cell–cell communication to determine their responses toward different antigens. These stimulatory and inhibitory receptors enable the immune system to react appropriately to foreign antigens and inhibit responses against autoantigens (Bardhan et al. 2016). Programmed cell death 1 (PD-1) is a major inhibitory receptor that is preferentially expressed on T and B cells. It is also reported to be expressed by other subsets such as natural killer cells, monocytes, and dendritic cells (He et al. 2015). PD-1 is a member of the CD28 superfamily that generates a negative signal upon interaction with its ligand. It conjugates with two ligands (PD-L1 and PD-L2), which are expressed widely by the components of the immune system as well as other cell types in the body (Bardhan et al. 2016). Interactions between PD-1 and its ligands play a key immunoregulatory role in T-lymphocyte activation and tolerance. The discovery of PD-1 and its ligands, PD-L1 and PD-L2, has changed our perspective of the immune system. The PD-1/PD-L-1 pathway plays a major role in maintaining tolerance, therefore protecting the body from auto-reactivity (Okazaki et al. 2013). PD-1 signals are also associated with the attenuation of immune responses against viruses, chronic bacterial infections, and tumors. Several strategies for blocking PD-1/PD-L1 interaction, promoting the responses of T-lymphocytes toward antigens, led to many basic and clinical studies. These will be discussed later in this chapter.

2 History of PD-1 and Its Ligands

2.1 Discovery

PD-1 was identified in 1992 by Honjo and colleagues at Kyoto University as a gene involved in inducing programmed cell death during T cell thymic selection. The group was studying two cell lines: a hematopoietic progenitor cell line (LyD9) that induces programmed cell death when deprived of interleukin IL-3, and a murine T cell hybridoma (2B4.11) that undergoes programmed cell death upon stimulation with phorbol 12-myristate 13-acetate and ionomycin (Ishida et al. 1992). Both cell lines induced de novo RNA and protein synthesis as they underwent apoptosis. The group performed subtractive hybridization to identify the gene responsible for the induction. They created two complementary DNA (cDNA) libraries subtracting collective mRNA of resting cells from collective mRNA of dying cells in LyD9 and 2B4.11 cell lines. Then they screened the resulting libraries for common genes by hybridization (Okazaki and Honjo 2007). PD-1 was predicted to be a type I transmembrane protein with a single IgV domain in the extracellular region based on the amino acid sequence. However, overexpression of PD-1 cDNA in these cell lines failed to induce apoptosis (Agata et al. 1996). The function of PD-1 remained unclear until the same group found that PD-1 negatively regulates immune responses, as PD-1-deficient mice spontaneously developed lupus-like arthritis and glomerulonephritis in 1999 (Nishimura et al. 1999).

PD-1 ligands were discovered through the integrated efforts of several research groups. Freeman's group at Harvard University identified a B7-like molecule (named as clone 129) by database searching and established a collaboration with the Genetics Institute Wyeth-Ayerst Research in Massachusetts to search for its receptor. At the same time, Honjo's group collaborated with the Genetics Institute to identify the ligand of PD-1. The Genetics Institute demonstrated an interaction between the two molecules as the engagement of PD-1 with the clone 129 molecule inhibited the proliferation and cytokine production of T-lymphocytes following stimulation with anti-CD3 antibodies (Freeman et al. 2000). This provided evidence that clone 129 is the ligand of PD-1, and this molecule was renamed as PD-L1 (Okazaki and Honjo 2007; Freeman et al. 2000). PD-L2 was later discovered as another ligand for PD-1 through the collaboration by the same groups (Latchman et al. 2001). At around the same time, other groups reported two molecules (B7-H1 and B7-DC) that were identical to PD-L1 and PD-L2, respectively, which were suspected to be co-stimulatory in regulating T cell activation (Dong et al. 1999; Tseng et al. 2001). PD-1 and its ligands were included as new members of the CD28 family (Fig. 1) (Okazaki and Honjo 2007).

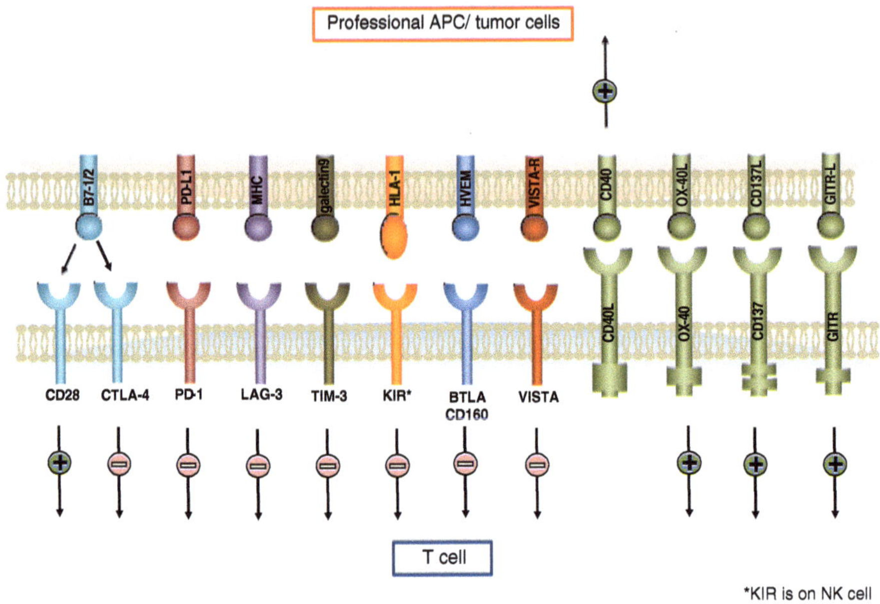

Fig. 1 CD28 family members that transduce positive or negative signals

2.2 *Structure*

PD-1 is a 50–55-kDa type I transmembrane glycoprotein composed of an IgV domain (Fig. 2). It belongs to the CD28/CTLA-4 subfamily of the Ig superfamily and shares 21–33% sequence identity with CTLA-4, CD28, and ICOS (Yamazaki et al. 2002). Unlike other CD28 family members, PD-1 lacks the membrane proximal cysteine residue required for homodimerization and exists as a monomer on the cell surface (Zhang et al. 2004). The cytoplasmic region of PD-1 contains two tyrosine residues, with the N-terminal tyrosine located in an immunoreceptor tyrosine-based inhibitory motif (ITIM) and C-terminal tyrosine in an immunoreceptor tyrosine-based switch motif (ITSM), unlike the C-terminal of CTLA-4 which contains no signaling motif (Fig. 2). Although the ITIM motif was originally thought to be responsible for the inhibitory function of PD-1, in vitro studies later showed that ITSM is more important for PD-1 activity (Okazaki et al. 2001). Studies also indicated that during evolution, only 60% of the amino acid sequence in the cytoplasmic region of PD-1 was conserved between humans and mice (Okazaki and Honjo 2007). However, the sequence surrounding ITSM was completely conserved between humans and mice, confirming its functional importance (Okazaki et al. 2001).

The ligands of PD-1 (PD-Ls) are type I transmembrane glycoproteins composed of typical IgC and IgV domains, which are representative features of B7 family molecules (Freeman 2008). The amino acid identity between PD-L1 and PD-L2 is

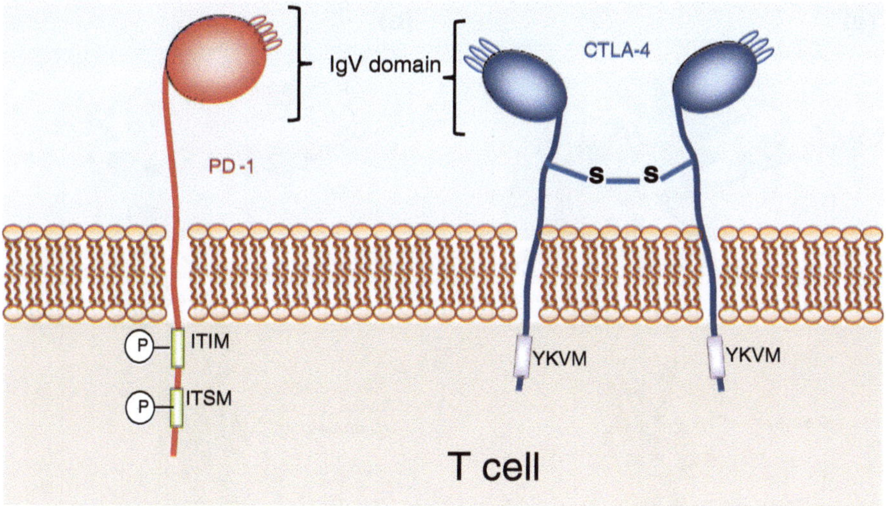

Fig. 2 Structure of PD-1 compared with that of CTLA-4

30–40%, while the amino acid identity between the PD-Ls and B7s is approximately 20% (Bardhan et al. 2016; Lazar-Molnar et al. 2008). The structures show a 1:1 receptor/ligand stoichiometry, with interactions primarily between the faces of the IgV domains. PD-1 is expressed on the cell surface as a monomer, unlike other members of the CD28 family: CTLA-4 forms covalent dimers, while B7-1 and CD28 form noncovalent dimers (Freeman 2008) (Fig. 2). An IgV domain is generally composed of approximately 120 amino acids organized into 9 parallel β-strands designated as ABCC′C″DEFG connected by loops (Fig. 3a, b). A disulfide bond connects the B and F strands folding IgV domain of PD-1 molecule into a two-layered sandwich structure (Freeman 2008). The IgV domains of PD-1 and its ligands form a binding interface via their front β-sheets. Residues from GFCC′ sheets as well as the ones from the CC′, CC″, and FG loops on PD-1 bind the frontal β-face (GFCC′) on PD-L1, and residues from the AGFC strands and FG loop on PD-L2 (Fig. 3a, b). This binding configuration hides the β-strands, which form a large hydrophobic surface area, from the aqueous environment. Eight of the 14 amino acids involved in binding to PD-1 are identical or highly conserved between PD-L1 and PD-L2. Face-to-face binding establishes an acute angle between PD-1 and PD-L1 or PD-L2, which shortens the distance between the distal ends of the IgV domain of PD-1 and IgC domain of PD-L1 or PD-L2 molecules (76 Å compared to 100 Å in CTLA-4/B7-1 complex) as shown in Fig. 3c. The size of the immunological synapse is determined by the length of T cell receptor (TCR)/major histocompatibility (MHC) complex (140 Å). To adjust to this size, PD-1 and its ligands have long segments connecting the IgV domain and IgC domain to the surface membrane, respectively (Fig. 3c) (Lazar-Molnar et al. 2008; Lin et al. 2008).

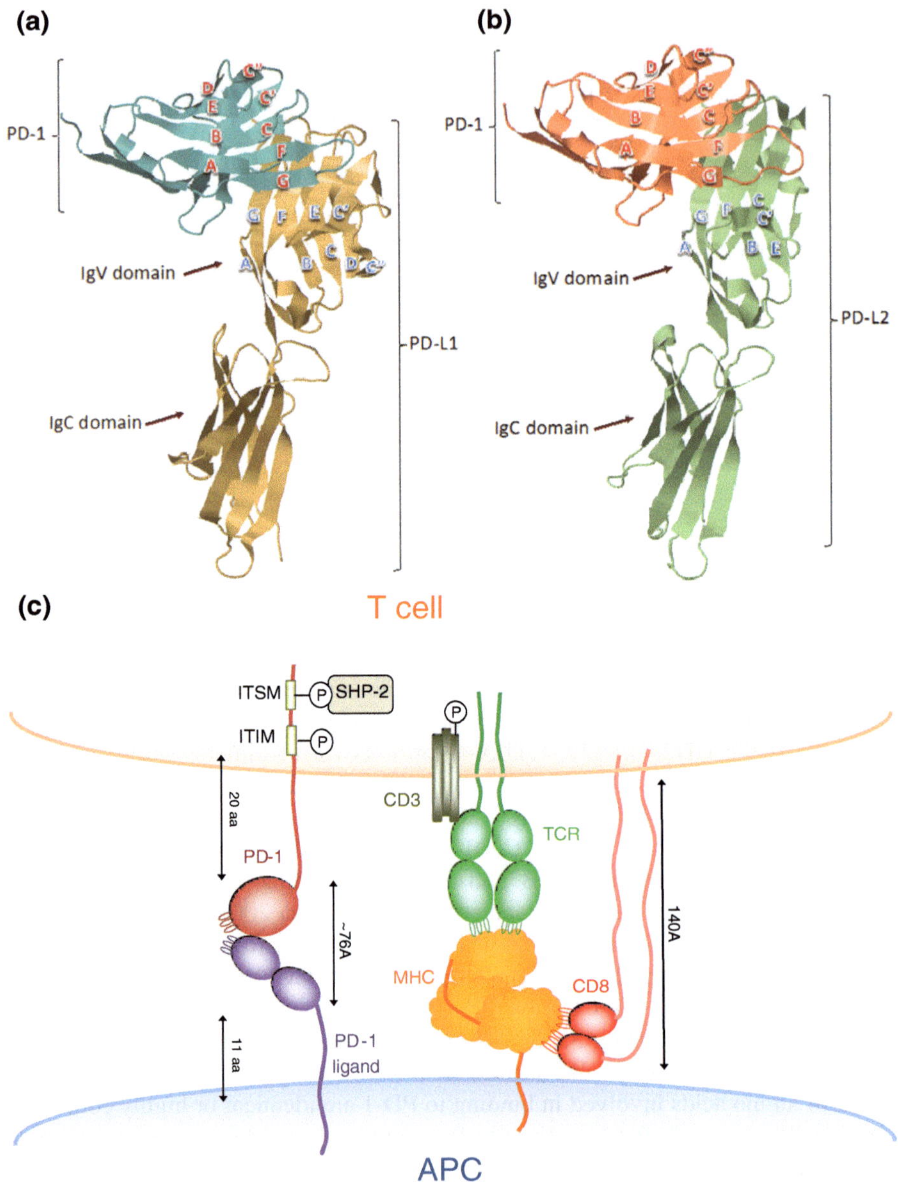

Fig. 3 Structure of PD-1/PD-L1, PD-L2 complexes

3 Molecular Functions

The inhibitory signal of PD-1 was initially analyzed in a B cell line (IIA1.6) by Okazaki et al. in 2001 (Okazaki et al. 2001). Upon BCR stimulation, tyrosine residues in both ITIM and ITSM of PD-1 were phosphorylated by Lyn. Phosphorylated tyrosine residues in ITSM recruited SH2-domain containing tyrosine phosphatase 2 (SHP-2), which in turn dephosphorylated BCR-proximal signaling molecules including Ig α/β and Syk. As a result, the activation of downstream molecules, including PLCγ2, PI3 K, vav, and ERK1/2, was reduced (Parry et al. 2005). A few years later, a similar mechanism was reported to inhibit TCR signaling (Fig. 4). The reports also stated that SHP-1 is involved in

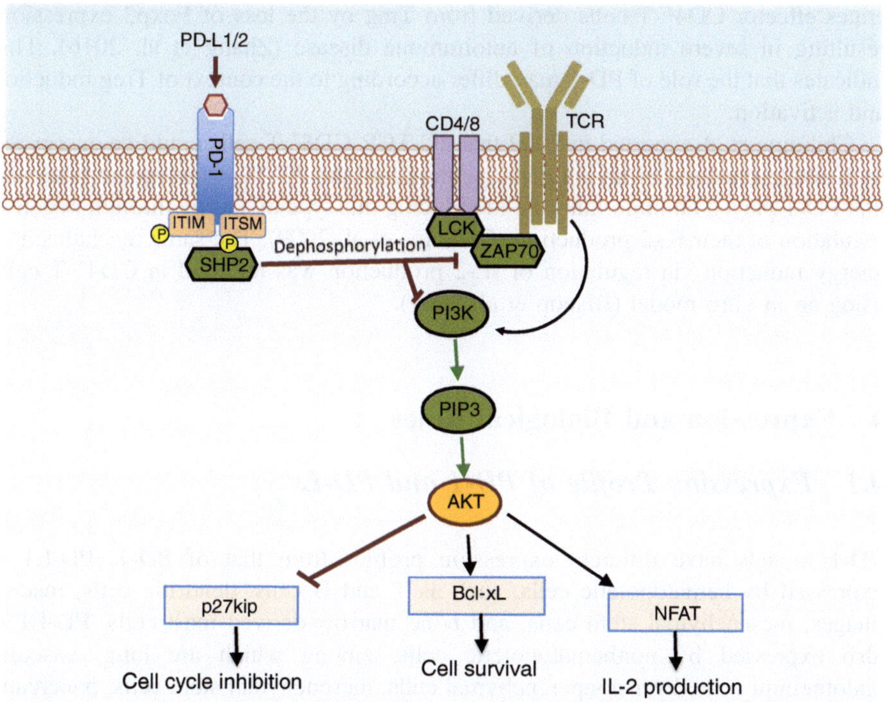

Fig. 4 PD-1 signaling pathway inhibits TCR signaling. Upon PD1/PD-L complex formation, the intracellular immunoreceptor tyrosine-based switch motif (ITSM) and immunoreceptor tyrosine-based inhibitory motif (ITIM) in PD-1 cytoplasmic tail become phosphorylated. SHP-2 can bind to phosphorylated ITSM and leads to overall inhibition of T cell receptor (TCR) signaling through dephosphorylation Zap-70, leading to (PI3 K)/Akt pathway shut down. This has many downstream effects, including downregulation of the Bcl-xL and NFAT pathways which affect cell survival and IL-2 production, respectively. It also results in the accumulation of p27kip1, an inhibitor of cyclin-dependent kinases, leading to cell cycle arrest, and thus blocks proliferation. Overall, PD-1 signaling causes T cells to become less proliferative and gradually lose their effector functions

attenuating TCR signaling through the PD-1 pathway, but to a lesser extent (Chemnitz et al. 2004; Sheppard et al. 2004).

Francisco et al. reported that PD-1 signaling was linked to the induction of regulatory T cells (iTregs). They showed that PD-L1$^{-/-}$ antigen-presenting cells reduced the ability to induce differentiation of naïve CD4$^+$ T cells to iTreg cells. In contrast, PD-L1-coated beads induced iTreg cells in vitro, indicating that PD-L1 itself regulates iTreg cell development and the successive expression of FoxP3. While inducing iTregs via PD-1 signaling, phospho-Akt, mTOR, S6, and ERK2 were downregulated, PTEN, on the other hand, was upregulated. This suggests that PD-1 signaling promoted the development of iTregs by blocking the Akt-mTOR signaling pathway (Francisco et al. 2009). In contrast, Zhang et al. demonstrated that PD-1-deficient Foxp3$^+$ Tregs suppress conventional T cell proliferation and activity more than PD-1-sufficient Foxp3$^+$ Tregs. However, PD-1 deficiency generates effector CD4$^+$ T cells derived from Treg by the loss of Foxp3 expression, resulting in severe induction of autoimmune disease (Zhang et al. 2016). This indicates that the role of PD-1 may differ according to the context of Treg induction and activation.

Chikuma et al. reported in 2009 that 2C-TCR CD8$^+$ T cells could be anergyzed by a single injection of an antigenic peptide into the 2C-TCR-transgenic mice and that PD-1 played an important role in inducing this type of anergy through negative regulation of their IL-2 production (Chikuma et al. 2009). The same mechanism of anergy induction via regulation of IL-2 production was reported in CD4$^+$ T cells using an in vitro model (Bishop et al. 2009).

4 Expression and Biological Roles

4.1 Expression Profile of PD-1 and PD-Ls

PD-1 ligands have different expression profiles from that of PD-1. PD-L1 is expressed by hematopoietic cells, such as T and B cells, dendritic cells, macrophages, mesenchymal stem cells, and bone marrow-derived mast cells. PD-L1 is also expressed by nonhematopoietic cells, among which are lung, vascular endothelium cells, liver nonparenchymal cells, mesenchymal stem cells, pancreatic islets, and keratinocytes (Yamazaki et al. 2002; Keir et al. 2008). In contrast, the PD-L2 expression is more restricted and limited to activated dendritic cells, macrophages, bone marrow-derived mast cells, and peritoneal B1 cells (Tseng et al. 2001; Zhong et al. 2007).

The fact that PD-1 ligands are expressed in various types of tissues including lymphoid and nonlymphoid tissues indicates the importance of this pathway in immune regulation. For example, PD-1/PD-Ls interaction regulates T cell maturation. PD-1 is expressed during the early stages of T-lymphocyte differentiation in the thymus, particularly in CD4$^-$ CD8$^-$ double-negative thymocytes (Nishimura et al.

1996; 2000). PD-L1 is expressed broadly in the thymus cortex and PD-L2 is expressed by medullary stromal cells in the thymus (Brown et al. 2003; Liang et al. 2003). PD-1 maintains a proper signaling threshold during thymic selection, as blockade of the interaction between PD-1 and PD-L1, but not PD-L2, enhanced negative selection, while overexpression of PD-1 attenuated positive selection (Keir et al. 2005; Blank et al. 2003). As another example, the PD-1/PD-L1 interaction was reported to prevent abortion by inhibiting immunity. During pregnancy, PD-L1 and PD-L2 are expressed on placental syncytiotrophoblasts and vascular endothelial cells, respectively (Guleria et al. 2005). PD-L1 continues to be differentially expressed throughout gestation (Holets et al. 2006). The interaction between PD-1 and its ligand is crucial for fetal–maternal tolerance by suppressing the maternal immune response to foreign antigens inherited from the father (Guleria et al. 2005).

The molecular mechanism regulating PD-1 expression is not fully understood but has been examined in several studies. Nakae et al. reported that signaling by TCR or B cell receptor (BCR) and/or tumor necrosis factor upregulate the PD-1 expression (Nakae et al. 2006). In 2008, Oestreich and colleagues demonstrated that PD-1 gene regulation occurs in part via the recruitment of NFATc1 to a regulatory element at the pdcd1 locus upon T cell activation. They also showed that an NFAT-specific inhibitor led to a sharp reduction in PD-1 expression (Oestreich et al. 2008). Other groups showed that interferon (IFN)-α mediates PD-1 expression on macrophages via the IFN-sensitive responsive element and STAT1/2 (Cho et al. 2008; Terawaki et al. 2011). During the late stages of an acute $CD8^+$ T cell effector response, the transcriptional repressor Blimp-1 is expressed and directly silences PD-1 expression through a process of chromatin reconfiguration, ultimately resulting in the loss of NFATc1 binding (Lu et al. 2014). On macrophages and monocytes, PD-1 expression correlates with increased IL-10 and decreased IL-12 levels in the blood of HIV-infected patients (Said et al. 2010; Cho et al. 2009). It was also recently demonstrated that following stimulation of macrophages with TLR ligands, PD-1 expression was induced by nuclear factor-κB (Bally et al. 2015). Thus, these results demonstrate that PD-1 expression may be differently controlled in different types of immune cells.

4.2 *Role in Autoimmunity*

The role of PD-1 was unknown for several years after the discovery of PD-1 molecules. Nishimura et al. first demonstrated the spontaneous development of typical lupus-like glomerulonephritis and destructive arthritis in PD-1-deficient B6 mice, leading to the conclusion that PD-1 is an immune inhibitory molecule. In the 2C TCR-transgenic mice, which recognize MHC H2-L^d, PD-1 deficiency induced overt graft-versus-host-like disease in a heterogeneric C57BL/6 and BALB/c background (autoreactive genetic background) (Nishimura et al. 1999). Disruption of PD-1 in BALB/c mice, but not in BALB/c RAG-$2^{-/-}$ mice, caused dilated cardiomyopathy with severely impaired contraction and sudden death by congestive heart failure (Nishimura et al. 2001). This was because of the accumulation of high-titer autoantibodies against cardiac troponin I. PD-1 deficiency also increases the frequency and early onset of type I diabetes in nonobese diabetic mice (Wang et al. 2005). Ansari et al. (2003) reported that blockade of PD-1 or PD-L1, but not PD-L2, in nonobese diabetic mice led to rapid aggressive diabetes regardless of age. This was associated with accelerated insulitis and proinflammatory cytokine production by T cells (Ansari et al. 2003). In studies using Murphy Roths Large mice, which are genetically predisposed to systemic autoimmunity, PD-1 deficiency caused the development of fatal myocarditis. In this model, infiltration of $CD8^+$, $CD4^+$ T cells, and myeloid cells was increased in the heart. High levels of autoantibodies against cardiac myosin were detected in the heart of PD-1-deficient Murphy Roths Large mice (Wang et al. 2010). In an experimental autoimmune encephalomyelitis (EAE) model after immunization with myelin oligodendrocyte glycoprotein, expression of PD-1 and PD-L1, but not PD-L2 was increased in the central nervous system (CNS). Blockade of PD-1 in this model resulted in rapid progression and more severe disease with increased CNS-infiltrated lymphocyte. The increased severity of the disease was linked to the higher frequency of IFN-γ–producing T cells, augmented delayed-type hypersensitivity responses, and higher serum levels of anti-myelin oligodendrocyte glycoprotein antibody (Salama et al. 2003). Interestingly, EAE was worsened by blockade of PD-L1 in BALB/c and SJL/J mice, but not in C57BL/6-background mice. PD-L2 blockade preferentially exerts the same severity effect in C57BL/6 mice. In contrast, in B10.S mice immunized with myelin proteolipid protein peptide, either PD-L1 or PD-L2 blockade markedly enhanced EAE severity (Salama et al. 2003; Zhu et al. 2006). These reports demonstrate that PD-L1 and PD-L2 differentially regulate the susceptibility and progress of autoimmune responses in a strain-dependent manner. However, the genes regulating this susceptibility remain largely unknown.

The significance of PD-1 was highlighted in human autoimmune disease development, exemplified by the fact that autoantibodies against PD-L1 were found in patients with rheumatoid arthritis. The authors assumed that the autoantibody against PD-L1 blocks the PD-1/PD-L1 interaction. The level of anti-PD-L1 autoantibodies in the serum of patients with rheumatoid arthritis was correlated with an active status of the disease, suggesting that blocking the PD-1/PD-L1 pathway contributes to disease development via aberrant T cell activation (Dong et al. 2003). Additionally, Prokunina et al. reported an association between a single-nucleotide polymorphism (SNP; G-to-A change in *Pdcd1* gene) and the development of systemic lupus erythematosus (SLE). The allele A of this SNP (known as PD1.3A located at position +7146 in the 4th intronic region of *Pdcd1*) was associated with the development of SLE in Europeans and Mexicans but not African Americans. The 4th intronic region of *Pdcd1* is an enhancer-like structure. This structure is composed of four imperfect tandem repeats containing binding sites for transcription factors such as RUNX1, Ebox–binding factors, and nuclear factor of κ in B cells 1, which are exclusively involved in hematopoietic differentiation and inflammation. The SNP PD1.3A very likely interrupts the binding of RUNX1, resulting in poor induction of PD-1 (Prokunina et al. 2002; Bertsias et al. 2009). Among the various SNPs in *Pdcd1*-related genes, PD1.3A and PD1.9 play important roles in regulating autoimmune disease. Other than SLE, PD1.3A is associated with other various autoimmune diseases such as type I diabetes, progressive multiple sclerosis, and rheumatoid arthritis (Nielsen et al. 2003; Kroner et al. 2005; Prokunina et al. 2004). Moreover, PD1.9 is related to Grave's disease and ankylosing spondylitis (Hayashi et al. 2008; Lee et al. 2006).

While previous studies demonstrated the significance of the PD-1/PD-L1 interaction in regulating autoimmune responses in humans, several therapeutic strategies for autoimmune disease have been developed in mouse models. Hirata et al. reported that overexpression of PD-L1 on dendritic cells reduced the severity of EAE (Hirata et al. 2005). Ding et al. showed that intravenous injection of recombinant adenovirus encoding the full-length PD-L1 gene partially prevented lupus-like nephritis (Ding et al. 2006). Other approaches targeting immunoinhibitory molecules and cells were tested using receptor-Fc fusion protein, ligand–Fc fusion protein, agonistic antibody, and a bispecific antibody. Although some methods were tested in clinical trials or are underway, none have been approved (van der Vlist et al. 2016).

4.3 Role in Viral Immunology

The role of PD-1 in viral immunity was highlighted in dysfunctional $CD8^+$ T cells during chronic viral infections. This state of dysfunction is referred to as $CD8^+$ T cell exhaustion (Wherry and Ahmed 2004). T cell exhaustion develops gradually by repeated TCR stimulation as cells lose their functional abilities starting from a loss of cytotoxicity, IL-2 production, proliferation, IFN-γ production, and ending with cell death (Fig. 5). PD-1 was first reported by Barber et al. to be transiently expressed and rapidly downregulated by virus-specific $CD8^+$ T cells after acute lymphocyte choriomeningitis virus (LCMV) infection. In the same study using a chronic model of LCMV infection, however, $CD8^+$ T cells continued to show high PD-1 expression in lymphoid and nonlymphoid tissues (Barber et al. 2006). Subsequently, the high PD-1 expression on virus-specific $CD8^+$ T cells was demonstrated in several chronic infection systems such as HIV (Day et al. 2006), simian immunodeficiency virus (Velu et al. 2009), hepatitis B virus (Boni et al. 2007), and hepatitis C virus (HCV) (Urbani et al. 2006). Several transcription factor-binding sites regulate the expression of PD-1, which is preferentially expressed on exhausted T cells. TCR-mediated calcium influx initiates *Pdcd1* transcription by activating NFATc1, which binds to the 5′-promoter region of the *Pdcd1* gene (at position -1160 relative to the transcription start site) (Oestreich et al. 2008). In virus-infected situations, IFN-α causes prolonged *Pdcd1* transcription in T cells by binding of the transcription factor IRF9 to the *Pdcd1* promoter (at position −1040 relative to the transcription start site) (Terawaki et al. 2011). Furthermore, demethylation of the *Pdcd1* promoter region (located 500–1500 base

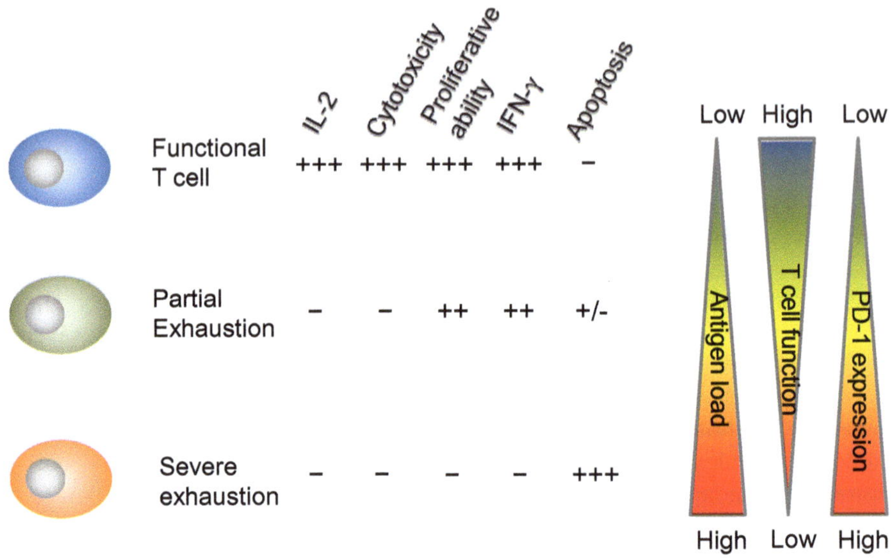

Fig. 5 T cell exhaustion stages. Activated T cells exhibit a wide range of exhausted states. First they lose the ability to produce IL-2, followed by reduction in cytotoxicity and proliferation. Finally, production of IFNgamma stops and apoptosis increases. During chronic infections, the extent of exhaustion and PD-1 expression correspond to the antigen load

pairs upstream of the initiation codon) has been reported to contribute to high PD-1 expression in exhausted $CD8^+$ T cells during chronic infections (Youngblood et al. 2011). While exhausted $CD8^+$ T cells express high Eomesodermin, which is regulated by transcription factor FoxO1, FoxO1 also binds to the *Pdcd1* promoter and enhances PD-1 expression (Staron et al. 2014). It has recently been suggested by Ahmed's group that coexpression of PD-1 and another immune inhibitory molecule, mucin-domain–containing molecule-3 (Tim-3), on T cells is strongly associated with the severity of their exhaustion. They reported the co-regulation of $CD8^+$ T cell exhaustion by Tim-3 and PD-1 during chronic LCMV infection as the majority (approximately 65–80%) of virus–specific $CD8^+$ T cells in lymphoid and nonlymphoid organs co-expressed Tim-3 and PD-1. PD-1/Tim-3 coexpression corresponded to more severe $CD8^+$ T cell exhaustion in terms of less proliferation and secretion of effector cytokines such as IFN-γ, tumor necrosis factor-α, and IL-2. Compared to type-1 cytokine production, $PD\text{-}1^+$ $Tim\text{-}3^+$ virus–specific $CD8^+$ T cells preferentially produced the suppressive cytokine IL-10 (Jin et al. 2010, 2011). However, how Tim-3 signaling regulates the severity of exhaustion remains largely unknown.

Because the interaction of PD-1 and its ligands plays an important role in $CD8^+$ T cell exhaustion during chronic viral infections, researchers developed viral-control strategies using PD-1 pathway-blocking antibodies. Iwai et al. first demonstrated the possibility of blocking the PD-1 pathway to clear the virus in the liver using a mouse model in which PD-$1^{-/-}$ mice cleared adenovirus more rapidly than wild-type mice (Iwai et al. 2002). PD-1 blockade exerts different effects in different viral models. For example in LCMV model; PD-1 blockade was associated with dramatic reduction in viral load (Barber et al. 2006). It caused rapid expansion of virus-specific $CD8^+$ T cells with improved functionality during simian immunodeficiency virus infection in nonhuman primates (Velu et al. 2009). PD-1 blockade is also reported to increase HIV-specific T cell proliferation with greater functionality (Day et al. 2006; D'Souza et al. 2007). In addition, combined blockade of Tim-3 and PD-1 in vivo synergistically rescued the exhausted $CD8^+$ T cell reduced viral levels in chronically infected mice (Jin et al. 2011). However, in the case of PD-1 blockade-resistant viral infection, virus-specific $CD8^+$ T cells were rescued from the exhausted state and then became exhausted again (Pauken et al. 2016). These results suggest that epigenetic control in T cells should be explored for viral therapies.

Golden-Mason et al. reported that in humans, $CD8^+$ T cell exhaustion was also correlated with high PD-1 expression during chronic infection with HCV. They observed significant upregulation of PD-1 on total and HCV-specific $CD8^+$ cytotoxic T-lymphocytes (CTLs) in the peripheral blood and livers of patients compared to in subjects with spontaneous HCV resolution or patients with the non-viral liver disease. They also reported increased levels of the senescence marker CD57 in PD-1 high HCV-specific CTLs and the recovery of the functional competence of these cells after blockade of PD-1/PD-L1 (Golden-Mason et al. 2007).

PD-1 was suggested to play an important role in the recovery of functional dysfunction of human $CD4^+$ T cells as well as $CD8^+$ T cells. D'Souza et al. found a direct correlation between PD-1 expression on exhausted HIV-specific $CD4^+$ T cells and plasma viral load. They also observed that PD-1 expression is upregulated on HIV-specific $CD4^+$ T cells in chronically infected subjects and that blockade of the PD-1 pathway enhances HIV-specific $CD4^+$ T cell function (D'Souza et al. 2007).

4.4 Role in Tumor Immunology

The results of various studies have indicated that PD-1 inhibitors can be applied in cancer immunotherapy because of the importance of the PD-1/PD-L1 pathway in regulating immune tolerance. Iwai et al. first developed a mouse model of PD-1 blockade cancer immunotherapy and demonstrated the importance of PD-L1 expression on tumor cells (Iwai et al. 2002). Subsequently, Honjo and his collaborators from the Ono and Medarex pharmaceutical company first developed a fully humanized monoclonal antibody against PD-1 in 2006 (nivolumab; also known as ONO4538, MDX-1106, or BMS-936558). Nivolumab was tested in phase I clinical trial in the US for patients with advanced metastatic melanoma, colorectal cancer, castrate-resistant prostate cancer, non-small-cell lung cancer (NSCLC), and renal-cell carcinoma (Brahmer et al. 2010; Topalian et al. 2012). The response rates were 18% for NSCLC, 28% for melanoma, and 27% for renal carcinoma. Grade 3 or 4 drug-related adverse events occurred in 14% of patients, which was much lower than those following treatment with ipilimumab, a monoclonal antibody against CTLA-4 (Topalian et al. 2012; Hodi et al. 2010). Numerous clinical trials have been conducted and reported. At least 500 clinical studies of PD-1 inhibitors have been performed using nine types of antibodies established by eight pharmaceutical companies for at least on 20 types of tumors, which are summarized in another review (Iwai et al. 2017).

Although many types of cancer immunotherapy have been developed such as peptide-vaccine therapy, dendritic cell-adjuvant therapy, and adoptive T cell therapy in the 2000s, the expected results were not obtained, except for immune checkpoint blockade therapy (Okazaki et al. 2013; Pardoll 2012). To determine why other immunotherapies were not successful, it is necessary to understand the types of tumor antigens evaluated. As shown in Table 1, tumor antigens are categorized into six main groups. Cancer-testis antigen, differentiation antigens, overexpressed antigens, and oncofetal antigens are self-antigens which are shared between cancer patients and are called "shared antigens". Mutated antigens (neo-antigens) and viral antigens are foreign antigens. As T cells that recognize self-antigens are depleted during thymic selection, the frequency of T cells with high-affinity TCRs to shared-tumor antigens should be very low in the body. In comparison, T cells in the peripheral blood can strongly react to foreign antigens such as neo-tumor antigens or virus-associated tumor antigens. Since the shared-tumor antigen MAGEA1 was identified by Tierry Boon's group (van der Bruggen et al. 1991), most cancer immunologists have focused on identifying shared-tumor antigens and targeting them using several immunotherapeutic methods (Coulie et al. 2014). However, because of the small number of shared-antigen-reactive T cells in our body, the immune response was minimal and obvious tumor inhibition was not observed in clinical studies. Importantly, in PD-1 blockade therapy, T cells with a large repertoire of TCRs against various tumor antigens including neo-antigens and virus antigens will be generated during therapy

Table 1 Types of cancer antigen

Antigen types	Self or non-self antigen?	Expression site	Examples
Cancer-Testis antigens	Self	Testis	MAGE, XAGE, NY-ESO-1
Differentiation antigens	Self	Tissues generating cancers	gp100, Melan, Mart-1, Tyrosinase, PSA, PAP
Overexpressed antigens	Self	Ubiquitously with low levels	HER2, MUC1 , PSMA, survivin, WT-1
Oncofetal antigens	Self	Fetus	CEA, AFP
Neo-antigen	Non-self	No	Different in each
Viral antigen	Non-self	Infected cells	Tax, Hbz, E6, E7

(Gros et al. 2016; Tumeh et al. 2014; Schumacher and Schreiber 2015). Therefore, PD-1 blockade therapy can generate T cells that are strongly reactive to neo-antigens on tumor cells, leading to stronger immune reactions against cancer.

Tumors grow while being edited by immune surveillance and undergo three phases: elimination, equilibrium, and escape (Fig. 6) (Schreiber et al. 2011). During the elimination phase, mutated abnormal cells proliferate, while most are eliminated by immune cells. In the equilibrium phase, few cells acquire the ability to escape from immune attack with accumulated mutations. In this phase, the tumor size appears stable. In the escape phase, cells escaping immune surveillance continue growing and become cancerous. Considering the high response rate in clinical studies to the PD-1 blockade, the PD-L1 expression is one of the most predominant escape mechanisms of human cancers. Indeed, several studies detected PD-L1 in various types of tumor tissues including melanoma, multiple myeloma, leukemia, glioblastoma, gastric, renal cell, bladder, hepatocellular, cutaneous, and breast cancers, and NSCLC (Thompson et al. 2004; Boland et al. 2013; Mittendorf et al. 2014; Velcheti et al. 2014; Cooper et al. 2015; Huang et al. 2015; Kakavand et al. 2015; Nduom et al. 2016) and that PD-1 is detected on tumor infiltrating lymphocytes (Ahmadzadeh et al. 2009; Sfanos et al. 2009). In addition, a very strong correlation was found between PD-L expression on tumor cells and poor prognosis in several types of cancer (Thompson et al. 2004; Ohigashi et al. 2005; Wu et al. 2006; Hamanishi et al. 2007; Nakanishi et al. 2007; Nomi et al. 2007). Thus, PD-1 blockade rejuvenates immune surveillance in cancer patients (Fig. 6).

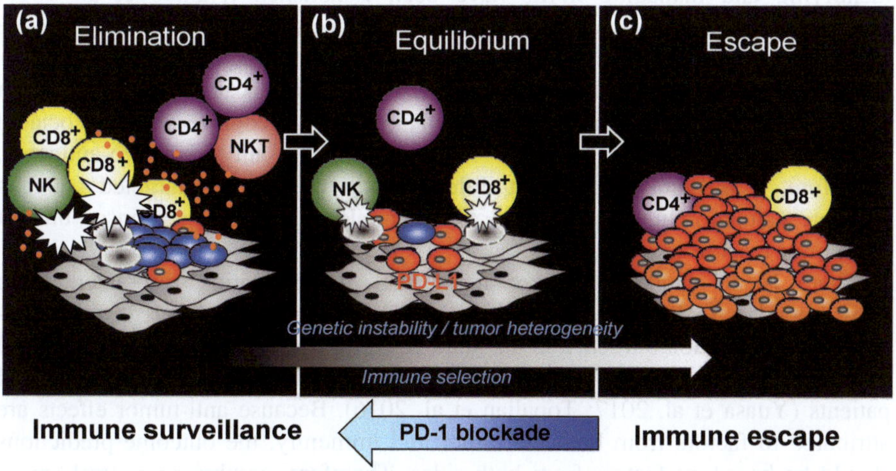

Fig. 6 Balance between immune surveillance and immune escape affects tumor growth. Most mutated abnormal cells (*blue*) are eliminated by immunity (**a**). Among the abnormal cells, some acquire mechanisms for escaping immune system attack (*red*) via genomic instability (**b**). One of the major mechanisms is a PD-L1 expression. Immune-escaping abnormal cells grow and become cancerous (**c**). PD-1/PD-L1 blockade reactivates immune surveillance

5 Issues and Future Directions in PD-1 Blockade Cancer Immunotherapy

Although the response rate of cancer therapy has been dramatically improved by the PD-1 blockade antibody, three main issues remain. (1) A fraction of patients undergoes autoimmune responses. Few strategies have been developed to reduce immune-related adverse events (irAEs) while ameliorating the anti-tumor effect. (2) More than half of patients are non-responders. There is no predictive biomarker for discriminating responders from non-responders to PD-1-blockade therapy. Additionally, the tumor inhibition mechanism by PD-1 blockade is not well understood. (3) Finally, the best combination therapy for improving therapeutic efficacy remains unclear.

In terms of irAEs, as described in the section "Role in autoimmunity", organs, where autoimmune responses occur through PD-1 signal blockade, depend on the genetic background of the patient. While the frequency of adverse events observed with PD-1 blockade therapy is lower than with conventional chemotherapy and CTLA-4 blockade therapy (Table 2) (Robert et al. 2015; Weber et al. 2015; Larkin et al. 2015; Brahmer et al. 2015; Borghaei et al. 2015; Motzer et al. 2015), there is no method for predicting which patients and organs exhibit irAEs. irAEs can be roughly classified into two groups according to the timing of onset: irAE within 2 months after treatment such as cutaneous, gastrointestinal, and hepatic responses and irAE occurring more than 2 months post-treatment such as pulmonary, endocrine, and renal responses. Based on these accumulated cases, guidelines for identifying and managing irAEs have been established (Naidoo et al. 2016; Champiat et al. 2016). Interestingly, steroids reduce the severity of autoimmune responses but do not always impair anti-tumor effects (Beck et al. 2006; Maker et al. 2006; Cousin and Italiano 2016). This might be because of differences in antigens targeted by immunity: T cells weakly recognize irAE-associated self-antigen, while they strongly recognize tumor antigens (neo-antigens). Therefore, it is highly likely that differences in the type of targeted antigens facilitate a situation in which steroids attenuate only autoimmune responses, but not anti-tumor responses.

There is no single biomarker for clearly discriminating responders from non-responders to PD-1 blockade therapy. PD-L1 expression levels in tumor tissues, tumor mutation burden, and the number of infiltrated lymphocyte are reported to be positively correlated with the response rate, but this is not the case for all patients (Yuasa et al. 2017; Topalian et al. 2016). Because anti-tumor effects are attributed to factors from both the tumor and immunity, the outcome predictions should be based on factors from both sides. Therefore, combining several candidates would improve the precision of predictions (Inoue et al. 2016). New findings regarding the mechanism leading to the identification of novel biomarkers for differentiating between responders and non-responders. Several novel approaches have been reported from the aspects of the microbiome and metabolome, which are profoundly related to immune regulation (Sivan et al. 2015; Saito et al. 2016;

Table 2 Grade 3–5 adverse events in conventional chemotherapy and immune checkpoint blockade therapy

Clinical study	Tumor	Drugs	Cases	Grade 3–5 adverse event related to treatment
CheckMate 066 (Robert et al. 2015)	Untreated metastatic melanoma without a BRAF mutation	Nivolumab	206	11.70%
		Dacarbazine	205	17.60%
CheckMate 037 (Motzer et al. 2015)	Advanced melanoma progressed after anti–CTLA-4	Nivolumab	268	9%
		Investigator's choice of chemotherapy	102	32%
KEYNOTE-006 (Robert et al. 2015)	Advanced melanoma	Pembrolizumab every 2 week	278	13.30%
		Pembrolizumab every 3 week	277	10.10%
		Ipilimumab	256	19.90%
CheckMate 067 (Larkin et al. 2015)	Untreated stage III or IV melanoma	Nivolumab alone	313	16.30%
		Ipilimumab alone	311	27.30%
		Nivolumab plus ipilimumab	313	55.00%
CheckMate 017 (Brahmer et al. 2015)	Advanced squamous-cell non-small-cell lung cancer	Nivolumab	131	7%
		Docetaxel	129	55%
CheckMate 057 (Borghaei et al. 2015)	Advanced nonsquamous non–small-cell lung cancer	Nivolumab	287	10%
		Docetaxel	268	54%
CheckMate 025 (Motzer et al. 2015)	Advanced clear-cell renal-cell carcinoma	Nivolumab	406	19%
		Everolimus	397	37%

[a]Dacarbazine 1000 mg/m^2 every 3 weeks or paclitaxel 175 mg/m^2 combined with carboplatin area under the curve 6 every 3 weeks

Kawamoto et al. 2012; Chang et al. 2015; Chamoto et al. 2017). Sivan et al. reported that a commensal microbe, *Bifidobacterium*, stimulates dendritic cells, resulting in improved CTL-mediated anti-tumor immunity (Sivan et al. 2015). Regulation of T follicular cells by PD-1 in the gut is important for selecting appropriate IgA-producing B cells, which maintain the proper balance between the microbiota and immune responses (Kawamoto et al. 2012). Therefore, the environment of the gut flora is an important factor regulating the immunity and sensitivity of PD-1 blockade therapy. In immune metabolism, Pearce's group demonstrated that deprivation of glucose in the tumor microenvironment by tumor cells dampened the glycolysis/mTOR pathway in killer T cells, which was recovered by the PD-1 blockade. We reported that the modulation of T cell metabolism by activating mitochondria enhances energy production in T cells and is synergistic with the anti-tumor activity with PD-1 blockade (Chamoto et al. 2017). It was previously reported that an indoleamine 2,3-dioxygenase inhibitor, which inhibits tryptophan metabolism, enhanced PD-1 blockade therapy in melanoma patients (Zakharia et al. 2017). These reports suggest that regulating the microbiota and metabolism can be used to augment the efficacy of PD-1 blockade therapy.

References

Agata Y et al (1996) Expression of the PD-1 antigen on the surface of stimulated mouse T and B lymphocytes. Int Immunol 8:765–772

Ahmadzadeh M et al (2009) Tumor antigen-specific CD8 T cells infiltrating the tumor express high levels of PD-1 and are functionally impaired. Blood 114:1537–1544

Ansari MJ et al (2003) The programmed death-1 (PD-1) pathway regulates autoimmune diabetes in nonobese diabetic (NOD) mice. J Exp Med 198:63–69

Bally AP et al (2015) NF-kappaB regulates PD-1 expression in macrophages. J Immunol 194:4545–4554

Barber DL et al (2006) Restoring function in exhausted CD8 T cells during chronic viral infection. Nature 439:682–687

Bardhan K, Anagnostou T, Boussiotis VA (2016) The PD1:PD-L1/2 pathway from discovery to clinical implementation. Front Immunol 7:550

Beck KE et al (2006) Enterocolitis in patients with cancer after antibody blockade of cytotoxic T-lymphocyte-associated antigen 4. J Clin Oncol 24:2283–2289

Bertsias GK et al (2009) Genetic, immunologic, and immunohistochemical analysis of the programmed death 1/programmed death ligand 1 pathway in human systemic lupus erythematosus. Arthritis Rheum 60:207–218

Bishop KD et al (2009) Depletion of the programmed death-1 receptor completely reverses established clonal anergy in CD4(+) T lymphocytes via an interleukin-2-dependent mechanism. Cell Immunol 256:86–91

Blank C et al (2003) Absence of programmed death receptor 1 alters thymic development and enhances generation of CD4/CD8 double-negative TCR-transgenic T cells. J Immunol 171:4574–4581

Boland JM et al (2013) Tumor B7-H1 and B7-H3 expression in squamous cell carcinoma of the lung. Clin Lung Cancer 14:157–163

Boni C et al (2007) Characterization of hepatitis B virus (HBV)-specific T-cell dysfunction in chronic HBV infection. J Virol 81:4215–4225

Borghaei H et al (2015) Nivolumab versus docetaxel in advanced nonsquamous non-small-cell lung cancer. N Engl J Med 373:1627–1639

Brahmer JR et al (2010) Phase I study of single-agent anti-programmed death-1 (MDX-1106) in refractory solid tumors: safety, clinical activity, pharmacodynamics, and immunologic correlates. J Clin Oncol 28:3167–3175

Brahmer J et al (2015) Nivolumab versus docetaxel in advanced squamous-cell non-small-cell lung cancer. N Engl J Med 373:123–135

Brown JA et al (2003) Blockade of programmed death-1 ligands on dendritic cells enhances T cell activation and cytokine production. J Immunol 170:1257–1266
Chang CH et al (2015) Metabolic competition in the tumor microenvironment is a driver of cancer progression. Cell 162:1229–1241
Champiat S et al (2016) Management of immune checkpoint blockade dysimmune toxicities: a collaborative position paper. Ann Oncol: Off J Eur Soc Med Oncol 27:559–574
Chamoto K et al (2017) Mitochondrial activation chemicals synergize with surface receptor PD-1 blockade for T cell-dependent antitumor activity. Proc Natl Acad Sci USA 114:E761–E770
Chemnitz JM, Parry RV, Nichols KE, June CH, Riley JL (2004) SHP-1 and SHP-2 associate with immunoreceptor tyrosine-based switch motif of programmed death 1 upon primary human T cell stimulation, but only receptor ligation prevents T cell activation. J Immunol 173:945–954
Chikuma S et al (2009) PD-1-mediated suppression of IL-2 production induces CD8 + T cell anergy in vivo. J Immunol 182:6682–6689
Cho HY, Lee SW, Seo SK, Choi IW, Choi I (2008) Interferon-sensitive response element (ISRE) is mainly responsible for IFN-alpha-induced upregulation of programmed death-1 (PD-1) in macrophages. Biochim Biophys Acta 1779:811–819
Cho HY et al (2009) Programmed death-1 receptor negatively regulates LPS-mediated IL-12 production and differentiation of murine macrophage RAW264.7 cells. Immunol Lett 127:39–47
Cooper WA et al (2015) PD-L1 expression is a favorable prognostic factor in early stage non-small cell carcinoma. Lung Cancer 89:181–188
Coulie PG, Van den Eynde BJ, van der Bruggen P, Boon T (2014) Tumour antigens recognized by T lymphocytes: at the core of cancer immunotherapy. Nat Rev Cancer 14:135–146
Cousin S, Italiano A (2016) Molecular Pathways: Immune Checkpoint Antibodies and their Toxicities. Clin Cancer Res 22:4550–4555
Day CL et al (2006) PD-1 expression on HIV-specific T cells is associated with T-cell exhaustion and disease progression. Nature 443:350–354
Ding H et al (2006) Delivering PD-1 inhibitory signal concomitant with blocking ICOS co-stimulation suppresses lupus-like syndrome in autoimmune BXSB mice. Clin Immunol 118:258–267
Dong H, Zhu G, Tamada K, Chen L (1999) B7-H1, a third member of the B7 family, co-stimulates T-cell proliferation and interleukin-10 secretion. Nat Med 5:1365–1369
Dong H et al (2003) Costimulating aberrant T cell responses by B7-H1 autoantibodies in rheumatoid arthritis. J Clin Invest 111:363–370
D'Souza M et al (2007) Programmed death 1 expression on HIV-specific CD4 + T cells is driven by viral replication and associated with T cell dysfunction. J Immunol 179:1979–1987
Francisco LM et al (2009) PD-L1 regulates the development, maintenance, and function of induced regulatory T cells. J Exp Med 206:3015–3029
Freeman GJ (2008) Structures of PD-1 with its ligands: sideways and dancing cheek to cheek. Proc Natl Acad Sci USA 105:10275–10276
Freeman GJ et al (2000) Engagement of the PD-1 immunoinhibitory receptor by a novel B7 family member leads to negative regulation of lymphocyte activation. J Exp Med 192:1027–1034
Golden-Mason L et al (2007) Upregulation of PD-1 expression on circulating and intrahepatic hepatitis C virus-specific CD8 + T cells associated with reversible immune dysfunction. J Virol 81:9249–9258
Gros A et al (2016) Prospective identification of neoantigen-specific lymphocytes in the peripheral blood of melanoma patients. Nat Med 22:433–438
Guleria I et al (2005) A critical role for the programmed death ligand 1 in fetomaternal tolerance. J Exp Med 202:231–237
Hamanishi J et al (2007) Programmed cell death 1 ligand 1 and tumor-infiltrating CD8 + T lymphocytes are prognostic factors of human ovarian cancer. Proc Natl Acad Sci U S A 104:3360–3365

Hayashi M, Kouki T, Takasu N, Sunagawa S, Komiya I (2008) Association of an A/C single nucleotide polymorphism in programmed cell death-ligand 1 gene with Graves' disease in Japanese patients. Eur J Endocrinol 158:817–822

He J, Hu Y, Hu M, Li B (2015) Development of PD-1/PD-L1 pathway in tumor immune microenvironment and treatment for non-small cell lung cancer. Sci Rep 5:13110

Hirata S et al (2005) Prevention of experimental autoimmune encephalomyelitis by transfer of embryonic stem cell-derived dendritic cells expressing myelin oligodendrocyte glycoprotein peptide along with TRAIL or programmed death-1 ligand. J Immunol 174:1888–1897

Hodi FS et al (2010) Improved survival with ipilimumab in patients with metastatic melanoma. N Engl J Med 363:711–723

Holets LM, Hunt JS, Petroff MG (2006) Trophoblast CD274 (B7-H1) is differentially expressed across gestation: influence of oxygen concentration. Biol Reprod 74:352–358

Huang Y et al (2015) The prognostic significance of PD-L1 in bladder cancer. Oncol Rep 33:3075–3084

Inoue H et al (2016) Intratumoral expression levels of PD-L1, GZMA, and HLA-A along with oligoclonal T cell expansion associate with response to nivolumab in metastatic melanoma. Oncoimmunology 5:e1204507

Ishida Y, Agata Y, Shibahara K, Honjo T (1992) Induced expression of PD-1, a novel member of the immunoglobulin gene superfamily, upon programmed cell death. EMBO J 11:3887–3895

Iwai Y et al (2002) Involvement of PD-L1 on tumor cells in the escape from host immune system and tumor immunotherapy by PD-L1 blockade. Proc Natl Acad Sci USA 99:12293–12297

Iwai Y, Hamanishi J, Chamoto K, Honjo T (2017) Cancer immunotherapies targeting the PD-1 signaling pathway. J Biomed Sci 24:26

Jin HT et al (2010) Cooperation of Tim-3 and PD-1 in CD8 T-cell exhaustion during chronic viral infection. Proc Natl Acad Sci USA 107:14733–14738

Jin HT, Ahmed R, Okazaki T (2011) Role of PD-1 in regulating T-cell immunity. Curr Top Microbiol Immunol 350:17–37

Kakavand H et al (2015) Tumor PD-L1 expression, immune cell correlates and PD-1 + lymphocytes in sentinel lymph node melanoma metastases. Mod Pathol 28:1535–1544

Kawamoto S et al (2012) The inhibitory receptor PD-1 regulates IgA selection and bacterial composition in the gut. Science 336:485–489

Keir ME, Latchman YE, Freeman GJ, Sharpe AH (2005) Programmed death-1 (PD-1):PD-ligand 1 interactions inhibit TCR-mediated positive selection of thymocytes. J Immunol 175:7372–7379

Keir ME, Butte MJ, Freeman GJ, Sharpe AH (2008) PD-1 and its ligands in tolerance and immunity. Annu Rev Immunol 26:677–704

Kroner A et al (2005) A PD-1 polymorphism is associated with disease progression in multiple sclerosis. Ann Neurol 58:50–57

Larkin J et al (2015) Combined nivolumab and ipilimumab or monotherapy in untreated melanoma. N Engl J Med 373:23–34

Latchman Y et al (2001) PD-L2 is a second ligand for PD-1 and inhibits T cell activation. Nat Immunol 2:261–268

Lazar-Molnar E et al (2008) Crystal structure of the complex between programmed death-1 (PD-1) and its ligand PD-L2. Proc Natl Acad Sci USA 105:10483–10488

Lee SH et al (2006) Association of the programmed cell death 1 (PDCD1) gene polymorphism with ankylosing spondylitis in the Korean population. Arthritis Res Therapy 8:R163

Liang SC et al (2003) Regulation of PD-1, PD-L1, and PD-L2 expression during normal and autoimmune responses. Eur J Immunol 33:2706–2716

Lin DY et al (2008) The PD-1/PD-L1 complex resembles the antigen-binding Fv domains of antibodies and T cell receptors. Proc Natl Acad Sci USA 105:3011–3016

Lu P et al (2014) Blimp-1 represses CD8 T cell expression of PD-1 using a feed-forward transcriptional circuit during acute viral infection. J Exp Med 211:515–527

Maker AV et al (2006) Intrapatient dose escalation of anti-CTLA-4 antibody in patients with metastatic melanoma. J Immunother 29:455–463

Mittendorf EA et al (2014) PD-L1 expression in triple-negative breast cancer. Cancer Immunol Res 2:361–370
Motzer RJ et al (2015) Nivolumab versus everolimus in advanced renal-cell carcinoma. N Engl J Med 373:1803–1813
Naidoo J et al (2016) Toxicities of the anti-PD-1 and anti-PD-L1 immune checkpoint antibodies. Ann Oncol: Off J Eur Soc Med Oncol 27:1362
Nakae S et al (2006) Mast cells enhance T cell activation: importance of mast cell costimulatory molecules and secreted TNF. J Immunol 176:2238–2248
Nakanishi J et al (2007) Overexpression of B7-H1 (PD-L1) significantly associates with tumor grade and postoperative prognosis in human urothelial cancers. Cancer Immunol Immunother 56:1173–1182
Nduom EK et al (2016) PD-L1 expression and prognostic impact in glioblastoma. Neuro Oncol 18:195–205
Nielsen C, Hansen D, Husby S, Jacobsen BB, Lillevang ST (2003) Association of a putative regulatory polymorphism in the PD-1 gene with susceptibility to type 1 diabetes. Tissue Antigens 62:492–497
Nishimura H et al (1996) Developmentally regulated expression of the PD-1 protein on the surface of double-negative (CD4-CD8-) thymocytes. Int Immunol 8:773–780
Nishimura H, Nose M, Hiai H, Minato N, Honjo T (1999) Development of lupus-like autoimmune diseases by disruption of the PD-1 gene encoding an ITIM motif-carrying immunoreceptor. Immunity 11:141–151
Nishimura H, Honjo T, Minato N (2000) Facilitation of beta selection and modification of positive selection in the thymus of PD-1-deficient mice. J Exp Med 191:891–898
Nishimura H et al (2001) Autoimmune dilated cardiomyopathy in PD-1 receptor-deficient mice. Science 291:319–322
Nomi T et al (2007) Clinical significance and therapeutic potential of the programmed death-1 ligand/programmed death-1 pathway in human pancreatic cancer. Clin Cancer Res 13:2151–2157
Oestreich KJ, Yoon H, Ahmed R, Boss JM (2008) NFATc1 regulates PD-1 expression upon T cell activation. J Immunol 181:4832–4839
Ohigashi Y et al (2005) Clinical significance of programmed death-1 ligand-1 and programmed death-1 ligand-2 expression in human esophageal cancer. Clin Cancer Res 11:2947–2953
Okazaki T, Honjo T (2007) PD-1 and PD-1 ligands: from discovery to clinical application. Int Immunol 19:813–824
Okazaki T, Maeda A, Nishimura H, Kurosaki T, Honjo T (2001) PD-1 immunoreceptor inhibits B cell receptor-mediated signaling by recruiting src homology 2-domain-containing tyrosine phosphatase 2 to phosphotyrosine. Proc Natl Acad Sci USA 98:13866–13871
Okazaki T, Chikuma S, Iwai Y, Fagarasan S, Honjo T (2013) A rheostat for immune responses: the unique properties of PD-1 and their advantages for clinical application. Nat Immunol 14:1212–1218
Pardoll DM (2012) The blockade of immune checkpoints in cancer immunotherapy. Nat Rev Cancer 12:252–264
Parry RV et al (2005) CTLA-4 and PD-1 receptors inhibit T-cell activation by distinct mechanisms. Mol Cell Biol 25:9543–9553
Pauken KE et al (2016) Epigenetic stability of exhausted T cells limits durability of reinvigoration by PD-1 blockade. Science 354:1160–1165
Probst HC, McCoy K, Okazaki T, Honjo T, van den Broek M (2005) Resting dendritic cells induce peripheral CD8 + T cell tolerance through PD-1 and CTLA-4. Nat Immunol 6:280–286
Prokunina L et al (2002) A regulatory polymorphism in PDCD1 is associated with susceptibility to systemic lupus erythematosus in humans. Nat Genet 32:666–669
Prokunina L et al (2004) Association of the PD-1.3A allele of the PDCD1 gene in patients with rheumatoid arthritis negative for rheumatoid factor and the shared epitope. Arthritis Rheum 50:1770–1773

Robert C et al (2015a) Nivolumab in previously untreated melanoma without BRAF mutation. N Engl J Med 372:320–330
Robert C et al (2015b) Pembrolizumab versus Ipilimumab in advanced melanoma. N Engl J Med 372:2521–2532
Said EA et al (2010) Programmed death-1-induced interleukin-10 production by monocytes impairs CD4 + T cell activation during HIV infection. Nat Med 16:452–459
Saito T et al (2016) Two FOXP3(+)CD4(+) T cell subpopulations distinctly control the prognosis of colorectal cancers. Nat Med 22:679–684
Salama AD et al (2003) Critical role of the programmed death-1 (PD-1) pathway in regulation of experimental autoimmune encephalomyelitis. J Exp Med 198:71–78
Schreiber RD, Old LJ, Smyth MJ (2011) Cancer immunoediting: integrating immunity's roles in cancer suppression and promotion. Science 331:1565–1570
Schumacher TN, Schreiber RD (2015) Neoantigens in cancer immunotherapy. Science 348:69–74
Sfanos KS et al (2009) Human prostate-infiltrating CD8 + T lymphocytes are oligoclonal and PD-1+. Prostate 69:1694–1703
Sheppard KA et al (2004) PD-1 inhibits T-cell receptor induced phosphorylation of the ZAP70/CD3zeta signalosome and downstream signaling to PKCtheta. FEBS Lett 574:37–41
Sivan A et al (2015) Commensal Bifidobacterium promotes antitumor immunity and facilitates anti-PD-L1 efficacy. Science 350:1084–1089
Staron MM et al (2014) The transcription factor FoxO1 sustains expression of the inhibitory receptor PD-1 and survival of antiviral CD8(+) T cells during chronic infection. Immunity 41:802–814
Terawaki S et al (2011) IFN-alpha directly promotes programmed cell death-1 transcription and limits the duration of T cell-mediated immunity. J Immunol 186:2772–2779
Thompson RH et al (2004) Costimulatory B7-H1 in renal cell carcinoma patients: Indicator of tumor aggressiveness and potential therapeutic target. Proc Natl Acad Sci USA 101:17174–17179
Topalian SL et al (2012) Safety, activity, and immune correlates of anti-PD-1 antibody in cancer. N Engl J Med 366:2443–2454
Topalian SL, Taube JM, Anders RA, Pardoll DM (2016) Mechanism-driven biomarkers to guide immune checkpoint blockade in cancer therapy. Nat Rev Cancer 16:275–287
Tseng SY et al (2001) B7-DC, a new dendritic cell molecule with potent costimulatory properties for T cells. J Exp Med 193:839–846
Tumeh PC et al (2014) PD-1 blockade induces responses by inhibiting adaptive immune resistance. Nature 515:568–571
Urbani S et al (2006) PD-1 expression in acute hepatitis C virus (HCV) infection is associated with HCV-specific CD8 exhaustion. J Virol 80:11398–11403
van der Bruggen P et al (1991) A gene encoding an antigen recognized by cytolytic T lymphocytes on a human melanoma. Science 254:1643–1647
van der Vlist M, Kuball J, Radstake TR, Meyaard L (2016) Immune checkpoints and rheumatic diseases: what can cancer immunotherapy teach us? Nat Rev Rheumatol 12:593–604
Velcheti V et al (2014) Programmed death ligand-1 expression in non-small cell lung cancer. Lab Invest 94:107–116
Velu V et al (2009) Enhancing SIV-specific immunity in vivo by PD-1 blockade. Nature 458:206–210
Wang J et al (2005) Establishment of NOD-Pdcd1-/- mice as an efficient animal model of type I diabetes. Proc Natl Acad Sci USA 102:11823–11828
Wang J et al (2010) PD-1 deficiency results in the development of fatal myocarditis in MRL mice. Int Immunol 22:443–452
Weber JS et al (2015) Nivolumab versus chemotherapy in patients with advanced melanoma who progressed after anti-CTLA-4 treatment (CheckMate 037): a randomised, controlled, open-label, phase 3 trial. Lancet Oncol 16:375–384
Wherry EJ, Ahmed R (2004) Memory CD8 T-cell differentiation during viral infection. J Virol 78:5535–5545

Wu C et al (2006) Immunohistochemical localization of programmed death-1 ligand-1 (PD-L1) in gastric carcinoma and its clinical significance. Acta Histochem 108:19–24

Yamazaki T et al (2002) Expression of programmed death 1 ligands by murine T cells and APC. J Immunol 169:5538–5545

Youngblood B et al (2011) Chronic virus infection enforces demethylation of the locus that encodes PD-1 in antigen-specific CD8(+) T cells. Immunity 35:400–412

Yuasa T, Masuda H, Yamamoto S, Numao N, Yonese J (2017) Biomarkers to predict prognosis and response to checkpoint inhibitors. Int J Clin Oncol 22:629–634

Zakharia Y, McWilliams R, Shaheen M et al (2017) Interim analysis of the phase 2 clinical trial of the IDO pathway inhibitor indoximod in combination with pembrolizumab for patients with advanced melanoma. Presented at: 2017 AACR Annual Meeting; 1–5 Apr 2017; San Washington, DC. Abstract CT117

Zhang X et al (2004) Structural and functional analysis of the costimulatory receptor programmed death-1. Immunity 20:337–347

Zhang B, Chikuma S, Hori S, Fagarasan S, Honjo T (2016) Nonoverlapping roles of PD-1 and FoxP3 in maintaining immune tolerance in a novel autoimmune pancreatitis mouse model. Proc Natl Acad Sci USA 113:8490–8495

Zhong X, Tumang JR, Gao W, Bai C, Rothstein TL (2007) PD-L2 expression extends beyond dendritic cells/macrophages to B1 cells enriched for V(H)11/V(H)12 and phosphatidylcholine binding. Eur J Immunol 37:2405–2410

Zhu B et al (2006) Differential role of programmed death-ligand 1 [corrected] and programmed death-ligand 2 [corrected] in regulating the susceptibility and chronic progression of experimental autoimmune encephalomyelitis. J Immunol 176:3480–3489

CTLA-4, an Essential Immune-Checkpoint for T-Cell Activation

Shunsuke Chikuma

Abstract The response of peripheral T lymphocytes (T cell) is controlled by multiple checkpoints to avoid unwanted activation against self-tissues. Two opposing costimulatory receptors, CD28 and CTLA-4, on T cells bind to the same ligands (CD80 and CD86) on antigen-presenting cells (APCs), and provide positive and negative feedback for T-cell activation, respectively. Early studies suggested that CTLA-4 is induced on activated T cells and binds to CD80/CD86 with much stronger affinity than CD28, providing a competitive inhibition. Subsequent studies by many researchers revealed the more complex mode of T-cell inhibition by CTLA-4. After T-cell activation, CTLA-4 is stored in the intracellular vesicles, and recruited to the immunological synapse formed between T cells and APCs, and inhibits further activation of T cells by blocking signals initiated by T-cell receptors and CD28. CTLA-4-positive cells can also provide cell-extrinsic regulation on other autoreactive T cells, and are considered to provide an essential regulatory mechanism for FoxP3+ regulatory T cells. Genetic deficiency of CTLA-4 leads to CD28-mediated severe autoimmunity in mice and humans, suggesting its function as a fundamental brake that restrains the expansion and activation of self-reactive T cells. In cancer, therapeutic approaches targeting CTLA-4 by humanized blocking antibodies has been demonstrated to be an effective immunotherapy by reversing T-cell tolerance against tumors. This chapter introduces CTLA-4 biology, including its discovery and mechanism of action, and discusses questions related to CTLA-4.

Contents

1 Introduction: CD28 and CTLA-4 as Checkpoints 100
2 Structure of CTLA-4 101

S. Chikuma (✉)
Department of Microbiology and Immunology, Keio University School of Medicine, 35 Shinanomachi, Shinjuku-Ku, Tokyo 160-8582, Japan
e-mail: schikuma@keio.jp

Current Topics in Microbiology and Immunology (2017) 410:99–126
DOI 10.1007/82_2017_61

Published Online: 13 September 2017

2.1 Discovery and Primary Structure of CTLA-4 ... 101
2.2 Alternatively Spliced Isoforms of CTLA-4 ... 102
3 Function of CTLA-4 ... 104
3.1 Cell-Intrinsic T-cell Inhibition by CTLA-4 ... 104
3.2 Cell-Extrinsic T-Cell Inhibition by CTLA-4 ... 105
4 Mechanisms of CTLA-4-Mediated Immune Regulation ... 107
4.1 The Biochemical Partners of CTLA-4 ... 107
4.2 Induction and Dynamic Localization of CTLA-4 to the Immunological Synapse... 110
4.3 Structure and Functional Relationship ... 112
5 Biological Roles of CTLA-4 ... 113
5.1 CTLA-4 Genetics in Autoimmunity: Mice to Humans ... 113
5.2 CTLA-4 in Infection ... 114
5.3 CTLA-4 in Cancer ... 115
References ... 118

1 Introduction: CD28 and CTLA-4 as Checkpoints

The activation and function of the peripheral T lymphocyte (T cell) reaction depends primarily on recognition of the antigen presented on the major histocompatibility complex (MHC) by the T-cell antigen receptor (TCR). This recognition, at the time of priming (the priming phase), causes the clonal expansion of T cells and the differentiation into effector T cells in secondary lymphoid organs. Recognition at the site of inflammation (the effector phase) results in an attack against infected tissue. The activation of self-reactive T cells can lead to an attack on self-tissue, leading to autoimmunity. To prevent this, there are many systems that have developed to keep self-reactive T cells in check.

The first signal, provided by TCR during recognition, does not cause activation of the T cells on its own. Full T-cell activation requires a second set of signals, called "costimulation," which is mainly provided by activated antigen-presenting cells (APCs). The best-characterized costimulatory system is the CD28 receptor on T cells, triggered by its ligands, CD80 or CD86 (previously called B7-1 and B7-2, respectively) on activated professional APCs (dendritic cells, macrophages, and B cells). CD80 and CD86 are upregulated on activated APCs by microbial "danger signals" so APCs presenting microbial antigens can efficiently stimulate T-cell activation (Banchereau and Steinman 1998; Akira et al. 2001). Successful engagement of CD28 leads to IL-2 production (June et al. 1987), induction of anti-apoptotic protein Bcl-xL (Boise et al. 1995), and stimulates glucose uptake by inducing glucose transporter and glycolysis (Frauwirth et al. 2002), as well as the cell-cycle progression necessary for the massive clonal expansion of antigen-specific T cells. In contrast, resting APCs do not express a high level of CD80/CD86. T cells that are stimulated in the absence of the CD28 signal fall into an unresponsive state called "clonal anergy" and become refractory to further stimulation by the same antigen (Schwartz 2003). CTLA-4, another receptor that is structurally similar to CD28, is induced on activated T cells and binds to CD80 and CD86 with greater avidity than CD28. All CTLA-4 KO mice showed massive lymphocyte proliferation in the lymph nodes and spleen, followed by an autoimmune attack against virtually all tissues by leucocytes, and premature death (Tivol et al. 1995; Waterhouse et al. 1995; Chambers et al. 1997b). The lethal phenotype is prevented by treating young CTLA-4 KO mice with antibody-depleting CD4+, but not CD8+ T cells (Chambers et al. 1997b). The conditional deletion of CTLA-4 only on CD4+ T-cells phenocopied germline CTLA-4 KO (Klocke et al. 2016), suggesting that the disease is CD4+ "helper T cell"-dependent. A recombinant soluble CTLA-4 (CTLA-4 Ig) blocking CD28 engagement by CD80/86 (Chambers et al. 1997b), CD28/CTLA-4 double knockout mice (Mandelbrot et al. 2001), or breeding CTLA-4 KO into CD80/CD86 double knockout mice (Triple KO; Mandelbrot et al. 1999) completely prevented lymphoproliferation and autoimmunity. Also, CTLA-4 KO remains healthy in mice genetically lacking autoreactive T cells (Waterhouse et al. 1997; Chambers et al. 1999; Bachmann et al. 2001; Greenwald et al. 2001; Gajewski et al. 2001). Conversely, by limiting the complexity of the T-cell repertoire by overexpressing TCRβ chain as a transgene, it was shown that CTLA-4 knockout disease is totally self-antigen-dependent (Ise et al. 2009). These data clearly suggest that CTLA-4 provides negative feedback on the CD28-mediated activation of polyclonal, self-reactive CD4 helper T cells.

Thus, T-cell activation against self is fundamentally prevented by two "checkpoints": 1. T cells in the absence of the CD28 signal (in the absence of CD80 and CD86 on APCs) become unresponsive. 2. Upon activation, CTLA-4 is induced, interacts with CD80/CD86, and prevents further activation of self-reactive T cells. After this discovery, several pairs of ligands and receptors were shown to have unique and functions in the immune system. It is now accepted that PD-1 and CTLA-4 blocks the T cell-mediated immune system at different levels, at the cell type and location (Chikuma 2016). CTLA-4 was also shown to be important for regulatory function. The dramatic phenotypes of CTLA-4 null mice attracted many researchers in immunology. It is now of particular importance that the artificial blockade of CTLA-4 was shown to break tolerance in the tumor setting, providing efficient immunotherapy against tumors (Leach et al. 1996). It turned out that the basis of T-cell inhibition by CTLA-4 was far more complex than previously expected. In this chapter, the history of discovery and basic research on CTLA-4 in terms of its biological roles and the mechanism of T-cell inactivation will be introduced, and the currently unsolved questions will be discussed.

2 Structure of CTLA-4

2.1 Discovery and Primary Structure of CTLA-4

CTLA-4 was originally isolated from a cDNA library derived from mouse cytotoxic T cells, hence the name cytotoxic T-cell-associated antigen-4 (Brunet et al. 1987). The CTLA-4 gene is encoded on human Ch2q33 and mouse Ch1C1. CTLA-4 consists of 4 exons and 3 introns, and encodes a protein of 223 amino acids (Ling et al. 1999). The protein is a type I transmembrane protein belonging to the immunoglobulin (Ig) superfamily, and it bears a single Ig-V (variable)-like domain on the extracellular portion. Like other receptors in the CD28 family, CTLA-4 is highly glycosylated at its extracellular domain. CTLA-4 forms a homodimer on activated T cells, which is mediated by the disulfide bond at the cysteine residue at amino acid position 120 (Linsley et al. 1995). There is a well-conserved motif (amino acid sequence: MYPPPY) on the extracellular domain that is shared by CD28, which is critical for binding its ligands CD80 and CD86 (Peach et al. 1994). CD80 and CD86 bind to CTLA-4 with stronger affinity than to CD28, which may account for CTLA-4 antagonizing to the activation signal. Dimerized CTLA-4 binds to two ligands (Linsley et al. 1995). Ostrov et al. (2000) demonstrated by solving the crystal structure of the extracellular portion that CTLA-4 shows a unique mode of dimerization with its ligand-binding domain distal to the dimerization interface. Stamper et al. (2001) and Schwartz et al. (2001) solved the crystal structures of CTLA-4 in complex with CD80 and CD86, respectively. Both groups concluded that each CTLA-4 dimer binds to two CD80 molecules, forming a lattice-like structure and providing very high avidity binding. The results suggest that CTLA-4 induction on the surface of activated T cells results in ligand sequestration of CD28. CTLA-4 also bears a short cytoplasmic tail that is almost 100% conserved among many species. The tail is reported to bind to various signaling molecules that mediate subcellular trafficking and the function of CTLA-4 (Fig. 1a, discussed later).

2.2 *Alternatively Spliced Isoforms of CTLA-4*

The splicing of messenger RNA of CTLA-4 might result in the expression of functionally important isoforms in the genetics of autoimmunity. For example, human resting blood lymphocytes express an alternatively spliced isoform of CTLA-4 lacking Exon3, encoding the entire transmembrane domain, resulting in a soluble isoform consisting solely of the extracellular domain (soluble CTLA-4: sCTLA-4; (Magistrelli et al. 1999; Oaks et al. 2000). Since the ectodomain of CTLA-4 can bind to its ligands (CD80 and CD86) with blocking activity, sCTLA-4 might be acting as a naturally occurring antagonist for CD28-mediated T-cell activation by CD80 and CD86. Indeed, Ueda et al. (2003) reported a CTLA-4 allelic variant associated with a lower mRNA level of sCTLA-4 linked with susceptibility to human autoimmune diseases. It was also reported that a naturally occurring CTLA-4 variant that is completely lacking its extracellular domain may determine autoimmune susceptibility in a particular mice strain. The "ligand-independent form" (referred to as liCTLA-4) can preferentially accumulate to the immunological synapse, and thereby provide a tonic inhibitory signal (Bour-Jordan et al. 2011 and Chikuma unpublished). Non-obese diabetes (NOD) mice that exhibit autoimmune juvenile diabetes express a reduced amount

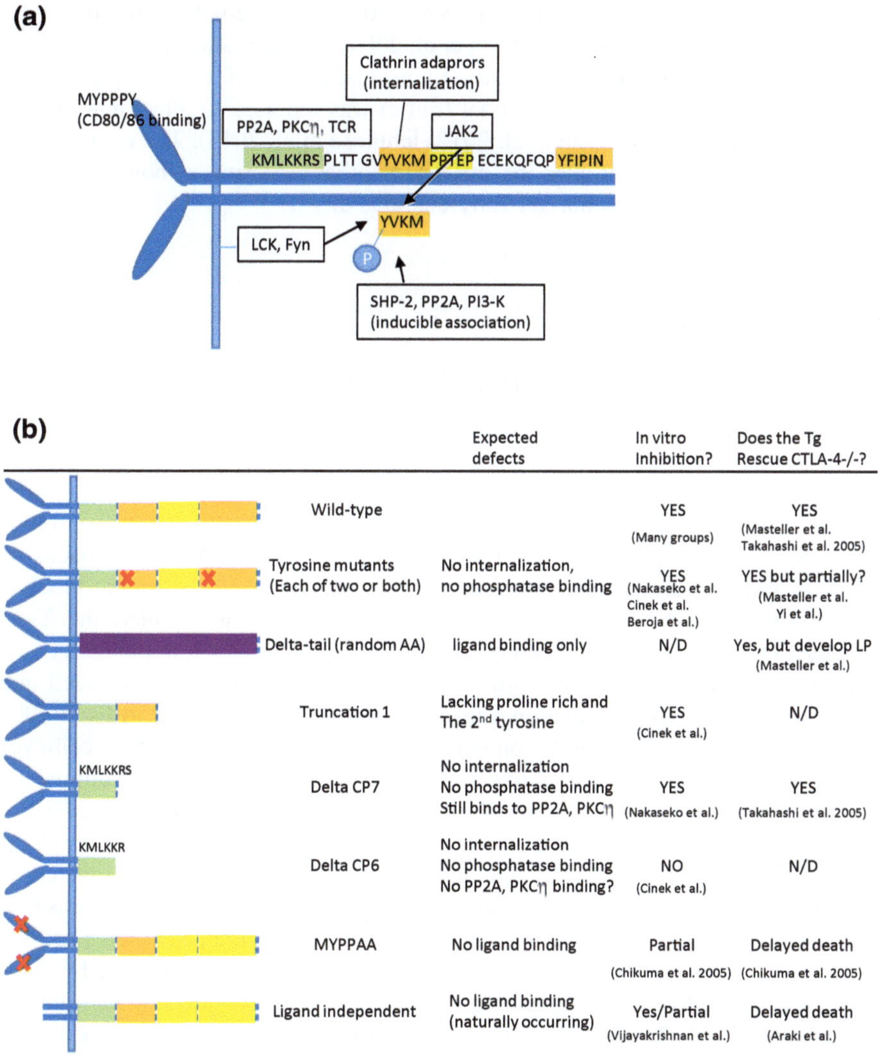

Fig. 1 A. schematic structure of CTLA-4. Note that the YVKM motif binds to the clathrin adaptor complex in its unphosphorylated form whereas the same motif binds to enzymes for signal transduction when phosphorylated, B. Summary of structure-functional studies. N/D: not done

of this isoform (Ueda et al. 2003), which accounts for their genetic susceptibility (Araki et al. 2009). A transgenic overexpression of liCTLA-4 or a mutant CTLA-4 that lacks the ligand-binding domain on T cells significantly retards the death of CTLA-4 KO mice, suggesting the function of a ligand-independent CTLA-4 signal (Vijayakrishnan et al. 2004; Chikuma et al. 2005; Araki et al. 2009). Interestingly, the activation status of T cells results in dynamic change in the ratio of transcripts encoding full-length, soluble, and ligand-independent forms of CTLA-4 transcripts, suggesting the unique role of each spliced isoform in different T-cell subpopulations (Oaks et al. 2000; Vijayakrishnan et al. 2004). A super-short form of CTLA-4 that encodes only the signal peptide (Exon1) and part of the cytoplasmic tail (Exon4; out of its original translational frame) was reported to affect autoimmune mice when overexpressed (Liu et al. 2012; Ichinose et al. 2013). The expression of this super-short form may alter the mRNA of other isoforms, although the exact function of this isoform is not yet fully elucidated.

3 Function of CTLA-4

3.1 Cell-Intrinsic T-cell Inhibition by CTLA-4

It was first reported that the monoclonal antibodies (mAb) that block the binding of CTLA-4 to CD80/CD86 augment T-cell responses when added to the co-culture of T cells, APCs, and anti-CD3 antibody (which work as a TCR ligand) in a soluble form (Walunas et al. 1994). The $F(ab)_2$ form of the same mAb, which lacks the ability to bind Fc receptor had similar activity, suggesting that genuine blocking activity of CTLA-4-CD80/CD86 binding in this culture augmented the T-cell response. In contrast to the blocking experiment, anti-CTLA-4 immobilized on the surface of plastics works to crosslink/stimulate CTLA-4 and inhibits T-cell activation, as well as IL-2 production and proliferation (Krummel and Allison 1995; Walunas et al. 1994). This inhibition was associated with inhibition of the cell-cycle progression and expression of activation markers CD69 and CD25 (Krummel and Allison 1996). A membrane-bound single-chain variable fragment (scFv) of anti-CTLA-4 antibody engineered to be expressed on a fibroblast cell line can ligate CTLA-4 in vitro and inhibits activation of co-cultured T cells, triggered by anti-CD3 and anti-CD28 on the same cell (Griffin et al. 2000). A transgenic expression of CTLA-4 scFv on B cells ameliorated autoimmune diabetes in NOD mice (Fife et al. 2006) in vivo. Importantly, this amelioration was observed in NOD mice lacking CD80/CD86, suggesting inhibition of the activation of islet-reactive diabetogenic T cells by B cells expressing the artificial CTLA-4 ligand directly, but not through inhibition of CD28 ligation. The strain also shows almost no FoxP3 + Tregs due to the absence of the CD28 signal necessary for Treg development/survival (Salomon et al. 2000; Tang et al. 2003; Tai et al. 2005). Thus, the CTLA-4 ligand-mediated inhibition of autoimmune diabetes in Tg mice is, at least in part independent of Tregs (Fife et al. 2006). An ectopic expression of CTLA-4-deficient primary mouse T cells or cell lines with CTLA-4 or its mutants reconstitutes ligand-mediated T-cell inactivation by CTLA-4 in many systems (Nakaseko et al. 1999; Chikuma et al. 2005; Cinek et al. 2000; Baroja et al. 2000; Vijayakrishnan et al. 2004). On the basis of these experiments, CTLA-4 was suggested to be a negative regulator for T-cell activation. Overexpression of CTLA-4 or its mutants to cell lines lacking endogenous CTLA-4 can be seen to mimic CTLA-4-mediated T-cell inhibition in vitro in many reports (Fig. 1b). Two groups reported that Jurkat cells (human-derived) expressing stable mouse CTLA-4 can be inhibited (in terms of IL-2 secretion) by immobilized anti-mouse CTLA-4 antibody. Subsequent mutational analysis showed that mutation of one or both tyrosine motifs to phenylalanine did not affect CTLA-4's ability to inhibit IL-2 production in the same system. Similar results were observed in a study, where CTLA-4-negative T cell clone was reconstituted with CTLA-4 by retroviral vectors (Nakaseko et al. 1999). Restimulated cells with wild-type CTLA-4 inhibited the activation, and the tyrosine mutants did not show any defects in the inhibition. In Nakaseko's study, CTLA-4 lacking almost all cytoplasmic regions but the membrane proximal to seven amino acids (KMLKKRS) showed suppressive activity. Although ligation of CTLA-4 alone does not cause changes in gene expression, it inhibits abundant gene expression by TCR/CD28 stimulation (Riley et al. 2002). This suppression is associated with inhibition of TCR-CD28-mediated biochemical events. It was reported that CTLA-4 ligation inhibits the phosphorylation of AKT (Parry et al. 2005), TCRζ (Lee et al. 1998), Erk (Calvo et al. 1997; Baroja et al. 2000; Chikuma et al. 2005), Jnk (Calvo et al. 1997), and Iκb (Pioli et al. 1999), all of which are triggered by TCR-CD28. Interestingly, some reports suggested that CTLA-4 can inhibit T-cell activation even without ligand engagement. In mice, the transgenic expression of CTLA-4 mutants that lack either the entire extracellular domain (Vijayakrishnan et al. 2004; Araki et al. 2009) or an essential ligand-binding motif (MYPPPY; Chikuma et al. 2005) inhibited T-cell proliferation and rescued the CTLA-4 null phenotype significantly, but not fully. The study clearly indicated the T-cell-intrinsic inhibition by CTLA-4. CTLA-4 also mediates the induction of T-cell anergy, a form of tolerance. Greenwald et al. (Greenwald et al. 2001) used a genuine genetic model in which CTLA-4 was lacking on T cells bearing single TCR with known specificity to show this. The CTLA-4 KO rag2 KO TCR Tg (classII MHC-restricted; CD4+) showed resistance to anergy induction in the transferred host when challenged by the cognate antigen, whereas CTLA-4 WT rag2 KO TCR Tg were tolerized. In contrast, CTLA-4 KO CD8+ T cells in the analogous system showed no defects in anergy induction. In a similar system, CTLA-4 on CD8+ T cells did not regulate their anergy induction (Frauwirth et al. 2000, 2001). In contrast, PD-1 deficiency in the same system had effects on CD8 + T cells which led to the augmentation of CD8+ T-cell activation and resistance to anergy (Chikuma et al. 2009), These results suggest that CTLA-4 and PD-1 sets the threshold for CD4+ and CD8+ T cells, respectively.

3.2 *Cell-Extrinsic T-Cell Inhibition by CTLA-4*

In contrast to the direct inhibition of T cells, CTLA-4 (or T-cell-expressing CTLA-4) can indirectly inhibit the activation of other T cells. This non-cell-autonomous tolerance mechanism was observed in lethally irradiated wild-type mice receiving mixed bone marrow from CTLA-4 wild-type and KO bone marrow when they did not develop the lymphoproliferative disease observed in CTLA-4 KO mice or recipients of CTLA-4 KO cells only (Bachmann et al. 1999). Similarly, Tivol and Gorski (2002) showed that the transfer of a mixture of Thy1+ splenocytes WT and CTLA-4 KO into Rag KO mice prevented inflammatory disease caused by the transfer of CTLA-4 KO cells only. In this setting, KO cells were eliminated in vivo, in the presence of WT T cells, suggesting that CTLA-4 on WT T cells can induce the deletion of KO T cells when the population coexist in vivo.

This non-cell-autonomous manner of CTLA-4-mediated inhibition is thought to be a major mechanism of tolerance by FoxP3+ Tregs. FoxP3+ Tregs are found to express high levels of CTLA-4 on the cell surface and intracellularly (Read et al. 2000; Takahashi et al. 2000; Salomon et al. 2000). FoxP3+ Tregs show a unique demethylation pattern on the CTLA-4 locus (Ohkura et al.), and FoxP3 can bind to CTLA-4 promoter (Wu et al. 2006) and drives its expression (Hori et al. 2003). FoxP3-deficient (germline KO or natural mutation; the so-called "scurfy" mutation) and CTLA-4 KO mice show a very similar phenotype with premature death by autoimmunity (Brunkow et al. 2001; Khattri et al. 2003; Fontenot et al. 2003). It has been, however difficult to determine if CTLA-4 expression on FoxP3+ Treg is critical for the regulatory function. For instance, FoxP3+ Tregs from CTLA-4 KO mice show suppression as efficiently as FoxP3+ Tregs from WT mice in vitro (Tang et al. 2004; Kataoka et al. 2005). To explore this question in vivo, Chikuma and Bluestone (2007) reconstituted RAG-1 KO recipients with bone marrow (BM) from CTLA-4KO (bred to a CD80/80 null background to avoid autoimmunity) mice and/or Foxp3-deficient mice. Control recipients, receiving either CTLA-4KO or Scurfy BM alone, rapidly developed interstitial pneumonitis, colitis, wasting, and dermatitis and died, which suggests that fatal autoimmunity occurred in the donor strain (Chikuma and Bluestone 2007). Next, the experiment was repeated using a 50:50 mixture of CTLA-4KO and Scurfy BM cells were tested to determine whether the addition of the CTLA-4+ FoxP3neg could overcome the pathology of and CTLA-4 or FoxP3-deficient cells in lymphocyte-null Rag1-KO recipients. Intriguingly, mixed BM chimeric recipients developed less severe autoimmunity and lived longer compared to CTLA-4KO or Scurfy BM alone, suggesting that the mixture of CTLA-4KO and Scurfy BM afforded some level of protection. The data suggest that CTLA-4 has a Treg-independent extrinsic function in the control of T-cell tolerance and homeostasis. Strikingly, the chimeric mice died with later kinetics, suggesting that CTLA-4 and Foxp3 must be expressed in cis on the same cell to fully prevent lethal autoimmunity and lymphoproliferation (Chikuma and Bluestone 2007). Wing et al. (2008) generated a mouse carrying a floxed CTLA-4 allele that allowed conditional knockout mice lacking CTLA-4 specifically on FoxP3+ Tregs when bred to mice expressing CRE under FoxP3 promotor. The resulting mice developed lethal autoimmunity, suggesting that the Treg-specific CTLA-4 expression is important for absolute maintenance of T-cell homeostasis. Again, mice showed much milder autoimmunity that affected limited organs and resulted in later death in comparison to the germline CTLA-4 KO (Wing et al. 2008). Conversely, Jain et al. (2010) showed that mice with CTLA-4 deleted only on activated conventional T cells (CTLA-4 on

FoxP3+ Tregs is intact) show significant, but not complete, protection from the lethal disease phenotype. Collectively, these results (Chikuma and Bluestone 2007; Wing et al. 2008; Jain et al. 2010) suggest that CTLA-4 on both activated conventional T cells and FoxP3+ Tregs is important for complete protection from death. As for the mechanism of cell-extrinsic regulation by CTLA-4, the wild-type T cells protecting autoimmunity appeared not to require inhibitory cytokines, such as TGFβ or IL-10 (Friedline et al. 2009). Instead, the CTLA-4 regulation of APC activation and function was proposed. Using two-photon microscopy, Tang et al. (2006) showed that FoxP3+ Tregs do not directly interact with conventional autoreactive cells in an inflammatory setting. Instead, Tregs interact with APC to down-modulate their ability to activate autoreactive conventional T cells. Using an in vitro co-culture system of conventional T cells, Tregs, and APCs, Wing et al. (2008) showed that the Tregs outcompeted with conventional T cells for binding to APCs. APCs cultured with FoxP3+ Tregs are attenuated in the expression of activation markers, CD80 and CD86, and lost the ability to stimulate conventional T cells (Wing et al. 2008). Qureshi et al. (2011) reported that the downregulation of CD80/86 by CTLA-4-expressing cells is through the CTLA-4-mediated capture of CD80/86 on APC and trans-endocytosis into CTLA-4+ T cells. The trans-endocytosed CD80/86 is degraded within CTLA-4+ cells, resulting in a reduction of molecules on the APC (Qureshi et al. 2011). This trans-endocytosis model is attractive to explain cell-extrinsic regulation by CTLA-4, especially because the model works in a FoxP3 null setting in an antigen-specific system (Wang et al. 2012). On the other hand, Tai et al. (2012) showed that the internalization-defective CTLA-4 transgene on CTLA-4 KO Tregs was functional in vivo, suggesting that trans-endocytosis may not be important. CTLA-4 binding to CD80/86 is also suggested in the transmission of "reverse signaling" that induces the production of an enzyme indolamine 2,3-dioxygenase (IDO) on APCs. IDO catalyzes the degradation of the amino acid tryptophan into the immune-inhibitory metabolite kynurenine, which is important in immuno-suppression in various immunologic settings, such as pregnancy, cancer, chronic infection, autoimmunity, and allergy (reviewed in Fallarino et al. 2003; Puccetti and Grohmann 2007). These hypotheses largely resulted from in vitro experiments using cells overexpressing CTLA-4 and/or CD80/CD86 so proof in the physiological system is less clear. CTLA-4-mediated cell-extrinsic inhibition is crucial in the maintenance of self-tolerance.

4 Mechanisms of CTLA-4-Mediated Immune Regulation

4.1 The Biochemical Partners of CTLA-4

The 33 amino-acid intracellular domain of CTLA-4 is 100% conserved among species, suggesting its importance in protection against lethal autoimmunity (Fig. 1a). The domain is composed of a lysine-rich membrane-proximal motif

(KMLKKRS) followed by two motifs containing tyrosine (YVKM and YFIP), with a proline-rich motif (PPTEP). The tail lacks a signaling module with enzymatic activity. Instead, the CTLA-4 tail interacts with many intracellular molecules, which control trafficking and signaling by CTLA-4.

4.1.1 SHP-2

Src-homology-containing tyrosine phosphatase-2 (SHP-2, also called Sh-PTP2 or Syp) is a cytosolic tyrosine phosphatase with a Src-homology 2 (SH2) domain. SHP-2 is involved in activation of the RAS-MAPK pathway, and transmits signals from cytokine receptor and co-inhibitory receptors (Lorenz 2009). SHP-2 interacts with CTLA-4 based on mutagenesis and immune coprecipitation studies. This association was suggested to be mediated by the SH2 of SHP-2 and the phosphotyrosine sequence YVKM within the CTLA-4 cytoplasmic tail when the motif is phosphorylated. This motif was not, however, a typical ITIM motif. The association of SHP-2 to CTLA-4 might be indirect or it may require other molecule(s) as adapter molecules (Schneider and Rudd 2000). The CTLA-4-SHP-2 interaction was proposed to cause tyrosine dephosphorylation of key signaling component required for T-\cell activation (Marengere et al. 1996). For instance, SHP-2 phosphatase activity can dephosphorylate the TCRζ chain(Lee et al. 1998) and the RAS regulator p52SHC(Marengere et al. 1996), and it has been suggested that the delivery of this enzyme by CTLA-4 to the specific region of the immunological synapse inhibits the TCR-CD3 complex-mediated signal transduction, leading to direct inhibition of T-cell activation (Lee et al. 1998). Similarly, Gab2, which binds directly to TCRζ signaling complex, recruits SHP-2 there and blocks proximal TCR signaling (Yamasaki et al. 2001).

4.1.2 PP2A

PP2A is a heterotrimeric serine-threonine phosphatase that dephosphorylates a wide variety of protein substrates involved in cellular activity. PP2A composes a very major fraction of cellular protein, and dephosphorylates many substrates (Lorenz 2009). In a yeast two-hybrid screen, Chuang et al. (2000) found that the catalytic subunit of PP2A associates with the cytoplasmic tail of CD28 and CTLA-4. Independently, Baroja et al. (2002) reported that the regulatory subunit (PP2AA) also interacted with CTLA-4. The catalytic and regulatory subunit of PP2A interacted with the lysine-rich KMLKKRS and YVKM motif within CTLA-4, respectively, suggesting tri-molecular interaction. Through mutagenesis of the KMLKKRS motif, Baroja et al. suggested that the PP2A interaction inhibits CTLA-4 suppressive activity. Parry et al. (Parry et al. 2005) showed that PP2A antagonizes phosphorylation-dependent AKT activation mediated by TCR and CD28, which was sensitive to ocadaic acid, an inhibitor of PP2A.

4.1.3 PKC-η(Eta)

Kong et al. (2014) reported that CTLA-4 associates with the η isoform of PKC. FoxP3+ Tregs contain a significant amount of phosphorylated PKC-η, which interacts with CTLA-4 at the immunological synapse. The mutation or truncation of the KMLKKRS motif greatly reduced PKC-η binding, suggesting the importance of this motif together with PP2A in CTLA-4 function. Germline PKC-η deficient mice demonstrated defects in conventional T-cell activations (Fu et al. 2011), but did not show any defects in the development of FoxP3+ Treg (Kong et al. 2014). They eventually exhibited lymphadenopathy, however, and their FoxP3+ Tregs were shown to have multiple defects in regulatory functions in vitro and in vivo. It appears that CTLA-4-PKC-η interaction is required for firm interaction of Tregs to APCs, by phosphorylation of PAK2, GIT2, two components of the focal adhesion complex, as well as a guanine nucleotide exchange factor, αPIX, at the immunological synapse. Mice with the Treg-specific deletion of PKC-η may result in Treg-specific defects in adhesion due to defective phosphorylation of the GIT2-aPIX-PAK complex.

4.1.4 Clathrin Adaptor Complex

Several groups reported that CTLA-4 interacts with AP1, and that AP2 subunits of the clathrin adaptor complex are primary components in the cell biology that induce the clathrin-mediated internalization of associated molecules (Chuang et al. 1997; Shiratori et al. 1997; Zhang and Allison 1997; Bradshaw et al. 1997; Schneider et al. 1999). AP1 and AP2 interact with the unphosphorylated form of Y201VKM, which may account for the preferential intracellular trafficking of CTLA-4.

4.1.5 Tyrosine Kinases

CTLA-4 phosphorylation and relocation to the immunological synapse are TCR-dependent. TCR-proximal kinase Lck and Fyn were shown to phosphorylate the Y201VKM motif (Chuang et al. 1999; Miyatake et al. 1998). Other tyrosine kinases, such as JAK2 (Chikuma et al. 2000) and Rlk (Schneider et al. 1998), were reported to directly bind to CTLA-4, and can phosphorylate Y201, suggesting that the broad extracellular signals that activate tyrosine kinases can stimulate CTLA-4 phosphorylation. Importantly, Lck and Fyn are membrane-associated src kinases, and their activity is TCR-dependent. The preferential recruitment of CTLA-4 to the IS may occur as a consequence of the trimolecular complex of LCK, CTLA-4, and TCR. The tyrosine motif is dispensable for CTLA-4-mediated inhibition in vitro. CTLA-4 KO mice overexpressing point-mutated CTLA-4 on tyrosine (YVKM → VKMs) develop late lymphoproliferative disease, suggesting the importance of this residue in CTLA-4 activation and function (Yi et al. 2004).

4.2 Induction and Dynamic Localization of CTLA-4 to the Immunological Synapse

CTLA-4 expression is restricted to T cells, but it shows a unique expression pattern, which is important for the inhibitory function. (1) The expression of CTLA-4 is activation-dependent. (2) CTLA-4 preferentially localizes to the intracellular vesicle compartment, (3) recycles to the T cell surface upon engagement of TCR, where it participates at the contact site of T cells and APC, called the "immunological synapse," (4) binds to CD80 and CD86, and is internalized from the immunological synapse to intracellular compartments. This unique pattern of CTLA-4 expression is critical for its specialized function as a negative regulatory molecule. Here, we focus on such dynamics of CTLA-4 mediated inhibition.

4.2.1 Induction of CTLA-4 on T Cells

In contrast to CD28's constitutive expression on the surface of all T cells, CTLA-4 expression is only found on activated T cells (Linsley et al. 1992). CTLA-4 on T cells is strongly induced by ConA and IL-2 (Brunet et al. 1987). TCR plus CD28 signals synergistically induce CTLA-4 expression by two mechanisms: enhanced transcription and an increase in the mRNA stability (Finn et al. 1997). A nucleotide sequence located within 335 bp upstream from the transcriptional start site of CTLA4 is sufficient for the induction (Perkins et al. 1996). FoxP3+ Tregs are poised to show demethylation on specific loci of CTLA-4, and an activated phenotype due to self-interactions during thymic development (Ohkura et al. 2012). CTLA-4 expression is sensitive to inhibition by cyclosporine, a calcineurin inhibitor. In addition, rapamycin, a small mTOR inhibitor that blocks the IL-2-mediated signaling cascade can control CTLA-4 expression. These results indicate that CTLA-4 upregulation is controlled by general T-cell activation signal. The effects of IL-2 and CD28 signaling were additive but independent, as the CD28 signal augmented CTLA4 expression in IL-2-deficient mice. In contrast, CTLA4 expression was not augmented by cytokines IL-4, IL-6, IL-7, or IL-12 (Alegre et al. 1996). Sodium butylate, an HDAC inhibitor, significantly augmented CTLA-4 expression, suggesting epigenetic silencing (Doyle et al. 2001). CD4+ CD25 + FoxP3+ Tregs showed the most abundant CTLA-4 expression in the steady-state condition, suggesting that these cells are constantly undergoing antigen recognition and activation (Takahashi et al. 2000; Read et al. 2000; Salomon et al. 2000). This concept is supported by the finding that Nur77, a proximal molecule expressed on recently activated cells, is expressed at its highest level(Zikherman et al. 2012).

4.2.2 CTLA-4 Storage

CTLA-4 is not primarily localized on the T-cell surface, but instead resides intracellularly in a region that overlaps the Golgi apparatus (Leung et al. 1995) and/or endocytic compartment(s) with perforin-containing secretory granules (Linsley et al. 1996; Iida et al. 2000). The transfer of 11 cytoplasmic residues, TTGVYVKMPPT, from the CTLA-4 cytoplasmic tail to CD28 conferred intracellular localization (Leung et al. 1995). Importantly, CTLA-4 expressed on the surface was also internalized, which explains its low levels of expression on the cell surface (Alegre et al. 1996). Consequently, on naïve, uninfected mice, CTLA-4 is hardly detected on the T-cell surface by highly sensitive multi-step fluorescent labeling and flow-cytometry detection, even in CD4+ CD25+ FoxP3+ Tregs, which have the most abundant CTLA-4 expression (Chikuma unpublished). CTLA-4 is likely transcribed and stored in the intracellular vesicle in the case of high-affinity antigen recognition.

4.2.3 CTLA-4 Localization and Inhibition at the Immunological Synapse

The dynamic movement of CTLA-4 to the contact site of APC and T cells immediately after transient calcium influx by TCR stimulation was first reported by Linsley et al. (1996). The immunological synapse has a structure composed of a central super-molecular cluster (c-SMAC) that includes TCR and signaling molecules, surrounded by peripheral SMAC containing adhesion molecules. Egen and Allison (2002) followed CTLA-4 recruitment to the immunological synapse by time-lapse microscopy. CTLA-4 is located at the uropod (the opposite site of T-cell movement), but rapidly relocated to the T-cell-APC contact site a few minutes after stimulation. Iida et al. (Iida et al. 2000) suggested that CTLA-4-containing secretory granules, also comprising perforin, rapidly relocates to the synapse upon TCR ligation. Chikuma et al. (2003) showed that CTLA-4, phosphorylated Lck, and TCR-ζ-chain form a trimolecular complex within the glycosphingolipid-enriched microdomain (also called the lipid raft) that is known to be enriched at the immunological synapse. The export of CTLA-4 to IS was dependent on TCR affinity, suggesting that CTLA-4 can preferentially inhibit CD4+ T cells that have higher affinity to antigens(Egen and Allison 2002). The CTLA-4 enrichment to the IS was dependent on CD80/CD86-binding (Pentcheva-Hoang et al. 2004), suggesting that the ligation of CTLA-4 by CD80/86 occurs before the formation of a mature immunological synapse. By observing the T-cell-APC contact surface using an artificial lipid bilayer as APC, Yokosuka et al. (2010) showed that CTLA-4 participated in the immunological synapse at a relatively later time point in mature synapse formation and sequestered CD28 out of the immunological synapse by competing for CD80/CD86 signaling. By collecting the lipid raft fraction by biochemical methods, Chikuma et al. (2003) showed that the TCR molecules within the raft decreased after the co-ligation of TCR and CTLA-4, suggesting that

CTLA-4 signal sequesters TCR from the lipid raft, enriched with Lck or other signaling molecules. The recruitment of phosphatases SHP-2 (Lee et al. 1998) and PP2A (Parry et al. 2005) to CTLA-4 causes dephosphorylation or the inhibition of phosphorylation of the TCR-ζ chain (Lee et al. 1998) and/or AKT (Parry et al. 2005), respectively; however, these phosphatases are not found to form a cluster with CTLA-4 to the immunological synapse (Yokosuka et al.). The association of phosphatases is likely very transient or happens at an earlier time point, when the CTLA-4 cluster is very small.

4.2.4 Internalization and Trans-Endocytosis

CTLA-4 binds to CD80/CD86 and brings them into the intracellular compartments for degradation (Qureshi et al. 2011), specifically at late time points of the T-APC interactions (Yokosuka et al. 2010). One key question is whether the CTLA-4 molecule trans-endocytosis and recycling is designed to move CTLA-4 into and out of the immunological synapse to regulate T-cell activation.

4.3 Structure and Functional Relationship

Given the importance of motifs used for molecular interaction (i.e., MYPPPY, KMLKKRS, YVKM) within CTLA-4, to date, many efforts have been made to understand their functional importance. Figure 2 summarizes many of the results. In vitro, CTLA-4-negative cells, reconstituted with mutants that are stably expressed on the plasma membrane show inhibitory activity when ligated by anti-CTLA-4 mAb or CD80/CD86. The critical point that makes it difficult to interpret the data is that CTLA-4 has two modes of inhibiting T-cell activation, namely, ligand competition and signaling. For example, deletion of the YVKM motif may abrogate phosphatase binding, but it augments cell surface expression and the inhibitory function of CTLA-4. Since the mutant CTLA-4 molecule lacking almost the entire cytoplasmic tail retains membrane-proximal KMLKKRS that is functionally able to protect the lethal phenotype of CTLA-4 KO mice. This motif binds to PP2A and PKC-η, which may be a minimal requirement of CTLA-4 function. Transgenic overexpression of this form of CTLA-4 completely (Takahashi et al. 2005) rescued the CTLA-4 KO phenotype whereas random amino acid substitution of the entire CTLA-4 tail only partially rescued it (Masteller et al. 2000) (see Fig. 2). Trans-endocytosis and degradation of CD80/86 by CTLA-4 is suggested to account for CTLA-4-mediated regulation (Qureshi et al.). Tailless CTLA-4 used in rescue experiments, which lacks the binding site for clathrin adaptors (YVKM), is functional when overexpressed on the T-cell compartment (Takahashi et al. 2005; Masteller et al. 2000) or the Treg compartment (Tai et al.), suggesting that endocytosis is not an absolute requirement for CTLA-4-mediated inhibition.

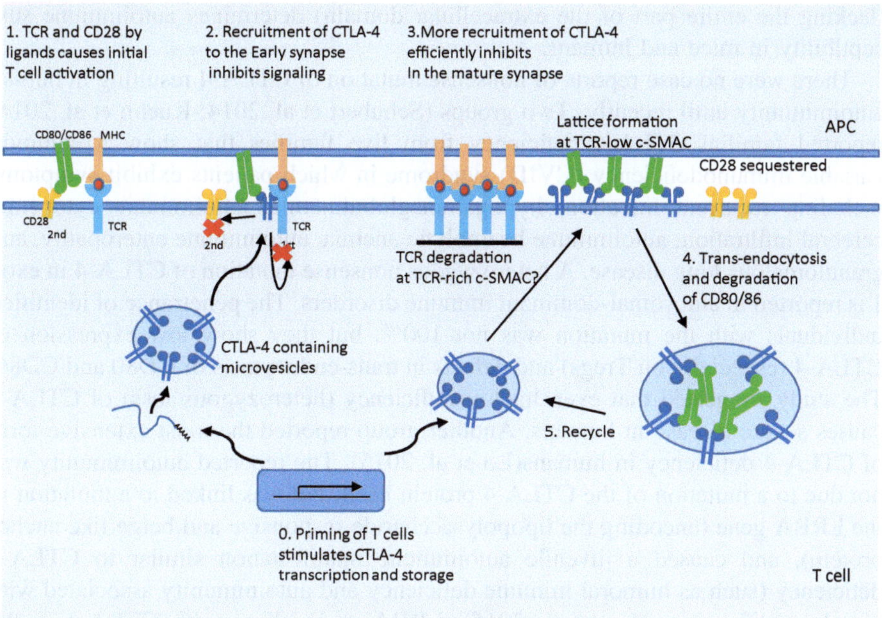

Fig. 2 A proposed model of spatiotemporal T-cell inhibition by CTLA-4

5 Biological Roles of CTLA-4

5.1 CTLA-4 Genetics in Autoimmunity: Mice to Humans

Germline CTLA-4 knockout mice show lymphoproliferation, and die from multi-organ failure and a cytokine storm, suggesting the indispensable role of CTLA-4 in regulating self-tolerance (Tivol et al. 1995; Waterhouse et al. 1995; Chambers et al. 1997a). In humans, single nucleotide polymorphisms in the regulatory region, Exon1, 3'untranslated region, have been suggested to be associated with various autoimmune diseases (Gough et al. 2005; Scalapino and Daikh 2008), including celiac disease (Djilali-Saiah et al. 1998), rheumatoid arthritis (Yanagawa et al. 2000), multiple sclerosis (Ligers et al. 1999), type I diabetes (Todd 1997), Graves' disease (Yanagawa et al. 1995), Hashimoto's thyroiditis, autoimmune Addison's disease (Donner et al. 1997), systemic lupus erythematosus (Hudson et al. 2002) to name a few. This clearly suggests that CTLA-4 insufficiency is a genetic factor in human autoimmunity. The alternative splicing of CTLA-4 mRNA is also suggested to determine autoimmune susceptibility. Ueda et al. (2003), through a comprehensive genetic association study, showed that the alternatively spliced form of the CTLA-4 locus that determines the relative amount of soluble CTLA-4 (sCTLA-4) ligand-independent liCTLA-4 versus ligand-binding CTLA-4 (lacking the entire part of the extracellular domain) determines autoimmune susceptibility in mice and humans.

There were no case reports of nonsense mutation of CTLA-4 resulting in human autoimmunity until recently. Two groups (Schubert et al. 2014; Kuehn et al. 2014) reported familial CTLA-4 deficiency from five families that show a common variable immunodeficiency (CVID) syndrome in which patients exhibit symptoms including recurrent infection, hypogammaglobulinemia, autoimmune cytopenia, cerebral infiltration, autoimmune heamolytic anemia, autoimmune enteropathy, and granulomatous lung disease. A heterozygous nonsense mutation of CTLA-4 in exon 1 is reported in autosomal-dominant immune disorders. The penetrance of identified individuals with the mutation was not 100%, but they show low expression of CTLA-4 (especially on Tregs) and defects in trans-endocytosis of CD80 and CD86. The study suggested that even haploinsufficiency (heterozygous loss) of CTLA-4 causes severe disease in humans. Another group reported the most extensive form of CTLA-4 deficiency in humans(Lo et al. 2015). The reported autoimmunity was not due to a mutation of the CTLA-4 protein itself, but was linked to a mutation in the LRBA gene (encoding the lipopolysaccharide-responsive and beige-like anchor protein), and caused a juvenile autoimmune manifestation similar to CTLA-4 deficiency (such as humoral immune deficiency and autoimmunity associated with lymphoproliferation) (Lo et al. 2015). LRBA co-localizes with CTLA-4 in the endosomal vesicles. Mutation/deficiency alters this functionality resulting in accelerated turnover of CTLA-4 by lysosomal degradation, leading to reduced CTLA-4 protein. Inhibiting the degradation by the lysosomal inhibitor chloroquine reduced CTLA-4 degradation and amelioration of autoimmunity. The authors concluded that chemicals that inhibit lysosomes (such as chroloquine) can be used as drugs to treat autoimmunity (by stabilizing CTLA-4 protein.) As described before, in mice, CD28-CTLA-4 double KO, CTLA-4/CD80/86 triple knockout, or CTLA-4 KO treated with CD28 antagonists (CTLA-4 Ig) do not develop autoimmunity, suggesting a key role of the CD28 signal. Patients with LRBA mutation who were treated with abatacept that blocks the CD28 signal show dramatic improvement in autoimmune symptoms (Lo et al. 2015), suggesting that excess CD28 signaling is the cause of the disease. Therefore, it would be beneficial to examine if individuals showing autoimmunity have mutation/polymorphism on the CTLA-4 locus, which should predict the effectiveness of abatacept therapy (Boussiotis 2014; Schubert et al. 2014).

5.2 *CTLA-4 in Infection*

Some reports suggest that CTLA-4 is critical for optimal T-cell response in antigen-specific immunity. CD4+ cells from CTLA-4 KO show spontaneous Th2-type skewing, even in the genetic absence of STAT6 (Bour-Jordan et al. 2003). Accordingly, CTLA-4 KO mice (partially rescued from lymphoproliferation by the tailless CTLA-4 transgene) show defects in controlling leishmania infection

(Masteller et al. 2000). Since CTLA-4 knockout mice experience early death by lymphoproliferation, it was difficult to experimentally address how CTLA-4 KO mice mount antigen-specific immunity to infection. To clarify this, CTLA-4 in infection was addressed using mice reconstituted by bone marrow from CTLA-4 KO and WT mice to protect them from autoimmunity-associated death (Bachmann et al. 1999; Homann et al. 2006). Bachmann et al. showed that although half of the T cells in these animals do not express CTLA-4 genetically, both CTLA-4 KO T cells and WT T cells responded to leishmania, lymphocytic choriomeningitis virus (LCMV), and mouse mammary tumor virus (MMTV) normally and contracted them equally. In addition, Homann et al. showed that both CTLA-4 KO T cells and WT T cells respond normally to invading viruses and decrease in number rapidly after elimination of the pathogen (Homann et al. 2006). These investigators proposed that CTLA-4 works mainly to inhibit autoreactive CD4+ helper T cells rather than by modulating/terminating ongoing immune responses against exogenous pathogens.

5.3 CTLA-4 in Cancer

The goal for cancer immune therapy is to enhance patients' immune systems in order to reject cancer. This concept was proposed by pioneer studies that showed that tumor cells mutated from healthy cells are recognized as "non-self" by T lymphocytes (De Plaen et al. 1988; Lurquin et al. 1989). The problem is that mutations in cancer cells may not always create a strong agonistic peptides epitope for T cells, and may instead stimulate tolerance rather than protective T cell responses. Therefore, enhancement of tumor recognition and activation by T cells through manipulation of the known pathway, such as costimulation, is beneficial. It was first shown that transfecting CD80, a CD28/CTLA-4 ligand, on a poorly immunogenic cancer cell line stimulated the mouse immune system for rejection upon tumor transplantation (Chen et al. 1992; Yang et al. 1995). The data demonstrated that recognition of antigens on a tumor can be augmented by additional signals mediated by CD28 and/or CTLA-4, leading to efficient T-cell activation and attack. Allison and colleagues demonstrated that a systemic administration with blocking anti-CTLA-4 mAb in mice boosted the anti-tumor response, resulting in the rejection of a transplanted tumors (Leach et al. 1996; Kwon et al. 1997). These reports established a milestone that the blockade of negative costimulatory molecules to their physiological ligand promotes tumor immunity. This and subsequent results led to the concept of immune-checkpoint blockade in cancer treatment. Subsequently, the CTLA-4 blockade was shown to be effective in combination with tumor vaccination in mice (van Elsas et al. 1999). Shrikant et al. (1999) used an antigen-specific tumor elimination mouse model to elucidate the mechanism of augmented tumor immunity by CTLA-4 blockade. They showed that tumor-specific CD8+ cells are generally anergic, but exhibit tumor attack upon administration of anti-CTLA-4 antibody in vivo. This re-activation of CD8+ T cells was dependent on CD4+ helper T cells and IL-2 produced by this population, suggesting that re-activation of the helper response indirectly boosts the killer-mediated anticancer responses (Shrikant et al. 1999). The result also supported the notion that CTLA-4 blockade not only directly augments effector CD8+ T cells, but also indirectly boosts immune responses by acting on helper T cells.

A fully humanized chimeric antibody named ipilimumab has been shown to be effective in melanoma in monotherapy (Hodi et al. 2010), in combination with chemotherapy (Robert et al. 2011) or with anti PD-1 antibody (Wolchok et al. 2013). Although manipulation of this pathway is attractive in cancer therapies, the immune-related adverse effects are often problematic. It was initially reported that 50–60% of patients receiving anti-CTLA-4 therapy showed adverse events (Hodi et al. 2010; Robert et al. 2011) affecting various organs (reviewed inMichot et al. 2016). Analogous to mouse studies, these data suggest that there are many autoreactive T cells in the periphery that need to be under continuous control by CTLA-4. Currently, patients with immune-related adverse events higher than grade 3 are primarily treated by steroids (reviewed in Michot et al. 2016). In the future, adequate management if these side effects are likely and it is expected that there will be an abrogation of adverse events without avoiding anticancer response. However, in case of subacute, life-threatening autoimmunity, given the example of human CTLA-4 deficiency, abatacept targeting the CD28 pathway might be important.

Conclusions and perspectives

CTLA-4 is no doubt important in the maintenance of T-cell homeostasis. The longstanding basic question of why CTLA-4 deficiency results in a lethal autoimmune phenotype in mice has not, however, been solved. An interesting observation was made by two groups. An induced deletion of CTLA-4 at adulthood resulted in milder autoimmunity compared to the original germline KO (Klocke et al. 2016). Another group developed a similar system deleting CTLA-4 in adult mice and did not cause autoimmunity (Paterson et al. 2015). Although there is a technical argument regarding the efficiency of the deletion in both models, the mice did not die, unlike germline KO mice. Similarly, it was suggested that successful anti-CTLA-4 treatment on adult cancer patients, at least in part rely on Fc-receptor-mediated depletion of CTLA-4+ T cells (Simpson et al. 2013), however, this treatment does not result in overt lymphoproliferation. An assumption based on these data is that CTLA-4 expression is particularly important during the neonatal period when polyclonal T cells undergo lymphopenia-driven expansion to seed the body (Fig. 3a) (Min et al. 2003). This form of T-cell proliferation does not require high-affinity interaction of TCR with the MHC-self-antigen complex, but is driven by the CD28 signal (Hagen et al. 2004). The neonatal period is also a time for development of self-reactive FoxP3+ Tregs that prevents activation of other T cells (Itoh et al. 1999) (Yang et al. 2015). Therefore, the loss of CTLA-4 from birth causes a CD28-dependent massive expansion of nearly the entire repertoire of low-affinity self-reactive T cells without proper regulation by FoxP3+ Tregs, resulting in lymphoproliferation and death, due to a cytokine storm (Fig. 3a). After

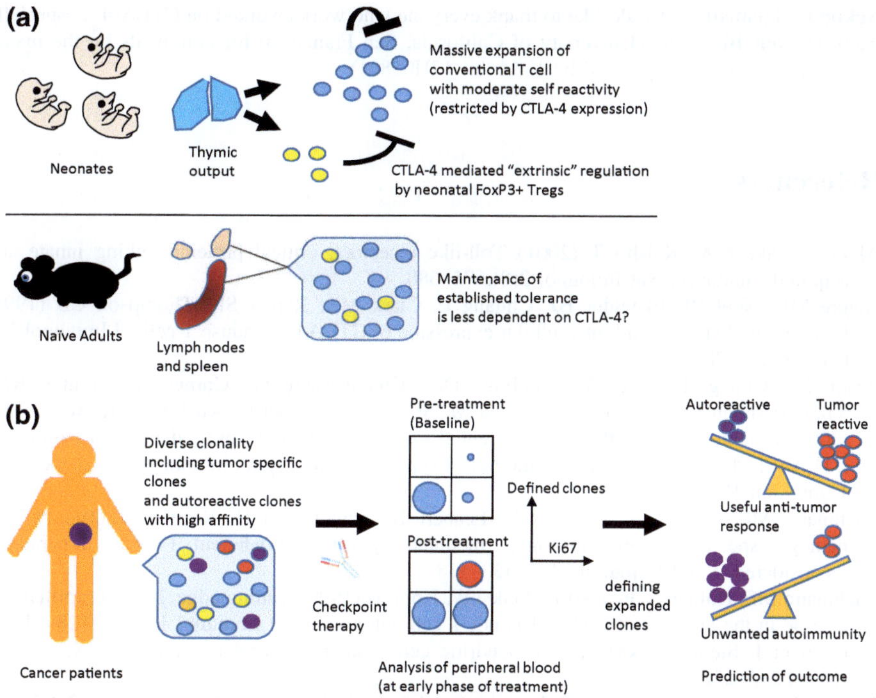

Fig. 3 A scheme representing the difference of CTLA-4-mediated inhibition between neonatal mice and naïve (uninfected) adult mice. A possible "immune monitoring" of cancer patients under checkpoint therapy to predict treatment outcome. The idea of Ki67-mediated detection of early-responding T-cell population is from (Huang et al. 2017) and (Kamphorst et al. 2017)

the initial wave of lymphocyte expansion ceases, CTLA-4-mediated inhibition is less important for maintaining the homeostatic condition of T cells (Klocke et al. 2016; Paterson et al. 2015).

In the cancer setting, high-affinity T cells receive the advantages of CTLA-4 blockade and regain their ability to fight, because these T cells recruit more CTLA-4 to the immunological synapse and are preferentially inhibited by CTLA-4 (Egen and Allison 2002). The problem is that the autoreactive memory T cells present in patients are also activated by this treatment. The balance and affinity of existing autoreactive/anticancer memory T cells will differ from person to person, which may determine the outcome of the treatment. Monitoring individuals for these clones will thus provide more efficient anticancer immunity with fewer treatment side effects (Fig. 3b). These approaches are currently in development, combining Ki67 (a proliferation marker) to monitor T cell re-activation before and after immune checkpoint therapy using patient peripheral blood samples (Huang et al. 2017; Kamphorst et al. 2017). It will hopefully become possible to detect and follow the activation of cancer/self-specific T-cell clones in the periphery during anti-CTLA-4 therapy. This will provide useful information when considering future cancer immune therapy (Fig. 3b).

Acknowledgements I would like to thank everyone who works/worked on CTLA-4. I especially thank Dr. Jeff Bluestone (University of California, San Francisco) for comments on the manuscript. This work was supported by Kakenhi (17H05801).

References

Akira S, Takeda K, Kaisho T (2001) Toll-like receptors: critical proteins linking innate and acquired immunity. Nat Immunol 2(8):675–680

Alegre ML, Noel PJ, Eisfelder BJ, Chuang E, Clark MR, Reiner SL, Thompson CB (1996) Regulation of surface and intracellular expression of CTLA4 on mouse T cells. J Immunol 157 (11):4762–4770

Araki M, Chung D, Liu S, Rainbow DB, Chamberlain G, Garner V, Hunter KM, Vijayakrishnan L, Peterson LB, Oukka M, Sharpe AH, Sobel R, Kuchroo VK, Wicker LS (2009) Genetic evidence that the differential expression of the ligand-independent isoform of CTLA-4 is the molecular basis of the Idd5.1 type 1 diabetes region in nonobese diabetic mice. J Immunol 183(8):5146–5157

Bachmann MF, Gallimore A, Jones E, Ecabert B, Acha-Orbea H, Kopf M (2001) Normal pathogen-specific immune responses mounted by CTLA-4-deficient T cells: a paradigm reconsidered. Eur J Immunol 31(2):450–458

Bachmann MF, Kohler G, Ecabert B, Mak TW, Kopf M (1999) Cutting edge: lymphoproliferative disease in the absence of CTLA-4 is not T cell autonomous. J Immunol 163(3):1128–1131

Banchereau J, Steinman RM (1998) Dendritic cells and the control of immunity. Nature 392 (6673):245–252

Baroja ML, Luxenberg D, Chau T, Ling V, Strathdee CA, Carreno BM, Madrenas J (2000) The inhibitory function of CTLA-4 does not require its tyrosine phosphorylation. J Immunol 164 (1):49–55

Baroja ML, Vijayakrishnan L, Bettelli E, Darlington PJ, Chau TA, Ling V, Collins M, Carreno BM, Madrenas J, Kuchroo VK (2002) Inhibition of CTLA-4 function by the regulatory subunit of serine/threonine phosphatase 2A. J Immunol 168(10):5070–5078

Boise LH, Minn AJ, Noel PJ, June CH, Accavitti MA, Lindsten T, Thompson CB (1995) CD28 costimulation can promote T cell survival by enhancing the expression of Bcl-XL. Immunity 3 (1):87–98

Bour-Jordan H, Esensten JH, Martinez-Llordella M, Penaranda C, Stumpf M, Bluestone JA (2011) Intrinsic and extrinsic control of peripheral T-cell tolerance by costimulatory molecules of the CD28/ B7 family. Immunol Rev 241(1):180–205

Bour-Jordan H, Grogan JL, Tang Q, Auger JA, Locksley RM, Bluestone JA (2003) CTLA-4 regulates the requirement for cytokine-induced signals in T(H)2 lineage commitment. Nat Immunol 4(2):182–188

Boussiotis VA (2014) Somatic mutations and immunotherapy outcome with CTLA-4 blockade in melanoma. N Engl J Med 371(23):2230–2232

Bradshaw JD, Lu P, Leytze G, Rodgers J, Schieven GL, Bennett KL, Linsley PS, Kurtz SE (1997) Interaction of the cytoplasmic tail of CTLA-4 (CD152) with a clathrin-associated protein is negatively regulated by tyrosine phosphorylation. Biochemistry 36(50):15975–15982

Brunet JF, Denizot F, Luciani MF, Roux-Dosseto M, Suzan M, Mattei MG, Golstein P (1987) A new member of the immunoglobulin superfamily–CTLA-4. Nature 328(6127):267–270

Brunkow ME, Jeffery EW, Hjerrild KA, Paeper B, Clark LB, Yasayko SA, Wilkinson JE, Galas D, Ziegler SF, Ramsdell F (2001) Disruption of a new forkhead/winged-helix protein, scurfin, results in the fatal lymphoproliferative disorder of the scurfy mouse. Nat Genet 27(1):68–73

Calvo CR, Amsen D, Kruisbeek AM (1997) Cytotoxic T lymphocyte antigen 4 (CTLA-4) interferes with extracellular signal-regulated kinase (ERK) and Jun NH2-terminal kinase

(JNK) activation, but does not affect phosphorylation of T cell receptor zeta and ZAP70. J Exp Med 186(10):1645–1653

Chambers CA, Cado D, Truong T, Allison JP (1997a) Thymocyte development is normal in CTLA-4-deficient mice. Proc Natl Acad Sci USA 94(17):9296–9301

Chambers CA, Kuhns MS, Allison JP (1999) Cytotoxic T lymphocyte antigen-4 (CTLA-4) regulates primary and secondary peptide-specific CD4(+) T cell responses. Proc Natl Acad Sci USA 96(15):8603–8608

Chambers CA, Sullivan TJ, Allison JP (1997b) Lymphoproliferation in CTLA-4-deficient mice is mediated by costimulation-dependent activation of CD4+ T cells. Immunity 7(6):885–895

Chen L, Ashe S, Brady WA, Hellstrom I, Hellstrom KE, Ledbetter JA, McGowan P, Linsley PS (1992) Costimulation of antitumor immunity by the B7 counterreceptor for the T lymphocyte molecules CD28 and CTLA-4. Cell 71(7):1093–1102

Chikuma S (2016) Basics of PD-1 in self-tolerance, infection, and cancer immunity. International journal of clinical oncology 21(3):448–455

Chikuma S, Abbas AK, Bluestone JA (2005) B7-independent inhibition of T cells by CTLA-4. J Immunol 175(1):177–181

Chikuma S, Bluestone JA (2007) Expression of CTLA-4 and FOXP3 in cis protects from lethal lymphoproliferative disease. Eur J Immunol 37(5):1285–1289

Chikuma S, Imboden JB, Bluestone JA (2003) Negative regulation of T cell receptor-lipid raft interaction by cytotoxic T lymphocyte-associated antigen 4. J Exp Med 197(1):129–135

Chikuma S, Murakami M, Tanaka K, Uede T (2000) Janus kinase 2 is associated with a box 1-like motif and phosphorylates a critical tyrosine residue in the cytoplasmic region of cytotoxic T lymphocyte associated molecule-4. J Cell Biochem 78(2):241–250

Chikuma S, Terawaki S, Hayashi T, Nabeshima R, Yoshida T, Shibayama S, Okazaki T, Honjo T (2009) PD-1-mediated suppression of IL-2 production induces CD8+ T cell anergy in vivo. J Immunol 182(11):6682–6689

Chuang E, Alegre ML, Duckett CS, Noel PJ, Vander Heiden MG, Thompson CB (1997) Interaction of CTLA-4 with the clathrin-associated protein AP50 results in ligand-independent endocytosis that limits cell surface expression. J Immunol 159(1):144–151

Chuang E, Fisher TS, Morgan RW, Robbins MD, Duerr JM, Vander Heiden MG, Gardner JP, Hambor JE, Neveu MJ, Thompson CB (2000) The CD28 and CTLA-4 receptors associate with the serine/threonine phosphatase PP2A. Immunity 13(3):313–322

Chuang E, Lee KM, Robbins MD, Duerr JM, Alegre ML, Hambor JE, Neveu MJ, Bluestone JA, Thompson CB (1999) Regulation of cytotoxic T lymphocyte-associated molecule-4 by Src kinases. J Immunol 162(3):1270–1277

Cinek T, Sadra A, Imboden JB (2000) Cutting edge: tyrosine-independent transmission of inhibitory signals by CTLA-4. J Immunol 164(1):5–8

De Plaen E, Lurquin C, Van Pel A, Mariame B, Szikora JP, Wolfel T, Sibille C, Chomez P, Boon T (1988) Immunogenic (tum-) variants of mouse tumor P815: cloning of the gene of tum-antigen P91A and identification of the tum- mutation. Proc Natl Acad Sci USA 85(7):2274–2278

Djilali-Saiah I, Schmitz J, Harfouch-Hammoud E, Mougenot JF, Bach JF, Caillat-Zucman S (1998) CTLA-4 gene polymorphism is associated with predisposition to coeliac disease. Gut 43(2):187–189

Donner H, Braun J, Seidl C, Rau H, Finke R, Ventz M, Walfish PG, Usadel KH, Badenhoop K (1997) Codon 17 polymorphism of the cytotoxic T lymphocyte antigen 4 gene in Hashimoto's thyroiditis and Addison's disease. J Clin Endocrinol Metab 82(12):4130–4132

Doyle AM, Mullen AC, Villarino AV, Hutchins AS, High FA, Lee HW, Thompson CB, Reiner SL (2001) Induction of cytotoxic T lymphocyte antigen 4 (CTLA-4) restricts clonal expansion of helper T cells. J Exp Med 194(7):893–902

Egen JG, Allison JP (2002) Cytotoxic T lymphocyte antigen-4 accumulation in the immunological synapse is regulated by TCR signal strength. Immunity 16(1):23–35

Fallarino F, Grohmann U, Hwang KW, Orabona C, Vacca C, Bianchi R, Belladonna ML, Fioretti MC, Alegre ML, Puccetti P (2003) Modulation of tryptophan catabolism by regulatory T cells. Nat Immunol 4(12):1206–1212

Fife BT, Griffin MD, Abbas AK, Locksley RM, Bluestone JA (2006) Inhibition of T cell activation and autoimmune diabetes using a B cell surface-linked CTLA-4 agonist. J Clin Invest 116 (8):2252–2261

Finn PW, He H, Wang Y, Wang Z, Guan G, Listman J, Perkins DL (1997) Synergistic induction of CTLA-4 expression by costimulation with TCR plus CD28 signals mediated by increased transcription and messenger ribonucleic acid stability. J Immunol 158(9):4074–4081

Fontenot JD, Gavin MA, Rudensky AY (2003) Foxp3 programs the development and function of CD4+ CD25+ regulatory T cells. Nat Immunol 4(4):330–336

Frauwirth KA, Alegre ML, Thompson CB (2000) Induction of T cell anergy in the absence of CTLA-4/B7 interaction. J Immunol 164(6):2987–2993

Frauwirth KA, Alegre ML, Thompson CB (2001) CTLA-4 is not required for induction of CD8(+) T cell anergy in vivo. J Immunol 167(9):4936–4941

Frauwirth KA, Riley JL, Harris MH, Parry RV, Rathmell JC, Plas DR, Elstrom RL, June CH, Thompson CB (2002) The CD28 signaling pathway regulates glucose metabolism. Immunity 16(6):769–777

Friedline RH, Brown DS, Nguyen H, Kornfeld H, Lee J, Zhang Y, Appleby M, Der SD, Kang J, Chambers CA (2009) CD4+ regulatory T cells require CTLA-4 for the maintenance of systemic tolerance. J Exp Med 206(2):421–434

Fu G, Hu J, Niederberger-Magnenat N, Rybakin V, Casas J, Yachi PP, Feldstein S, Ma B, Hoerter JA, Ampudia J, Rigaud S, Lambolez F, Gavin AL, Sauer K, Cheroutre H, Gascoigne NR (2011) Protein kinase C eta is required for T cell activation and homeostatic proliferation. Sci Signal 4 (202):ra84

Gajewski TF, Fallarino F, Fields PE, Rivas F, Alegre ML (2001) Absence of CTLA-4 lowers the activation threshold of primed CD8+ TCR-transgenic T cells: lack of correlation with Src homology domain 2-containing protein tyrosine phosphatase. J Immunol 166(6):3900–3907

Gough SC, Walker LS, Sansom DM (2005) CTLA4 gene polymorphism and autoimmunity. Immunol Rev 204:102–115

Greenwald RJ, Boussiotis VA, Lorsbach RB, Abbas AK, Sharpe AH (2001) CTLA-4 regulates induction of anergy in vivo. Immunity 14(2):145–155

Griffin MD, Hong DK, Holman PO, Lee KM, Whitters MJ, O'Herrin SM, Fallarino F, Collins M, Segal DM, Gajewski TF, Kranz DM, Bluestone JA (2000) Blockade of T cell activation using a surface-linked single-chain antibody to CTLA-4 (CD152). J Immunol 164(9):4433–4442

Hagen KA, Moses CT, Drasler EF, Podetz-Pedersen KM, Jameson SC, Khoruts A (2004) A role for CD28 in lymphopenia-induced proliferation of CD4 T cells. J Immunol 173(6):3909–3915

Hodi FS, O'Day SJ, McDermott DF, Weber RW, Sosman JA, Haanen JB, Gonzalez R, Robert C, Schadendorf D, Hassel JC, Akerley W, van den Eertwegh AJ, Lutzky J, Lorigan P, Vaubel JM, Linette GP, Hogg D, Ottensmeier CH, Lebbe C, Peschel C, Quirt I, Clark JI, Wolchok JD, Weber JS, Tian J, Yellin MJ, Nichol GM, Hoos A, Urba WJ (2010) Improved survival with ipilimumab in patients with metastatic melanoma. N Engl J Med 363(8):711–723

Homann D, Dummer W, Wolfe T, Rodrigo E, Theofilopoulos AN, Oldstone MB, von Herrath MG (2006) Lack of intrinsic CTLA-4 expression has minimal effect on regulation of antiviral T-cell immunity. J Virol 80(1):270–280

Hori S, Nomura T, Sakaguchi S (2003) Control of regulatory T cell development by the transcription factor Foxp3. Science 299(5609):1057–1061

Huang AC, Postow MA, Orlowski RJ, Mick R, Bengsch B, Manne S, Xu W, Harmon S, Giles JR, Wenz B, Adamow M, Kuk D, Panageas KS, Carrera C, Wong P, Quagliarello F, Wubbenhorst B, D'Andrea K, Pauken KE, Herati RS, Staupe RP, Schenkel JM, McGettigan S, Kothari S, George SM, Vonderheide RH, Amaravadi RK, Karakousis GC, Schuchter LM, Xu X, Nathanson KL, Wolchok JD, Gangadhar TC, Wherry EJ (2017) T-cell invigoration to tumour burden ratio associated with anti-PD-1 response. Nature 545(7652):60–65

Hudson LL, Rocca K, Song YW, Pandey JP (2002) CTLA-4 gene polymorphisms in systemic lupus erythematosus: a highly significant association with a determinant in the promoter region. Hum Genet 111(4–5):452–455

Ichinose K, Zhang Z, Koga T, Juang YT, Kis-Toth K, Sharpe AH, Kuchroo V, Crispin JC, Tsokos GC (2013) Brief report: increased expression of a short splice variant of CTLA-4 exacerbates lupus in MRL/lpr mice. Arthritis Rheum 65(3):764–769

Iida T, Ohno H, Nakaseko C, Sakuma M, Takeda-Ezaki M, Arase H, Kominami E, Fujisawa T, Saito T (2000) Regulation of cell surface expression of CTLA-4 by secretion of CTLA-4-containing lysosomes upon activation of CD4+ T cells. J Immunol 165(9):5062–5068

Ise W, Kohyama M, Nutsch KM, Lee HM, Suri A, Unanue ER, Murphy TL, Murphy KM (2009) CTLA-4 suppresses the pathogenicity of self antigen-specific T cells by cell-intrinsic and cell-extrinsic mechanisms. Nat Immunol 11(2):129–135

Itoh M, Takahashi T, Sakaguchi N, Kuniyasu Y, Shimizu J, Otsuka F, Sakaguchi S (1999) Thymus and autoimmunity: production of CD25+ CD4+ naturally anergic and suppressive T cells as a key function of the thymus in maintaining immunologic self-tolerance. J Immunol 162(9):5317–5326

Jain N, Nguyen H, Chambers C, Kang J (2010) Dual function of CTLA-4 in regulatory T cells and conventional T cells to prevent multiorgan autoimmunity. Proc Natl Acad Sci USA 107 (4):1524–1528

June CH, Ledbetter JA, Gillespie MM, Lindsten T, Thompson CB (1987) T-cell proliferation involving the CD28 pathway is associated with cyclosporine-resistant interleukin 2 gene expression. Mol Cell Biol 7(12):4472–4481

Kamphorst AO, Pillai RN, Yang S, Nasti TH, Akondy RS, Wieland A, Sica GL, Yu K, Koenig L, Patel NT, Behera M, Wu H, McCausland M, Chen Z, Zhang C, Khuri FR, Owonikoko TK, Ahmed R, Ramalingam SS (2017) Proliferation of PD-1+ CD8 T cells in peripheral blood after PD-1-targeted therapy in lung cancer patients. Proc Natl Acad Sci USA

Kataoka H, Takahashi S, Takase K, Yamasaki S, Yokosuka T, Koike T, Saito T (2005) CD25(+) CD4(+) regulatory T cells exert in vitro suppressive activity independent of CTLA-4. Int Immunol 17(4):421–427

Khattri R, Cox T, Yasayko SA, Ramsdell F (2003) An essential role for Scurfin in CD4+ CD25 + T regulatory cells. Nat Immunol 4(4):337–342

Klocke K, Sakaguchi S, Holmdahl R, Wing K (2016) Induction of autoimmune disease by deletion of CTLA-4 in mice in adulthood. Proc Natl Acad Sci USA 113(17):E2383–E2392

Kong KF, Fu G, Zhang Y, Yokosuka T, Casas J, Canonigo-Balancio AJ, Becart S, Kim G, Yates JR 3rd, Kronenberg M, Saito T, Gascoigne NR, Altman A (2014) Protein kinase C-eta controls CTLA-4-mediated regulatory T cell function. Nat Immunol 15(5):465–472

Krummel MF, Allison JP (1995) CD28 and CTLA-4 have opposing effects on the response of T cells to stimulation. J Exp Med 182(2):459–465

Krummel MF, Allison JP (1996) CTLA-4 engagement inhibits IL-2 accumulation and cell cycle progression upon activation of resting T cells. J Exp Med 183(6):2533–2540

Kuehn HS, Ouyang W, Lo B, Deenick EK, Niemela JE, Avery DT, Schickel JN, Tran DQ, Stoddard J, Zhang Y, Frucht DM, Dumitriu B, Scheinberg P, Folio LR, Frein CA, Price S, Koh C, Heller T, Seroogy CM, Huttenlocher A, Rao VK, Su HC, Kleiner D, Notarangelo LD, Rampertaap Y, Olivier KN, McElwee J, Hughes J, Pittaluga S, Oliveira JB, Meffre E, Fleisher TA, Holland SM, Lenardo MJ, Tangye SG, Uzel G (2014) Immune dysregulation in human subjects with heterozygous germline mutations in CTLA4. Science 345(6204):1623–1627

Kwon ED, Hurwitz AA, Foster BA, Madias C, Feldhaus AL, Greenberg NM, Burg MB, Allison JP (1997) Manipulation of T cell costimulatory and inhibitory signals for immunotherapy of prostate cancer. Proc Natl Acad Sci USA 94(15):8099–8103

Leach DR, Krummel MF, Allison JP (1996) Enhancement of antitumor immunity by CTLA-4 blockade. Science 271(5256):1734–1736

Lee KM, Chuang E, Griffin M, Khattri R, Hong DK, Zhang W, Straus D, Samelson LE, Thompson CB, Bluestone JA (1998) Molecular basis of T cell inactivation by CTLA-4. Science 282(5397):2263–2266

Leung HT, Bradshaw J, Cleaveland JS, Linsley PS (1995) Cytotoxic T lymphocyte-associated molecule-4, a high-avidity receptor for CD80 and CD86, contains an intracellular localization motif in its cytoplasmic tail. J Biol Chem 270(42):25107–25114

Ligers A, Xu C, Saarinen S, Hillert J, Olerup O (1999) The CTLA-4 gene is associated with multiple sclerosis. J Neuroimmunol 97(1–2):182–190

Ling V, Wu PW, Finnerty HF, Sharpe AH, Gray GS, Collins M (1999) Complete sequence determination of the mouse and human CTLA4 gene loci: cross-species DNA sequence similarity beyond exon borders. Genomics 60(3):341–355

Linsley PS, Bradshaw J, Greene J, Peach R, Bennett KL, Mittler RS (1996) Intracellular trafficking of CTLA-4 and focal localization towards sites of TCR engagement. Immunity 4(6):535–543

Linsley PS, Greene JL, Tan P, Bradshaw J, Ledbetter JA, Anasetti C, Damle NK (1992) Coexpression and functional cooperation of CTLA-4 and CD28 on activated T lymphocytes. J Exp Med 176(6):1595–1604

Linsley PS, Nadler SG, Bajorath J, Peach R, Leung HT, Rogers J, Bradshaw J, Stebbins M, Leytze G, Brady W et al (1995) Binding stoichiometry of the cytotoxic T lymphocyte-associated molecule-4 (CTLA-4). A disulfide-linked homodimer binds two CD86 molecules. J Biol Chem 270(25):15417–15424

Liu SM, Sutherland AP, Zhang Z, Rainbow DB, Quintana FJ, Paterson AM, Sharpe AH, Oukka M, Wicker LS, Kuchroo VK (2012) Overexpression of the Ctla-4 isoform lacking exons 2 and 3 causes autoimmunity. J Immunol 188(1):155–162

Lo B, Zhang K, Lu W, Zheng L, Zhang Q, Kanellopoulou C, Zhang Y, Liu Z, Fritz JM, Marsh R, Husami A, Kissell D, Nortman S, Chaturvedi V, Haines H, Young LR, Mo J, Filipovich AH, Bleesing JJ, Mustillo P, Stephens M, Rueda CM, Chougnet CA, Hoebe K, McElwee J, Hughes JD, Karakoc-Aydiner E, Matthews HF, Price S, Su HC, Rao VK, Lenardo MJ, Jordan MB (2015) AUTOIMMUNE DISEASE. Patients with LRBA deficiency show CTLA4 loss and immune dysregulation responsive to abatacept therapy. Science 349(6246):436–440

Lorenz U (2009) SHP-1 and SHP-2 in T cells: two phosphatases functioning at many levels. Immunol Rev 228(1):342–359

Lurquin C, Van Pel A, Mariame B, De Plaen E, Szikora JP, Janssens C, Reddehase MJ, Lejeune J, Boon T (1989) Structure of the gene of tum- transplantation antigen P91A: the mutated exon encodes a peptide recognized with Ld by cytolytic T cells. Cell 58(2):293–303

Magistrelli G, Jeannin P, Herbault N, Benoit De Coignac A, Gauchat JF, Bonnefoy JY, Delneste Y (1999) A soluble form of CTLA-4 generated by alternative splicing is expressed by nonstimulated human T cells. Eur J Immunol 29(11):3596–3602

Mandelbrot DA, McAdam AJ, Sharpe AH (1999) B7-1 or B7-2 is required to produce the lymphoproliferative phenotype in mice lacking cytotoxic T lymphocyte-associated antigen 4 (CTLA-4). J Exp Med 189(2):435–440

Mandelbrot DA, Oosterwegel MA, Shimizu K, Yamada A, Freeman GJ, Mitchell RN, Sayegh MH, Sharpe AH (2001) B7-dependent T-cell costimulation in mice lacking CD28 and CTLA4. J Clin Invest 107(7):881–887

Marengere LE, Waterhouse P, Duncan GS, Mittrucker HW, Feng GS, Mak TW (1996) Regulation of T cell receptor signaling by tyrosine phosphatase SYP association with CTLA-4. Science 272(5265):1170–1173

Masteller EL, Chuang E, Mullen AC, Reiner SL, Thompson CB (2000) Structural analysis of CTLA-4 function in vivo. J Immunol 164(10):5319–5327

Michot JM, Bigenwald C, Champiat S, Collins M, Carbonnel F, Postel-Vinay S, Berdelou A, Varga A, Bahleda R, Hollebecque A, Massard C, Fuerea A, Ribrag V, Gazzah A, Armand JP, Amellal N, Angevin E, Noel N, Boutros C, Mateus C, Robert C, Soria JC, Marabelle A, Lambotte O (2016) Immune-related adverse events with immune checkpoint blockade: a comprehensive review. Eur J Cancer 54:139–148

Min B, McHugh R, Sempowski GD, Mackall C, Foucras G, Paul WE (2003) Neonates support lymphopenia-induced proliferation. Immunity 18(1):131–140
Miyatake S, Nakaseko C, Umemori H, Yamamoto T, Saito T (1998) Src family tyrosine kinases associate with and phosphorylate CTLA-4 (CD152). Biochem Biophys Res Commun 249 (2):444–448
Nakaseko C, Miyatake S, Iida T, Hara S, Abe R, Ohno H, Saito Y, Saito T (1999) Cytotoxic T lymphocyte antigen 4 (CTLA-4) engagement delivers an inhibitory signal through the membrane-proximal region in the absence of the tyrosine motif in the cytoplasmic tail. J Exp Med 190(6):765–774
Oaks MK, Hallett KM, Penwell RT, Stauber EC, Warren SJ, Tector AJ (2000) A native soluble form of CTLA-4. Cell Immunol 201(2):144–153
Ohkura N, Hamaguchi M, Morikawa H, Sugimura K, Tanaka A, Ito Y, Osaki M, Tanaka Y, Yamashita R, Nakano N, Huehn J, Fehling HJ, Sparwasser T, Nakai K, Sakaguchi S (2012) T cell receptor stimulation-induced epigenetic changes and Foxp3 expression are independent and complementary events required for Treg cell development. Immunity 37(5):785–799
Ostrov DA, Shi W, Schwartz JC, Almo SC, Nathenson SG (2000) Structure of murine CTLA-4 and its role in modulating T cell responsiveness. Science 290(5492):816–819
Parry RV, Chemnitz JM, Frauwirth KA, Lanfranco AR, Braunstein I, Kobayashi SV, Linsley PS, Thompson CB, Riley JL (2005) CTLA-4 and PD-1 receptors inhibit T-cell activation by distinct mechanisms. Mol Cell Biol 25(21):9543–9553
Paterson AM, Lovitch SB, Sage PT, Juneja VR, Lee Y, Trombley JD, Arancibia-Carcamo CV, Sobel RA, Rudensky AY, Kuchroo VK, Freeman GJ, Sharpe AH (2015) Deletion of CTLA-4 on regulatory T cells during adulthood leads to resistance to autoimmunity. J Exp Med 212 (10):1603–1621
Peach RJ, Bajorath J, Brady W, Leytze G, Greene J, Naemura J, Linsley PS (1994) Complementarity determining region 1 (CDR1)- and CDR3-analogous regions in CTLA-4 and CD28 determine the binding to B7-1. J Exp Med 180(6):2049–2058
Pentcheva-Hoang T, Egen JG, Wojnoonski K, Allison JP (2004) B7-1 and B7-2 selectively recruit CTLA-4 and CD28 to the immunological synapse. Immunity 21(3):401–413
Perkins D, Wang Z, Donovan C, He H, Mark D, Guan G, Wang Y, Walunas T, Bluestone J, Listman J, Finn PW (1996) Regulation of CTLA-4 expression during T cell activation. J Immunol 156(11):4154–4159
Pioli C, Gatta L, Frasca D, Doria G (1999) Cytotoxic T lymphocyte antigen 4 (CTLA-4) inhibits CD28-induced IkappaBalpha degradation and RelA activation. Eur J Immunol 29(3):856–863
Puccetti P, Grohmann U (2007) IDO and regulatory T cells: a role for reverse signalling and non-canonical NF-kappaB activation. Nat Rev Immunol 7(10):817–823
Qureshi OS, Zheng Y, Nakamura K, Attridge K, Manzotti C, Schmidt EM, Baker J, Jeffery LE, Kaur S, Briggs Z, Hou TZ, Futter CE, Anderson G, Walker LS, Sansom DM (2011) Trans-endocytosis of CD80 and CD86: a molecular basis for the cell-extrinsic function of CTLA-4. Science 332(6029):600–603
Read S, Malmstrom V, Powrie F (2000) Cytotoxic T lymphocyte-associated antigen 4 plays an essential role in the function of CD25(+)CD4(+) regulatory cells that control intestinal inflammation. J Exp Med 192(2):295–302
Riley JL, Mao M, Kobayashi S, Biery M, Burchard J, Cavet G, Gregson BP, June CH, Linsley PS (2002) Modulation of TCR-induced transcriptional profiles by ligation of CD28, ICOS, and CTLA-4 receptors. Proc Natl Acad Sci U S A 99(18):11790–11795
Robert C, Thomas L, Bondarenko I, O'Day S, Weber J, Garbe C, Lebbe C, Baurain JF, Testori A, Grob JJ, Davidson N, Richards J, Maio M, Hauschild A, Miller WH Jr, Gascon P, Lotem M, Harmankaya K, Ibrahim R, Francis S, Chen TT, Humphrey R, Hoos A, Wolchok JD (2011) Ipilimumab plus dacarbazine for previously untreated metastatic melanoma. N Engl J Med 364 (26):2517–2526
Salomon B, Lenschow DJ, Rhee L, Ashourian N, Singh B, Sharpe A, Bluestone JA (2000) B7/CD28 costimulation is essential for the homeostasis of the CD4+ CD25+ immunoregulatory T cells that control autoimmune diabetes. Immunity 12(4):431–440

Scalapino KJ, Daikh DI (2008) CTLA-4: a key regulatory point in the control of autoimmune disease. Immunol Rev 223:143–155

Schneider H, Martin M, Agarraberes FA, Yin L, Rapoport I, Kirchhausen T, Rudd CE (1999) Cytolytic T lymphocyte-associated antigen-4 and the TCR zeta/CD3 complex, but not CD28, interact with clathrin adaptor complexes AP-1 and AP-2. J Immunol 163(4):1868–1879

Schneider H, Rudd CE (2000) Tyrosine phosphatase SHP-2 binding to CTLA-4: absence of direct YVKM/YFIP motif recognition. Biochem Biophys Res Commun 269(1):279–283

Schneider H, Schwartzberg PL, Rudd CE (1998) Resting lymphocyte kinase (Rlk/Txk) phosphorylates the YVKM motif and regulates PI 3-kinase binding to T-cell antigen CTLA-4. Biochem Biophys Res Commun 252(1):14–19

Schubert D, Bode C, Kenefeck R, Hou TZ, Wing JB, Kennedy A, Bulashevska A, Petersen BS, Schaffer AA, Gruning BA, Unger S, Frede N, Baumann U, Witte T, Schmidt RE, Dueckers G, Niehues T, Seneviratne S, Kanariou M, Speckmann C, Ehl S, Rensing-Ehl A, Warnatz K, Rakhmanov M, Thimme R, Hasselblatt P, Emmerich F, Cathomen T, Backofen R, Fisch P, Seidl M, May A, Schmitt-Graeff A, Ikemizu S, Salzer U, Franke A, Sakaguchi S, Walker LS, Sansom DM, Grimbacher B (2014) Autosomal dominant immune dysregulation syndrome in humans with CTLA4 mutations. Nat Med 20(12):1410–1416

Schwartz JC, Zhang X, Fedorov AA, Nathenson SG, Almo SC (2001) Structural basis for co-stimulation by the human CTLA-4/B7-2 complex. Nature 410(6828):604–608

Schwartz RH (2003) T cell anergy. Annu Rev Immunol 21:305–334

Shiratori T, Miyatake S, Ohno H, Nakaseko C, Isono K, Bonifacino JS, Saito T (1997) Tyrosine phosphorylation controls internalization of CTLA-4 by regulating its interaction with clathrin-associated adaptor complex AP-2. Immunity 6(5):583–589

Shrikant P, Khoruts A, Mescher MF (1999) CTLA-4 blockade reverses CD8+ T cell tolerance to tumor by a CD4+ T cell- and IL-2-dependent mechanism. Immunity 11(4):483–493

Simpson TR, Li F, Montalvo-Ortiz W, Sepulveda MA, Bergerhoff K, Arce F, Roddie C, Henry JY, Yagita H, Wolchok JD, Peggs KS, Ravetch JV, Allison JP, Quezada SA (2013) Fc-dependent depletion of tumor-infiltrating regulatory T cells co-defines the efficacy of anti-CTLA-4 therapy against melanoma. J Exp Med 210(9):1695–1710

Stamper CC, Zhang Y, Tobin JF, Erbe DV, Ikemizu S, Davis SJ, Stahl ML, Seehra J, Somers WS, Mosyak L (2001) Crystal structure of the B7-1/CTLA-4 complex that inhibits human immune responses. Nature 410(6828):608–611

Tai X, Cowan M, Feigenbaum L, Singer A (2005) CD28 costimulation of developing thymocytes induces Foxp3 expression and regulatory T cell differentiation independently of interleukin 2. Nat Immunol 6(2):152–162

Tai X, Van Laethem F, Pobezinsky L, Guinter T, Sharrow SO, Adams A, Granger L, Kruhlak M, Lindsten T, Thompson CB, Feigenbaum L, Singer A (2012) Basis of CTLA-4 function in regulatory and conventional CD4(+) T cells. Blood 119(22):5155–5163

Takahashi S, Kataoka H, Hara S, Yokosuka T, Takase K, Yamasaki S, Kobayashi W, Saito Y, Saito T (2005) In vivo overexpression of CTLA-4 suppresses lymphoproliferative diseases and thymic negative selection. Eur J Immunol 35(2):399–407

Takahashi T, Tagami T, Yamazaki S, Uede T, Shimizu J, Sakaguchi N, Mak TW, Sakaguchi S (2000) Immunologic self-tolerance maintained by CD25(+)CD4(+) regulatory T cells constitutively expressing cytotoxic T lymphocyte-associated antigen 4. J Exp Med 192 (2):303–310

Tang Q, Adams JY, Tooley AJ, Bi M, Fife BT, Serra P, Santamaria P, Locksley RM, Krummel MF, Bluestone JA (2006) Visualizing regulatory T cell control of autoimmune responses in nonobese diabetic mice. Nat Immunol 7(1):83–92

Tang Q, Boden EK, Henriksen KJ, Bour-Jordan H, Bi M, Bluestone JA (2004) Distinct roles of CTLA-4 and TGF-beta in CD4+ CD25+ regulatory T cell function. Eur J Immunol 34 (11):2996–3005

Tang Q, Henriksen KJ, Boden EK, Tooley AJ, Ye J, Subudhi SK, Zheng XX, Strom TB, Bluestone JA (2003) Cutting edge: CD28 controls peripheral homeostasis of CD4 + CD25 + regulatory T cells. J Immunol 171(7):3348–3352

Tivol EA, Borriello F, Schweitzer AN, Lynch WP, Bluestone JA, Sharpe AH (1995) Loss of CTLA-4 leads to massive lymphoproliferation and fatal multiorgan tissue destruction, revealing a critical negative regulatory role of CTLA-4. Immunity 3(5):541–547

Tivol EA, Gorski J (2002) Re-establishing peripheral tolerance in the absence of CTLA-4: complementation by wild-type T cells points to an indirect role for CTLA-4. J Immunol 169 (4):1852–1858

Todd JA (1997) Genetics of type 1 diabetes. Pathol Biol (Paris) 45(3):219–227

Ueda H, Howson JM, Esposito L, Heward J, Snook H, Chamberlain G, Rainbow DB, Hunter KM, Smith AN, Di Genova G, Herr MH, Dahlman I, Payne F, Smyth D, Lowe C, Twells RC, Howlett S, Healy B, Nutland S, Rance HE, Everett V, Smink LJ, Lam AC, Cordell HJ, Walker NM, Bordin C, Hulme J, Motzo C, Cucca F, Hess JF, Metzker ML, Rogers J, Gregory S, Allahabadia A, Nithiyananthan R, Tuomilehto-Wolf E, Tuomilehto J, Bingley P, Gillespie KM, Undlien DE, Ronningen KS, Guja C, Ionescu-Tirgoviste C, Savage DA, Maxwell AP, Carson DJ, Patterson CC, Franklyn JA, Clayton DG, Peterson LB, Wicker LS, Todd JA, Gough SC (2003) Association of the T-cell regulatory gene CTLA4 with susceptibility to autoimmune disease. Nature 423(6939):506–511

van Elsas A, Hurwitz AA, Allison JP (1999) Combination immunotherapy of B16 melanoma using anti-cytotoxic T lymphocyte-associated antigen 4 (CTLA-4) and granulocyte/macrophage colony-stimulating factor (GM-CSF)-producing vaccines induces rejection of subcutaneous and metastatic tumors accompanied by autoimmune depigmentation. J Exp Med 190(3):355–366

Vijayakrishnan L, Slavik JM, Illes Z, Greenwald RJ, Rainbow D, Greve B, Peterson LB, Hafler DA, Freeman GJ, Sharpe AH, Wicker LS, Kuchroo VK (2004) An autoimmune disease-associated CTLA-4 splice variant lacking the B7 binding domain signals negatively in T cells. Immunity 20(5):563–575

Walunas TL, Lenschow DJ, Bakker CY, Linsley PS, Freeman GJ, Green JM, Thompson CB, Bluestone JA (1994) CTLA-4 can function as a negative regulator of T cell activation. Immunity 1(5):405–413

Wang CJ, Kenefeck R, Wardzinski L, Attridge K, Manzotti C, Schmidt EM, Qureshi OS, Sansom DM, Walker LS (2012) Cutting edge: cell-extrinsic immune regulation by CTLA-4 expressed on conventional T cells. J Immunol 189(3):1118–1122

Waterhouse P, Bachmann MF, Penninger JM, Ohashi PS, Mak TW (1997) Normal thymic selection, normal viability and decreased lymphoproliferation in T cell receptor-transgenic CTLA-4-deficient mice. Eur J Immunol 27(8):1887–1892

Waterhouse P, Penninger JM, Timms E, Wakeham A, Shahinian A, Lee KP, Thompson CB, Griesser H, Mak TW (1995) Lymphoproliferative disorders with early lethality in mice deficient in Ctla-4. Science 270(5238):985–988

Wing K, Onishi Y, Prieto-Martin P, Yamaguchi T, Miyara M, Fehervari Z, Nomura T, Sakaguchi S (2008) CTLA-4 control over Foxp3 + regulatory T cell function. Science 322 (5899):271–275

Wolchok JD, Kluger H, Callahan MK, Postow MA, Rizvi NA, Lesokhin AM, Segal NH, Ariyan CE, Gordon RA, Reed K, Burke MM, Caldwell A, Kronenberg SA, Agunwamba BU, Zhang X, Lowy I, Inzunza HD, Feely W, Horak CE, Hong Q, Korman AJ, Wigginton JM, Gupta A, Sznol M (2013) Nivolumab plus ipilimumab in advanced melanoma. N Engl J Med 369(2):122–133

Wu Y, Borde M, Heissmeyer V, Feuerer M, Lapan AD, Stroud JC, Bates DL, Guo L, Han A, Ziegler SF, Mathis D, Benoist C, Chen L, Rao A (2006) FOXP3 controls regulatory T cell function through cooperation with NFAT. Cell 126(2):375–387

Yamasaki S, Nishida K, Hibi M, Sakuma M, Shiina R, Takeuchi A, Ohnishi H, Hirano T, Saito T (2001) Docking protein Gab2 is phosphorylated by ZAP-70 and negatively regulates T cell receptor signaling by recruitment of inhibitory molecules. J Biol Chem 276(48):45175–45183

Yanagawa T, Gomi K, Nakao EI, Inada S (2000) CTLA-4 gene polymorphism in Japanese patients with rheumatoid arthritis. J Rheumatol 27(12):2740–2742

Yanagawa T, Hidaka Y, Guimaraes V, Soliman M, DeGroot LJ (1995) CTLA-4 gene polymorphism associated with Graves' disease in a Caucasian population. J Clin Endocrinol Metab 80(1):41–45
Yang G, Hellstrom KE, Hellstrom I, Chen L (1995) Antitumor immunity elicited by tumor cells transfected with B7-2, a second ligand for CD28/CTLA-4 costimulatory molecules. J Immunol 154(6):2794–2800
Yang S, Fujikado N, Kolodin D, Benoist C, Mathis D (2015) Immune tolerance. Regulatory T cells generated early in life play a distinct role in maintaining self-tolerance. Science 348 (6234):589–594
Yi LA, Hajialiasgar S, Chuang E (2004) Tyrosine-mediated inhibitory signals contribute to CTLA-4 function in vivo. Int Immunol 16(4):539–547
Yokosuka T, Kobayashi W, Takamatsu M, Sakata-Sogawa K, Zeng H, Hashimoto-Tane A, Yagita H, Tokunaga M, Saito T (2010) Spatiotemporal basis of CTLA-4 costimulatory molecule-mediated negative regulation of T cell activation. Immunity 33(3):326–339
Zhang Y, Allison JP (1997) Interaction of CTLA-4 with AP50, a clathrin-coated pit adaptor protein. Proc Natl Acad Sci USA 94(17):9273–9278
Zikherman J, Parameswaran R, Weiss A (2012) Endogenous antigen tunes the responsiveness of naive B cells but not T cells. Nature 489(7414):160–164

Tim-3, Lag-3, and TIGIT

Nicole Joller and Vijay K. Kuchroo

Abstract Co-inhibitory receptors play a key role in regulating T cell responses and maintaining immune homeostasis. Their inhibitory function prevents autoimmune responses but also restricts the ability of T cells to mount effective immune responses against tumors or persistent pathogens. T cells express a module of co-inhibitory receptors, which display great diversity in expression, structure, and function. Here, we focus on the co-inhibitory receptors Tim-3, Lag-3, and TIGIT and how they regulate T cell function, maintenance of self-tolerance, their role in regulating ongoing T cell responses at peripheral tissues, and their synergistic effects in regulating autoimmunity and antitumor responses.

Contents

1 Introduction ... 128
2 Tim-3 ... 129
2.1 Signaling ... 131
2.2 Tim-3 in Innate Cells ... 131
2.3 Role of Tim-3 in Effector Cells ... 132
2.4 Role of Tim-3 in Regulatory Cells ... 133
3 Lag-3 ... 134
3.1 Signaling ... 134
3.2 Role of Lag-3 in Effector Cells ... 134
3.3 Role of Lag-3 in Regulatory Cells ... 135
4 TIGIT ... 135
4.1 Signaling and Direct Inhibition ... 137
4.2 Indirect Inhibition ... 139

N. Joller
Institute for Experimental Immunology, University of Zurich, Zurich, Switzerland
e-mail: nicole.joller@immunology.uzh.ch

V.K. Kuchroo (✉)
Harvard Medical School and Brigham & Women's Hospital, Evergrande Center for Immunologic Diseases, Boston, MA, USA
e-mail: vkuchroo@evergrande.hms.harvard.edu

Current Topics in Microbiology and Immunology (2017) 410:127–156
DOI 10.1007/82_2017_62

Published Online: 13 September 2017

5 Role in Diseases and Therapy 140
5.1 T Cell Exhaustion 140
5.2 Cancer and Checkpoint Inhibitors 142
5.3 Autoimmunity 144
6 Conclusion 146
References 147

1 Introduction

Initiation of adaptive T cell responses requires two signals: T cell receptor (TCR) stimulation through antigen recognition in the context of MHC and co-stimulation through interaction of co-receptors on T cells with their ligands on APCs. Many of these receptors are members of the B7 family that either positively or negatively contribute to TCR signaling and thus fine-tune the threshold for T cell activation. Positive co-stimulatory molecules promote T cell proliferation and effector function. The most prominent example of such a molecule is CD28, which allows for proper T cell activation upon recognition of its ligands CD80 or CD86 expressed by mature antigen presenting cells (APCs) (Esensten et al. 2016). Signaling via co-inhibitory receptors counteracts TCR-driven T cell activation and promotes functional inactivation of T cells leading to anergy and a state of tolerance. The best studied co-inhibitory receptor is CTLA-4, which outcompetes CD28 for its ligands and actively delivers inhibitory signals to the T cell to dampen T cell activation (Schildberg et al. 2016). At steady state, co-inhibitory receptors are critical for maintaining immune homeostasis as they counterbalance co-stimulatory signals and prevent excessive effector T cell activation, which would lead to autoimmunity. In addition to regulating effector T cell responses directly, co-inhibitory receptors like CTL-4 also dampen T cells responses indirectly by promoting the suppressive function of regulatory T cells (Tregs) (Wing et al. 2008). Through interaction with their ligands, co-inhibitory molecules can further regulate the ability of APCs to prime effector T cells. For instance, binding of B7 by CTLA-4 leads to back signaling into the APC, resulting in reduced cytokine production and induction of IDO and conditions DCs to downmodulate co-stimulatory ligands (Fallarino et al. 2003; Dejean et al. 2009; Grohmann et al. 2002). Co-inhibitory receptors thus regulate T cell responses at three different levels, namely by directly inhibiting effector T cell activation, by promoting the suppressive function of Tregs, and by modulating APC function to prevent T cell activation.

The strict control of T cell responses through co-inhibitory molecules is of utmost importance to a functioning immune system, as dysregulation of T cell responses results in pathology. Failure to keep T cell responses in check causes excessive immune activation and autoimmunity (Linsley et al. 1991; Waterhouse et al. 1995).

On the other hand, excessive inhibition of T cell responses predisposes to cancer and allows for pathogen persistence. The importance of co-inhibitory receptors in cancer and chronic viral infection is highlighted by the fact that in these settings, co-inhibitory pathways are being targeted clinically to improve antitumor and antiviral T cell responses (Pauken and Wherry 2015; Chen and Mellman 2013).

T cells express a diverse repertoire of co-inhibitory receptors, which changes depending on the activation status of the T cell and its environment. Which of these receptors regulate T cell responses in any given setting not only depends on the expression pattern of the co-inhibitory receptor but also on where their ligands are expressed. Furthermore, triggers in the tissue microenvironment may dictate the expression and persistence of co-inhibitory molecules on T cells. While CTLA-4 primarily acts as a global switch during the early priming phase in lymphoid organs, other co-inhibitory receptors predominantly regulate effector T cell responses within tissue where effector T cell responses are being executed. This chapter will focus on the co-inhibitory receptors Tim-3, Lag-3, and TIGIT and how they regulate T cell function, especially during ongoing T cell responses.

2 Tim-3

T cell immunoglobulin-3 (Tim-3) is a type I transmembrane protein originally identified as a specific marker for Th1 and Tc1 cells and its expression is regulated by the Th1 transcription factor T-bet (Monney et al. 2002; Anderson et al. 2010) together with another transcription factor NFIL3 (Zhu et al. 2015). Tim-3 is further expressed on Tregs, NK cells, monocytes, macrophages, and DCs. The discovery of Tim-3 also led to the identification of the Tim family of genes, of which 3 proteins (Tim-1, Tim-3, and Tim-4) are conserved between mouse and humans (Meyers et al. 2005). They share a common structure consisting of an N-terminal IgV domain, a mucin stalk containing potential O-linked glycosylation sites, a type I transmembrane domain, and a cytoplasmic tail, which does not harbor any classical inhibitory signaling motifs but contains five conserved tyrosine residues (Meyers et al. 2005).

Initial efforts to determine the ligands of Tim-3 identified the C-Type lectin galectin-9 and a second protein, which was recently characterized as Ceacam1 (Fig. 1a) (Huang et al. 2015; Zhu et al. 2005). Galectin-9 binds to the N-linked sugar moieties in the Tim-3 IgV domain and this interaction triggers cell death in Th1 and Tc1 cells (Zhu et al. 2005; Kang et al. 2015). Ceacam-1 is co-expressed with Tim-3 on T cells and their *cis*-interaction in required for the inhibitory function of Tim-3, which is compromised in the absence of Ceacam-1 (Huang et al. 2015). Furthermore, the Tim-3—Ceacam-1 *trans*-interaction also suppresses T cell function. While Ceacam-1 and galectin-9 bind to different regions of the IgV domain of Tim-3, both ligands induce phosphorylation of the same two tyrosine residues required for functional activity of Tim-3 (Huang et al. 2015; Rangachari et al. 2012).

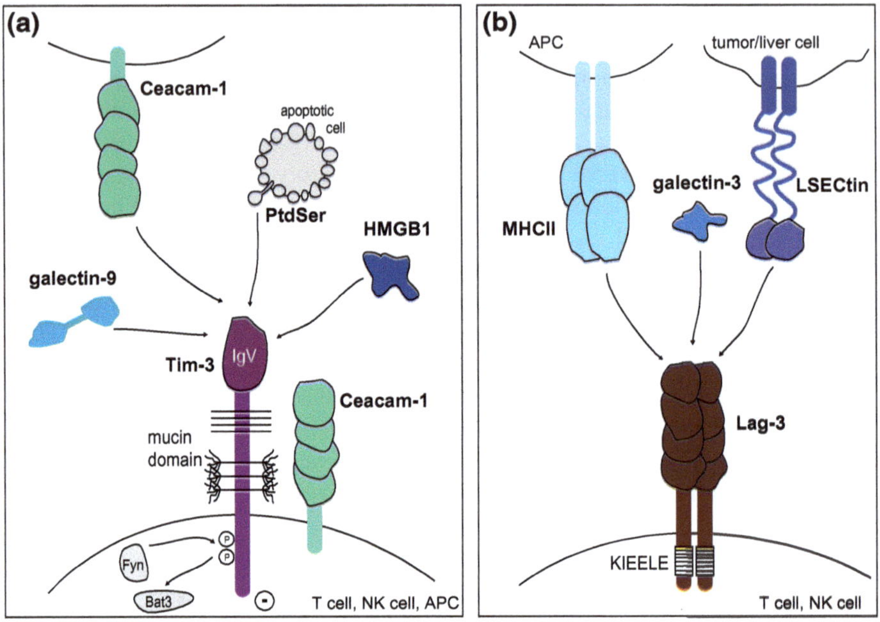

Fig. 1 Tim-3 and Lag-3 pathways. **a** The Tim-3 pathway. Tim-3 is composed of an extracellular IgV domain, a mucin stalk with N- and O-linked glycosylation sites, and an intracellular tail with conserved tyrosine residues. It is expressed on T cells, NK cells, and APCs and binds to cell surface receptors (Ceacam-1 and phosphatidyl serine (PtdSer)) and soluble ligands (galectin-9 and HMGB1). Ligand binding triggers phosphorylation of two conserved tyrosine residues and release of Bat3 from the cytoplasmic tail of Tim-3, allowing Tim-3 to exert its inhibitory function. **b** The Lag-3 pathway. Lag-3 is composed of four extracellular Ig-like domains, a transmembrane domain, and a cytoplasmic tail containing a unique KIEELE motif. It is expressed on T cells and NK cells and binds to MHC class II on APCs, galectin-3, and LSECtin on tumor cells or liver cells

In addition to galectin-9 and Ceacam-1, Tim-3 has been reported to bind two additional ligands, phosphatidyl serine (PtdSer) and high mobility group protein B1 (HMGB1), which mainly play a role in the action of Tim-3 in innate cells (Fig. 1a) (Chiba et al. 2012). PtdSer is a nonprotein ligand, which is shared between the different Tim family members and is expressed on apoptotic cells (Cao et al. 2007; Santiago et al. 2007a, 2007b). Binding of PtdSer by Tim-3 expressing phagocytic cells can mediate uptake of apoptotic cells (DeKruyff et al. 2010; Nakayama et al. 2009) and perhaps contribute to initiating Tim-3-mediated inhibitory function. Tim-3 binding to HMGB1 has been shown to promote inhibitory function by blocking the transport of nucleic acids to endosomes and thereby interfering with nucleic acid sensing and danger signaling pathways in DCs (Chiba et al. 2012). Whether interactions of Tim-3 and PtdSer or HMGB1 take place in T cells and whether such contacts have functional consequences is still unknown. However, one can imagine that binding of apoptotic cells to Tim-3 in DCs could promote Tim-3 dependent inhibitory function and thus indirectly dampen T cell responses.

2.1 Signaling

Tim-3 acts as a negative regulator of Th1 and CTL responses. However, as mentioned above, the Tim-3 tail does not contain classical inhibitory signaling motifs. Instead, it harbors five conserved tyrosine residues, two of which are phosphorylated and important for binding to the intracellular adapter protein Bat3 (HLA-B associated transcript 3). Phosphorylation of the two tyrosines in the Tim-3 tail promotes downstream inhibitory signals (Rangachari et al. 2012; Lee et al. 2011). In the absence of binding of Tim-3 ligands, Bat3 is bound to the unphosphorylated cytoplasmic tail of Tim-3, recruits Lck, and preserves or may even promote T cell signaling (Rangachari et al. 2012). The interaction of Tim-3 with its ligands (galectin-9 or Ceacam1) induces an intracellular calcium flux and phosphorylation of the two critical tyrosine residues (Y256 and Y263), which releases Bat3 from the cytoplasmic tail of Tim-3 (Zhu et al. 2005; Rangachari et al. 2012). The release of Bat3 allows for the binding of SH2 domain-containing Src kinases and promotion of subsequent negative regulation of TCR signaling (Huang et al. 2015; Rangachari et al. 2012). Since Fyn and Bat3 bind to the same domain of the Tim-3 tail, it is possible that a switch between Tim-3/Bat3 and Tim-3/Fyn triggers the switch of Tim-3 being permissive to TCR signaling to Tim-3 inhibiting upstream TCR signaling (Rangachari et al. 2012). Much less is known about the downstream signaling events of Tim-3 in innate cells and it will be important to determine how the different Tim-3 ligands affect its function depending on the cellular context.

2.2 Tim-3 in Innate Cells

Tim-3 is highly expressed on innate cells and regulates their function in several ways. In human, all mature resting $CD56^{dim}$ NK cells express Tim-3 and its expression on NK cells is induced upon cytokine stimulation (Gleason et al. 2012; Ndhlovu et al. 2012), however, the role of Tim-3 on NK cells is controversial. High expression of Tim-3 is found on effector NK cells that display high cytotoxicity and produce IFN-γ. In this context, ligation of Tim-3 by galectin-9 promotes IFN-γ production by the NK cells and antibody blockade of Tim-3 results in impaired IFN-γ production (Gleason et al. 2012). In contrast, in advanced melanoma patients, Tim-3 marks functionally exhausted NK cells with impaired IFN-γ secretion and cytotoxicity (da Silva et al. 2014). Moreover, in this context Tim-3 expression levels on NK cells correlate with poor prognosis (da Silva et al. 2014; Xu et al. 2015). Furthermore, Tim-3 blockade reverses the exhausted phenotype and restores NK function (da Silva et al. 2014; Xu et al. 2015), suggesting that, similar to its role in $CD8^+$ CTLs, Tim-3 not only marks exhausted NK cells but also contributes to their dysfunction in cancer settings.

DCs and macrophages also constitutively express high levels of Tim-3, which seems to act as a negative regulator of their function. In DCs, Tim-3 binding to

HMGB1 inhibits DC activation by interfering with nucleic acid sensing as described above (Chiba et al. 2012). Tim-3 expression on macrophages down-modulates responses to TLR4 stimulation and has a dampening effect during sepsis (Yang et al. 2013). Tim-3 can also suppress the immune response indirectly by promoting generation of myeloid-derived suppressor cells (MDSC) in a Tim-3/galecin-9 dependent manner. Transgenic overexpression of Tim-3 or galectin-9 promotes expansion of MDSCs leading to decreased adaptive immunity as exemplified in accelerated tumor growth and decreased autoimmunity (Dardalhon et al. 2010). Loss of Tim-3 results in reversing the MDSC phenotype further supporting that the interaction between Tim-3/galectin-9 is promoting generation of MDSC (Dardalhon et al. 2010).

Reverse signaling on macrophages through the Tim-3/galectin-9 interaction also seems to have a protective effect in infections with intracellular pathogens. In a mouse tuberculosis model, treatment with Tim-3-Fc fusion protein reduced the bacterial burden in macrophages. This treatment was effective in both WT and Tim-$3^{-/-}$, but not galectin-$9^{-/-}$ macrophages. Similarly, Tim-3 transgenic but not Tim$3^{-/-}$ $CD4^{+}$ T cells controlled mycobacterial replication in galectin-9 expressing macrophages (Jayaraman et al. 2010; Sada-Ovalle et al. 2012). The interaction of Tim-3 expressing T cells with galectin-9 expressing macrophages thus controls microbial replication through reverse signaling. This suggests that Tim-3 expressed on effector T cells might directly interact with galectin-9 on macrophages/DCs to control intracellular pathogen, but in return the galectin-9/Tim-3 interaction might inhibit or delete Tim-3 bearing T cells, providing an effective mechanism by which to control effector T cells.

2.3 Role of Tim-3 in Effector Cells

Tim-3 was originally identified as a receptor expressed on Th1 and Tc1 cells, where it acts as a negative regulator of type 1 immunity (Monney et al. 2002). Tim-3 blocking antibodies were shown to exacerbate experimental autoimmune encephalomyelitis (EAE), a mouse model for human multiple sclerosis (MS) (Monney et al. 2002). Furthermore, blockade of the Tim-3 pathway accelerated Th1-driven diabetes in non-obese diabetic mice (Sanchez-Fueyo et al. 2003). In contrast, activation of Tim-3 by administration of its ligand galectin-9 dampened Th1 responses through induction of cell death in Tim-3^{+} Th1 cells and ameliorated EAE (Zhu et al. 2005).

Tim-3 also plays an important role in the induction of T cell tolerance. Loss of Tim-3 abrogates the induction of antigen-specific tolerance (Sabatos et al. 2003). Furthermore, anti-Tim-3 treatment prevents tolerance induction as its administration accelerated disease in a model of acute GVHD and negatively affected maternal–fetal tolerance resulting in increased risk of miscarriage (Oikawa et al. 2006; Wang et al. 2015).

In addition to expression on IFN-γ expressing Th1 and $CD8^+$ T cells, Tim-3 is also transiently expressed on T cells upon activation, but stable expression is only observed upon sustained stimulation (Zhu et al. 2005; Sanchez-Fueyo et al. 2003). Interestingly, recent studies have investigated the role of Tim-3 in settings of chronic antigenic stimulation such as chronic infections or cancer. Tim-3 is indeed highly expressed on exhausted T cells in HIV-infected patients, where increased frequencies of Tim-3 expressing $CD4^+$ T cells correlated with disease progression (Jones et al. 2008). Similarly, in mice chronic infection with lymphocytic choriomeningitis virus (LCMV) clone 13 induced exhausted $CD8^+$ T cells that co-express Tim-3 and PD-1. These $PD\text{-}1^+Tim\text{-}3^+$ double positive cells are functionally more deeply impaired than those expressing PD-1 alone (Jin et al. 2010). Tim-3 is also highly expressed on tumor-infiltrating lymphocytes and parallel to its expression pattern during chronic viral infection, Tim-3 is usually co-expressed with PD-1 and marks the most highly dysfunctional T cell subset (Sakuishi et al. 2010). Importantly, in both settings blockade of Tim-3 alone or in combination with PD-1 restores functionality in these cells, resulting in viral control or tumor regression suggesting that Tim-3 is involved in actively enforcing T cell exhaustion (Jin et al. 2010; Sakuishi et al. 2010; Baitsch et al. 2011; Ngiow et al. 2011).

2.4 *Role of Tim-3 in Regulatory Cells*

Under steady-state conditions, Tim-3 is barely expressed on $Foxp3^+$ Tregs. In contrast, Tim-3 is unregulated on Tregs during an active immune response and is highly expressed on allograft and tissue infiltrating Tregs including tumor-infiltrating Tregs (Gupta et al. 2012; Sakuishi et al. 2013; Yan et al. 2013; Gao et al. 2012). In in vitro suppression assays, $Tim\text{-}3^+$ Tregs display enhanced suppressive capacity compared to their $Tim\text{-}3^-$ counterparts (Gupta et al. 2012; Sakuishi et al. 2013; Gautron et al. 2014). Furthermore, Tim-3 expressing Tregs show increased expression of molecules associated with the suppressive function of Tregs including co-inhibitory receptors such as CTLA-4, Lag-3, and PD-1 as well as higher secretion of suppressive cytokines such as IL-10 and TGF-β (Gupta et al. 2012; Gautron et al. 2014). In tumor settings, the majority of tumor-infiltrating Tregs expresses Tim-3 and seems to represent a specialized tissue resident Treg subset with enhanced suppressive activity. Importantly, tumor-resident $Tim\text{-}3^+$ Tregs may play a role in promoting effector T cell dysfunction observed in tumor-infiltrating lymphocytes, as their depletion restores functionality to effector T cells (Sakuishi et al. 2013). In an allograft rejection model, graft infiltrating $Tim\text{-}3^+$ Tregs arise from activated, proliferating $Tim\text{-}3^-$ Tregs. Nevertheless, despite their superior suppressive function, $Tim\text{-}3^+$ Tregs were inferior compared to their $Tim\text{-}3^-$ counterparts in prolonging graft survival in an adoptive transfer model as $Tim\text{-}3^+$ Tregs were more short-lived and prone to undergo apoptosis (Gupta et al. 2012). These data suggest that parallel to its function in Th1 cells, Tim-3 might promote cell death in Tim-3 expressing Tregs.

3 Lag-3

Lag-3 (CD223) is expressed on activated $CD4^+$ and $CD8^+$ effector T cells, $CD4^+Foxp3^+$ Treg, Tr1 cells, B cells, plasmacytoid DCs, and a subset of NK cells (Triebel et al. 1990; Huang et al. 2004; Kisielow et al. 2005; Workman et al. 2009; Gagliani et al. 2013). Lag-3 is an immunoglobulin superfamily member composed of four extracellular Ig-like domains and a type I transmembrane domain and hence structurally resembles the CD4 co-receptor (Huard et al. 1995). Like CD4, Lag-3 binds MHC class II, but with higher affinity (Huard et al. 1995). Recently, two additional binding partners for Lag-3 have been described, LSECtin and galectin-3 (Kouo et al. 2015; Xu et al. 2014). LSECtin, a member of the DC-sign family, is expressed in the liver and on tumor cells, while galectin-3 is a soluble lectin expressed in a wide variety of cell types including tumor cells (Fig. 1b).

3.1 *Signaling*

Following TCR engagement, Lag-3 associates with CD3 in the TCR complex and crosslinking of Lag-3 together with CD3 negatively regulates signal transduction leading to reduced T cell proliferation and cytokine production (Hannier et al. 1998). However, the molecular aspects of the inhibitory effects of Lag-3 are still largely unknown, because of lack of a definable motif in the cytoplasmic tail. The cytoplasmic tail of Lag-3 does not contain any inhibitory motifs that are shared with other inhibitory receptors. As a consequence, the exact signaling pathway utilized by Lag-3 is still unclear. Nevertheless, the Lag-3 cytoplasmic tail contains three regions that are conserved between human and mouse and are thus likely involved in signal transduction. The first region contains a serine-phosphorylation site, the second region a single lysine residue within a unique "KIEELE" motif, and the third region contains glutamic acid-proline (EP) repeats (Workman et al. 2002). Among these three regions, the KIEELE motif was shown to be essential for signal transduction and the inhibitory function of Lag-3 (Workman et al. 2002). However, which binding partners interact with Lag-3 and mediate signal transduction is still unknown.

3.2 *Role of Lag-3 in Effector Cells*

Lag-3 acts as a negative regulator of T cell activation as blockade of Lag-3 or Lag-3 deficiency induces enhanced T cell proliferation and cytokine production (Workman et al. 2004; Workman and Vignali 2003). Lag-3 deficient OVA-specific $CD4^+$ T cells show uncontrolled expansion upon immunization with their cognate antigen (Workman and Vignali 2003). Similarly, increased proliferation of Lag-3 deficient donor T cells causes more severe acute GVHD (Sega et al. 2014).

On $CD8^+$ T cells, Lag-3 expression is induced by T cell activation and, like in $CD4^+$ T cells, blockade of Lag-3 improves CTL proliferation and effector function (Grosso et al. 2007). Importantly, Lag-3 is also highly expressed on exhausted $CD8^+$ T cells in both chronic viral infections and cancer (Blackburn et al. 2009; Richter et al. 2010). As Lag-3 blockade during chronic LCMV infection synergizes with PD-1 blockade to reverse exhaustion and improve viral control, Lag-3 seems to functionally contribute to $CD8^+$ T cell exhaustion (Blackburn et al. 2009). Lag-3 is also co-expressed with PD-1 in tumor-infiltrating lymphocytes in various human tumors and mouse tumor models. While Lag-3 blockade alone might not necessarily be able to reverse the exhausted phenotype in $CD8^+$ T cells, it synergizes with PD-1 blockade to improve tumor control or regression (Woo et al. 2012). Lag-3 hence functionally contributes to T cell exhaustion in both chronic viral infections and cancer.

3.3 Role of Lag-3 in Regulatory Cells

In addition to its expression in effector cells, Lag-3 has been reported to be highly expressed in regulatory IL-10 producing Tr1 cells and $Foxp3^+$ Tregs. In fact, together with CD49b, Lag-3 was shown to identify IL-10 producing Tr1 cells in both mice and humans (Gagliani et al. 2013). While Lag-3 expression correlates with IL-10 levels (Burton et al. 2014), it has not been addressed whether Lag-3 directly contributes to the suppressive function of Tr1 cells.

In Tregs, loss of Lag-3 reduced the suppressive function of Tregs, while forced expression of Lag-3 conferred effector T cells with suppressive capacity (Huang et al. 2004). In line with these results, tumor-infiltrating $Lag\text{-}3^+$ Treg display an activated phenotype and produce high amounts of IL-10 and TGF-β1 (Camisaschi et al. 2010). Furthermore, Lag-3 crosslinking of MHC II on DCs was shown to tolerize DCs and thus suppress the priming of effector T cell responses (Liang et al. 2008). Lag-3 thus plays an important role in dampening immune responses by functionally contributing to immune suppression by regulatory T cells.

4 TIGIT

Several groups simultaneously identified TIGIT (T cell immunoglobulin and ITIM domain; also called VSig9, Vstm3, or WUCAM) by bioinformatic analysis as a novel member of the CD28 family (Boles et al. 2009; Levin et al. 2011; Stanietsky et al. 2009; Yu et al. 2009). It acts as a co-inhibitory receptor and is expressed on NK cells and T cells, specifically activated, memory, and follicular T helper cells as well as on a subset of regulatory T cells (Boles et al. 2009; Levin et al. 2011; Stanietsky et al. 2009; Yu et al. 2009; Joller et al. 2011, 2014). TIGIT is composed of one extracellular IgV domain, a type I transmembrane region, and a cytoplasmic

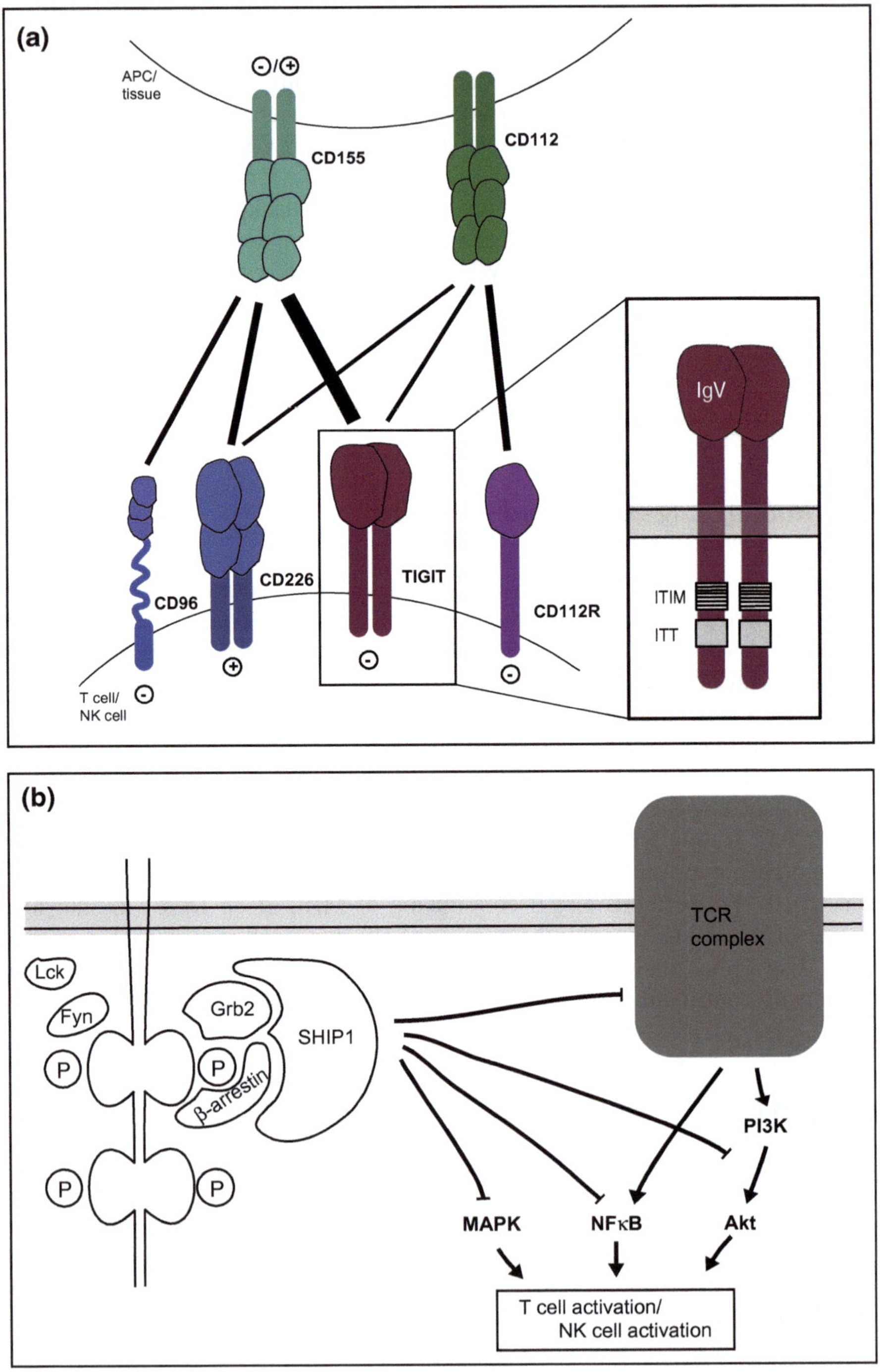
(a)
APC/
tissue
-/+
CD155
CD112
IgV
ITIM
ITT
CD96
CD226
TIGIT
CD112R
T cell/
NK cell
(b)
TCR
complex
Lck
Fyn
Grb2
SHIP1
P
β-arrestin
PI3K
MAPK
NFκB
Akt
T cell activation/
NK cell activation

◀**Fig. 2** TIGIT pathway. **a** TIGIT forms part of a complex network where TIGIT, the co-stimulatory receptor CD226, and the co-inhibitory receptors CD96 and CD112R are expressed on T and NK cells and their ligands CD155 and CD112 are expressed on APCs and in tissue. TIGIT is composed of an extracellular IgV domain and a cytoplasmic tail containing an ITIM and ITT-like motif. **b** Upon ligand binding, the ITIM and ITT-like motifs in the TIGIT tail are phosphorylated and recruit SHIP1 via the adaptor proteins Grb2 or β-arrestin. SHIP1 inhibits signaling through the MAPK, NFκB, and Akt pathways, thus inhibiting activation

tail containing an ITAM and an immunoglobulin tail tyrosine (ITT)-like motif (Boles et al. 2009; Levin et al. 2011; Stanietsky et al. 2009; Yu et al. 2009; Stengel et al. 2012).

TIGIT is part of a complex ligand/receptor network in which it binds with high affinity to CD155 (PVR) and weakly interacts with CD112 (PVRL2, nectin-2) (Fig. 2a) (Levin et al. 2011; Stanietsky et al. 2009; Yu et al. 2009). Both of these ligands are expressed on APCs and a variety of non-hematopoietic cell types including tumor cells and are shared with CD226 (DNAM-1), the co-stimulatory receptor of this network (Bottino et al. 2003; Casado et al. 2009; Mendelsohn et al. 1989). We, in fact, predicted the existence of an inhibitory molecule that parallels CD226, in our in vivo blocking studies (Dardalhon et al. 2005), which was later identified as TIGIT. CD226 binds the two ligands with about 10 times lower affinity than TIGIT, which can inhibit the interaction between CD226 and CD155 in a dose-dependent manner (Levin et al. 2011; Stanietsky et al. 2009; Yu et al. 2009; Lozano et al. 2012). In addition to ligand competition, TIGIT can also directly bind to CD226 in *cis* and disrupt its homodimerization and hence its co-stimulatory function (Johnston et al. 2014). The network is completed by the co-inhibitory receptors CD96 (Tactile), an additional binding partner for CD155, and CD112R, which interacts with CD112 (Fig. 2a) (Chan et al. 2014; Fuchs et al. 2004; Zhu et al. 2016).

4.1 Signaling and Direct Inhibition

Although initial studies suggested that TIGIT only inhibits immune responses in a cell extrinsic manner, subsequent studies clearly demonstrated direct, cell intrinsic inhibitory functions of TIGIT (Levin et al. 2011; Stanietsky et al. 2009; Joller et al. 2011; Lozano et al. 2012). In NK cells, TIGIT directly inhibits NK cytotoxicity, granule polarization, and cytokine secretion (Stanietsky et al. 2009, 2013; Li et al. 2014; Liu et al. 2013). In T cells, TIGIT engagement inhibits their proliferation, cytokine production, and TCR signaling in a cell intrinsic manner (Levin et al. 2011; Joller et al. 2011; Lozano et al. 2012; Inozume et al. 2016).

The cytoplasmic tail of TIGIT contains an ITIM and an ITT-like motif, which are highly conserved between mouse and human and mediate its cell intrinsic inhibitory function (Boles et al. 2009; Levin et al. 2011; Stanietsky et al. 2009; Yu et al. 2009; Stengel et al. 2012). However, there is still some debate as to which of the two motifs is important for signaling and the inhibitory function of TIGIT. In mouse, the two motifs seem redundant as phosphorylation of the tyrosine residue in either the ITIM motif (Y233) or the ITT-like motif (Y227) is sufficient for signal transduction and the inhibitory activity of TIGIT is only abolished when both tyrosine residues are mutated (Stanietsky et al. 2013). Which of the two motifs is important for TIGIT signaling in human cells is still unclear as contradictory reports have been published claiming an essential role for either the ITIM motif (Y231) (Stanietsky et al. 2009) or the ITT-like motif (Y225) (Li et al. 2014; Liu et al. 2013). These differences might stem from the experimental system used as both groups performed their studies in TIGIT overexpressing cell lines. Hence, it will be important to study the relative contribution of ITIM versus ITT-like motifs under physiological conditions in primary human cells.

Engagement of TIGIT in NK cells induces the phosphorylation of the tyrosine residues in its ITIM and ITT-like motifs through the Src-family kinases Fyn and Lck. This allows for binding of the adaptor proteins Grb2 (growth factor receptor-bound protein 2) and β-arrestin 2, which in turn recruit SHIP1 (SH2 domain-containing inositol-5-phosphatase 1) to the cytoplasmic tail of TIGIT (Li et al. 2014; Liu et al. 2013). SHIP1 recruitment prematurely terminates PI3K (phosphoinositide 3-kinase), MAPK (mitogen-activated protein kinase), and NF-κB (nuclear factor-κB) signaling and results in NK cell inhibition marked by reduced cytotoxicity and cytokine secretion (Fig. 2b) (Li et al. 2014; Liu et al. 2013). In T cells, TIGIT signaling has not been studied at the protein level. Nevertheless, TIGIT engagement downregulates transcription of central components of the TCR signaling pathway as well as the TCR complex itself, thereby inhibiting productive T cell activation (Joller et al. 2011). In addition to its inhibitory effects on the TCR signaling pathway, TIGIT engagement promotes T cell survival through the induction of anti-apoptotic molecules (e.g., Bcl-xL) and receptors for pro-survival cytokines such as IL-2, IL-7, and IL-15 (Joller et al. 2011). TIGIT thus not only inhibits T cell activation but also promotes T cell survival and maintenance.

4.1.1 TIGIT in Effector Cells

The direct inhibitory function of TIGIT was first described in NK cells (Stanietsky et al. 2009). Here, TIGIT inhibits NK cytotoxicity and IFN-γ secretion and this TIGIT-mediated inhibition is dominant over co-activation through CD226 (Stanietsky et al. 2009; Stanietsky et al. 2013). In human NK cells, expression of TIGIT correlated with reduced cytokine production, degranulation, and cytotoxicity (Wang et al. 2015). Importantly, TIGIT expression levels on NK cells determine the effectiveness of inhibition of their cytotoxicity (Sarhan et al. 2016). TIGIT-mediated inhibition of NK function thus is an important mechanism for determining the threshold for NK activation. Furthermore, engagement of TIGIT through CD155 plays an important role in dampening NK-mediated immunopathology, as shown in a murine model of acute viral hepatitis (Bi et al. 2014).

In effector T cells, TIGIT can directly inhibit T cell activation and proliferation. Stimulation of T cells in the absence of APCs but in the presence of an agonistic anti-TIGIT antibody inhibits proliferation of mouse as well as human T cells (Levin et al. 2011; Joller et al. 2011). Furthermore, TIGIT stimulation inhibits IFN-γ production in human $CD4^+$ T cells (Lozano et al. 2012). This T cell intrinsic inhibitory role of TIGIT was also confirmed in vivo, as mice with a $CD4^+$ T cell specific loss of TIGIT developed augmented T cell responses marked by increased levels of the pro-inflammatory cytokines IFN-γ and IL-17 upon immunization (Joller et al. 2011). Similarly, stimulation of $CD8^+$ T cell with CD155 expressing melanoma cells inhibits their IFN-γ production in a T cell intrinsic manner, as signaling through CD155 was not required to mediate the effect (Inozume et al. 2016). As seen for $CD4^+$ T cells, in vivo, selective loss of TIGIT on T cells results in increased IFN-γ secretion by $CD8^+$ T cells (Johnston et al. 2014). TIGIT thus acts as a cell intrinsic inhibitor by dampening effector cell activation, by inhibiting their proliferation and thereby limiting the effector cell pool, and finally by reducing effector cell function and cytokine production.

4.1.2 TIGIT in Regulatory Cells

In addition to effector cells, TIGIT is highly expressed in regulatory cells, where it promotes their suppressive function. In Tr1 cells ($CD4^+Foxp3^-IL\text{-}10^+$), induction of the regulatory cytokine IL-10 correlates with TIGIT expression (Burton et al. 2014). TIGIT is also a direct target of Foxp3 and is expressed in a subset of predominantly natural $CD4^+Foxp3^+$ Tregs (Joller et al. 2014; Zhang et al. 2013). $TIGIT^+$ Tregs express higher levels of Treg signature genes, such as Foxp3, CD25, and CTLA-4 and show enhanced demethylation in Treg-specific demethylated regions (TSDR) leading to higher lineage stability (Joller et al. 2014; Fuhrman et al. 2015). Interestingly, while $TIGIT^+$ Tregs are highly suppressive and stable, CD226 expression in Tregs is associated with lineage instability and decreased suppressive capacity (Fuhrman et al. 2015). Engagement of TIGIT on Tregs directly induces expression of the suppressive mediators IL-10 and Fgl2 (Joller et al. 2014). TIGIT-dependent induction of Fgl2 confers superior suppressive capacity to $TIGIT^+$ Tregs and most importantly enables them to selectively suppress pro-inflammatory Th1 and Th17 responses but not Th2 responses (Joller et al. 2014). In regulatory cells, TIGIT thus contributes to lineage stability and enhances their inhibitory function through the direct induction of suppressive mediators.

4.2 Indirect Inhibition

Early studies suggested that TIGIT indirectly suppresses immune responses via interacting with CD155 on DCs (Yu et al. 2009). TIGIT engages CD155 as a homodimer, where each TIGIT molecule binds one CD155 molecule, thus assembling into a heterotetramer with a TIGIT homodimeric core (Stengel et al. 2012). TIGIT binding induces phosphorylation of the ITIM motif in the CD155 cytoplasmic tail resulting in recruitment of SHP-2 and phosphorylation of Erk (Yu et al. 2009; Coyne et al. 2007). Overall, this leads to a decrease in IL-12p40 while increasing IL-10 production in DCs (Yu et al. 2009). TIGIT thus inhibits T cell responses indirectly by inducing tolerogenic DCs through engagement of CD155.

More importantly, TIGIT is also able to indirectly dampen immune responses by enhancing and modulating the suppressive function of Tregs. TIGIT expressing Tregs displays an activated phenotype and increased suppressive capacity compared to their TIGIT^{-} counterparts (Joller et al. 2014). TIGIT engagement in Tregs induces suppressive mediators, which dampen T cell proliferation and mediate pleiotropic immunosuppressive effects (Chan et al. 2003). TIGIT-dependent induction of Fgl2 also confers Tregs with the ability to selectively dampen pro-inflammatory type 1 and type 17, but not type 2 immune responses (Joller et al. 2014). TIGIT^{+} Tregs thus potently suppress activation and proliferation of effector T cells in general, but also shape the effector response by shifting the balance away from pro-inflammatory Th1/Th17 towards more anti-inflammatory, IL-10-dominated responses.

5 Role in Diseases and Therapy

5.1 T Cell Exhaustion

The main function of co-inhibitory receptors is to maintain tolerance under homeostatic conditions and to dampen excessive immune responses in order to prevent immunopathology. In chronic infections, the pathogens and therefore also the antigens persist and constitute a source of persistent antigenic stimulation. This chronic activation is counterbalanced by an upregulation of co-inhibitory receptors, which serve to keep the effector response in check and prevent immunopathology. At the same time, this persistent high expression of co-inhibitory receptors can constrain effector T cells to a degree where they become dysfunctional or exhausted and are no longer able to promote pathogen clearance. Exhausted T cells show impaired effector function (cytokine production, cytotoxicity) and are marked by the sustained expression of multiple inhibitory receptors (reviewed in (Wherry and Kurachi 2015)). PD-1 was the first inhibitory receptor identified to selectively mark exhausted T cells and actively contribute to the dysfunctional state, as its blockade was able to restore function in virus-specific T cells (Barber et al. 2006). It has since become clear that several other co-inhibitory receptors are co-expressed with PD-1 and synergistically act to curb T cell responsiveness and function.

Chronic infection with LCMV serves as the prototypic model for studying T cell exhaustion (Moskophidis et al. 1993). Here, Lag-3, Tim-3, and more recently also TIGIT were shown to be co-expressed with PD-1 on exhausted virus-specific CD8^{+} T cells (Table 1.) (Jin et al. 2010; Blackburn et al. 2009; Richter et al. 2010;

Johnston et al. 2014). The extent of exhaustion seems to correlate with the number of co-inhibitory receptors expressed and while the expression of a single co-inhibitory receptor is not indicative of exhaustion, co-expression of multiple inhibitory receptors is a hallmark of exhausted T cells (Fig. 3.) (Doering et al. 2012). However, it is not clear whether the co-inhibitory molecules are co-expressed on the same dysfunctional T cell or the co-inhibitory molecules are progressively acquired in the population, which leads to the loss of effector functions. The co-expression of inhibitory molecules also bears functional relevance as blockade of a single co-inhibitory receptor does not or only poorly reverse exhaustion, while co-blockade results in synergistic effects. This concept was demonstrated for PD-1 and Lag-3 (Blackburn et al. 2009), PD-1 and Tim-3 (Jin et al. 2010), and PD-1 and TIGIT (Johnston et al. 2014), where simultaneous targeting of the two pathways was able to synergistically reverse exhaustion and improve T cell function (Table 1.). Similar to their cooperative action in enforcing T cell exhaustion during chronic LCMV infection in mice, inhibitory receptors are also found to be co-expressed in chronic viral infections in humans such as HIV, HBV, or HCV infection (Fromentin et al. 2016; Jones et al. 2008; Golden-Mason et al. 2009; McMahan et al. 2010; Wu et al. 2012; Nebbia et al. 2012). In these chronic viral infections, Tim-3 marks virus-specific dysfunctional $CD8^+$ T cells and Tim-3 blockade improves their function (Jones et al. 2008; Golden-Mason et al. 2009; McMahan et al. 2010; Wu et al. 2012; Nebbia et al. 2012). Similarly, TIGIT expression correlates with parameters of HIV disease progression and combinational blockade of TIGIT and PD-L1 restore $CD8^+$ T cell function (Chew et al. 2016). Collectively, these data suggest that Tim-3, Lag-3, and TIGIT act in a cooperative manner with PD-1 and have nonredundant functions. Their divers expression patterns, binding partners, and signaling motifs all contribute to their synergistic effects. However, whether co-blockade of Tim-3, Lag-3, or TIGIT will have synergistic effect with each other independent of PD-1 is currently being

Table 1 Expression pattern and therapeutic effect on Tim-3, Lag-3, and TIGIT in chronic infections and cancer

		Expression			Therapeutic targeting reverses exhaustion		
		Tim-3	Lag-3	TIGIT	Tim-3	Lag-3	TIGIT
Chronic infection	LCMV	x	x	x	With α-PD-1	With α-PD-1	With α-PD-1
	HIV	x	x	x	x		With α-PD-L1
Cancer	$CD8^+$ TILs	x	x	x	With α-PD-1	with α-PD-1	With α-PD-1
	$CD4^+$ TILs	Not tested	x	Not tested	With α-TIGIT		With α-Tim-3
	Treg TILs	x	Not tested	x			

tested. As such their therapeutic targeting alone or in combination bears the potential of gradual or site specific restoration of function in exhausted T cells.

5.2 Cancer and Checkpoint Inhibitors

Chronic antigen exposure is a key feature shared between persistent infections and cancer. Indeed, tumor-specific T cells resemble the dysfunctional effector T cells present in settings of chronic infection in that they acquire expression of multiple co-inhibitory receptors during tumor progression, leading to their functional exhaustion (Fig. 3). Exhausted tumor-specific T cells express high levels of CTLA-4 and PD-1 and recent immunotherapeutic advances have aimed at targeting

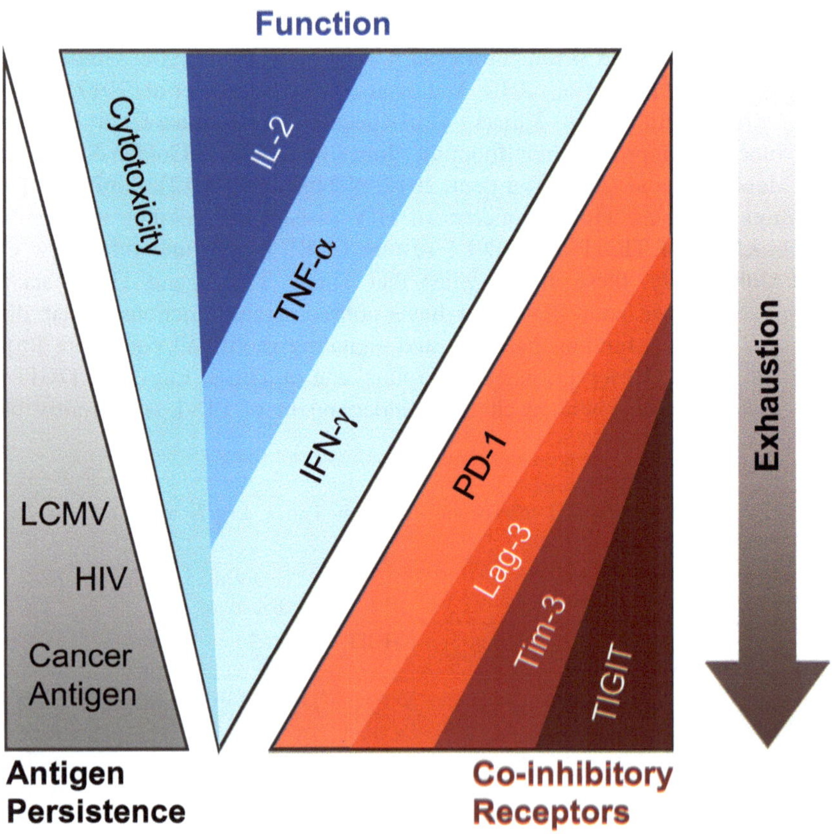

Fig. 3 Co-inhibitory receptors in chronic infections and cancer. Antigen persistence drives T cells into a state of exhaustion/dysfunction characterized by hierarchical loss of cytokine production as well as impairment of cytotoxicity. As T cells enter the state of T cell exhaustion they progressively express PD-1 and upregulate Lag-3, Tim-3, and TIGIT

these co-inhibitory receptors to reverse their dysfunctional phenotype, reinvigorate tumor-specific T cell responses, and promote tumor elimination. The success of therapies targeting CTLA-4 (Ipilimumab) and PD-1 (Prembrolizumab and Nivolumab) has marked a major breakthrough in cancer therapy (Couzin-Frankel 2013; Gravitz 2013). Despite their successes, response rates for these therapies only range around 20–30% (Hodi et al. 2010; Topalian et al. 2012) and at present, combinatorial approaches are being explored to improve their efficacy. More recently, the list of co-inhibitory receptors expressed on TILs has been extended to include Tim-3, Lag-3, and TIGIT, which might represent additional therapeutic targets for cancer immunotherapy.

Expression of Tim-3 was found on functionally exhausted cells in a broad spectrum of both murine tumor models and cancer patients (melanoma, non-small cell lung cancer, follicular B cell non-Hodgkin lymphoma) (Sakuishi et al. 2010; Gao et al. 2012; Fourcade et al. 2010; Yang et al. 2012). Tim-3 positively correlates with cancer severity and poor prognosis in different cancer settings (Gao et al. 2012; Yang et al. 2012) and identifies a highly exhausted population of $CD8^+$ TILs, which fail to produce IL-2, TNF-α, or IFN-γ (Sakuishi et al. 2010). Similar to observations in chronic infections, Tim-3 is co-expressed with other co-inhibitory receptors, most notably PD-1. While blockade of Tim-3 alone only shows minor effects, co-blockade of Tim-3 with PD-1 is superior at improving antitumor effector function and suppressing tumor growth than blockade of either pathway alone (Sakuishi et al. 2010; Ngiow et al. 2011; Zhou et al. 2011). Similarly, Lag-3 is co-expressed with PD-1 on dysfunctional $CD4^+$ as well as $CD8^+$ TILs and co-blockade of both pathways shows synergistic effects in improving antitumor immunity (Woo et al. 2012; Matsuzaki et al. 2010). TIGIT was shown to negatively regulate antitumor responses as TIGIT deficiency results in significantly delayed tumor growth (Kurtulus et al. 2015). Like Tim-3 and Lag-3, TIGIT is highly expressed on human and murine TILs and shows synergistic effects with PD-1 (Johnston et al. 2014; Kurtulus et al. 2015; Chauvin et al. 2015). In murine tumors, $CD8^+$ $TIGIT^+$ TILs co-express PD-1, Tim-3, and Lag3, where TIGIT marks the most dysfunctional T cell population among the $CD8^+$ TILs (Kurtulus et al. 2015). Importantly, TIGIT was shown to not only synergize with PD-1 but also with Tim-3 as their co-blockade synergistically improved antitumor immunity (Kurtulus et al. 2015). Thus, in both chronic infections and in cancer PD-1, Tim-3, Lag-3, and TIGIT are co-expressed on highly dysfunctional effector T cells and their cooperative action seems to functionally contribute to T cell exhaustion (Table 1).

In addition to their inhibitory function on tumor-infiltrating effector cells, co-inhibitory receptors also play a role in dampening antitumor responses through their action on Tregs. Indeed, tumor-infiltrating Tregs have been shown to express high levels of co-inhibitory receptors. While Tregs in peripheral lymphoid tissues express only moderate levels of TIGIT and are mostly negative for Tim-3, the majority of tumor-infiltrating Tregs express TIGIT and Tim-3 (Yan et al. 2013; Gao et al. 2012; Kurtulus et al. 2015; Sakuishi et al. 2013). Both Tim-3^+ and $TIGIT^+$ Tregs have been shown to possess superior suppressive capacity in vitro and express high levels Treg signature genes including *Foxp3* (Sakuishi et al. 2013;

Gautron et al. 2014; Joller et al. 2014; Sakuishi et al. 2013; Gupta et al. 2012). Furthermore, $TIGIT^+$ and $Tim\text{-}3^+$ Tregs both display increased production of suppressive mediators such as IL-10, perforin, or TGF-β, which contribute to their superior suppression (Sakuishi et al. 2013; Joller et al. 2014; Kurtulus et al. 2015). Highly suppressive Tregs are indeed a main driver in actively suppressing antitumor responses (Nishikawa and Sakaguchi 2010) and we found that loss of TIGIT on Tregs but not on $CD8^+$ TILs was able to delay tumor growth and restore $CD8^+$ T cell function (Kurtulus et al. 2015). Hence, expression of TIGIT, and possibly also Tim-3, on Tregs seems to play a dominant role in restraining antitumor responses and may actively promote the dysfunctional phenotype observed in $CD8^+$ TILs.

Taken together, enhanced expression of co-inhibitory receptors directly impairs effector T cell responses and their concerted action synergistically contributes to the dysfunctional T cell phenotype observed within the tumor microenvironment. In addition, these co-inhibitory receptors also contribute to enhanced suppression through Tregs present in the tumor tissue, further dampening the antitumor response. The success of cancer immunotherapy targeting checkpoint inhibitors has demonstrated that the enhanced expression of co-inhibitory receptors on both effector and regulatory T cells represents a central obstacle for tumor elimination. The synergistic effects of co-blockade of several co-inhibitory receptors seen in preclinical cancer models and on patient-derived samples suggest that combination therapy might greatly improve the low response rates observed in current monotherapies and Tim-3, Lag-3, and TIGIT represent the most promising new targets. As these novel checkpoint inhibitors and their ligands show distinct expression patterns, personalized combination therapy bears the potential of yielding optimal results depending on the type of cancer and the tissue affected.

5.3 *Autoimmunity*

Despite the striking success of therapies targeting the co-inhibitory receptors CTLA-4 and PD-1 in several cancer indications, a sizable portion of patients (10 −20%) shows significant side effects, in particular autoimmune syndromes, including colitis, pneumonitis, skin disorders, and hepatitis (Callahan et al. 2016; Robert et al. 2015). Similarly, in certain settings of chronic infections, interference with co-inhibitory pathways results in severe immune-mediated tissue damage that can have detrimental consequences (Frebel et al. 2012; Hafalla et al. 2012; Lazar-Molnar et al. 2010; Vaccari et al. 2012). Indeed, co-inhibitory receptors play a central role in maintaining immune homeostasis and their loss, most notably of CTLA-4 or PD-1, results in spontaneous, severe autoimmunity with loss of CTLA-4 and a milder form of tissue inflammation with loss of PD-1 (Tivol et al. 1995; Waterhouse et al. 1995; Nishimura et al. 1999; Nishimura et al. 2001). Many co-inhibitory pathways, including the Tim-3 and CD226/TIGIT pathways, have been genetically linked to susceptibility to autoimmune diseases and their function in regulating immune responses has been intensively studied in the context of autoimmune diseases (Kasagi et al. 2011; Qu et al. 2009; Wang et al. 2014; Song et al. 2011; Hafler et al. 2009). However, deficiency in Tim-3, Lag-3, or TIGIT alone does not predispose to autoimmunity unless the mice are on a permissive background (Joller et al. 2011; Bettini et al. 2011; Okazaki et al. 2011; Lee and Goverman 2013). While loss of Lag-3 was shown to accelerate type 1 diabetes in NOD mice (Bettini et al. 2011; Okazaki et al. 2011), the function of Tim-3 and TIGIT has most intensively been studied in the context of CNS autoimmunity. Loss or blockade of either Tim-3 or TIGIT resulted in enhanced pro-inflammatory T cell responses leading to exacerbated EAE (Monney et al. 2002; Sanchez-Fueyo et al. 2003; Levin et al. 2011; Joller et al. 2011). The inhibitory function of the receptors is achieved through their direct inhibitory action on effector T cells as well as their ability to indirectly dampen immune responses by promoting Treg-mediated suppression and production of regulatory cytokines. An important common feature among these co-inhibitory receptors is their association with the immunoregulatory cytokine IL-10. In Tregs, IL-10 is almost exclusively found within the Tim-3^+, Lag-3^+, and $TIGIT^+$ Treg subsets and TIGIT ligation is able to directly induce IL-10 production in Tregs (Sakuishi et al. 2013; Camisaschi et al. 2010; Joller et al. 2014). Furthermore, in Tr1 cells induced to mediate antigen-specific tolerance, Tim-3, Lag-3, and TIGIT correlate with the expression of IL-10 (Burton et al. 2014). In DCs, ligation of CD155 by TIGIT induces IL-10 production and promotes a tolerogenic phenotype (Yu et al. 2009). Finally, dysfunctional $CD8^+$ TILs found in melanoma co-express Tim-3, Lag-3, and TIGIT and show enhanced IL-10 production (Singer et al. 2016; Tirosh et al. 2016). These observations suggest that expression of Tim-3, Lag-3, and TIGIT might be co-regulated to ensure optimal T cell regulation through their cooperative function.

In contrast to CTLA-4 or PD-1, Tim-3 and TIGIT do not act as global inhibitors of immune responses but specifically target certain aspects of the immune response (Fig. 4.). Tim-3 is predominantly expressed on Th1 but not Th2 cells and interaction with its ligands galectin-9 or Ceacam-1 triggers cell death in Th1 and Tc1 cells, thereby specifically dampening Th1 responses (Huang et al. 2015; Zhu et al. 2005; Kang et al. 2015). The specificity of Tim-3-mediated inhibition is also demonstrated by the fact that Tim-3-deficiency regulates Th1- but not Th17-driven EAE (Lee and Goverman 2013). Similarly, TIGIT and its co-stimulatory counterpart CD226 have differential effects on different types of immune responses. While CD226 promotes Th1 and Th17 responses, TIGIT selectively inhibits production of IFN-γ and IL-17 but enhances Th2 cytokines (Burton et al. 2014; Yu et al. 2009; Joller et al. 2011, 2014; Dardalhon et al. 2005; Lozano et al. 2012, 2013). TIGIT selectively inhibits pro-inflammatory Th1 and Th17 responses through its action on multiple cell types. In DCs, TIGIT ligation of CD155 inhibits IL-12 production and thus interferes with polarization of naïve $CD4^+$ T helper cells into Th1 cells during priming (Yu et al. 2009). In effector T cells, loss of TIGIT results in upregulation of the Th1 master transcription factor T-bet and enhanced production of IFN-γ and IL-17 (Joller et al. 2011; Lozano et al. 2012). Finally in Tregs, TIGIT directly induces production of the suppressive mediator Fgl2, which enables $TIGIT^+$ Tregs to selectively suppress Th1 and Th17 responses (Joller et al. 2014). TIGIT and

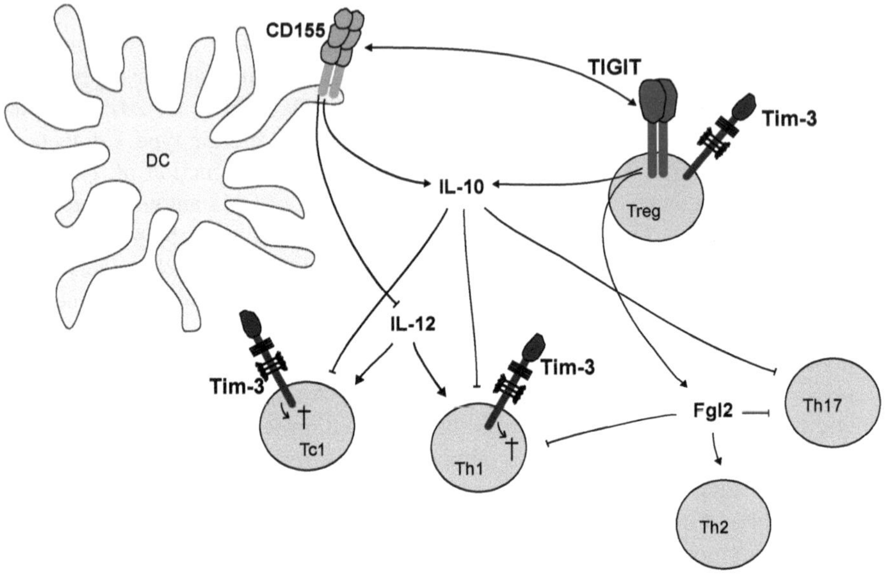

Fig. 4 The Tim-3 and TIGIT pathways specifically inhibit pro-inflammatory responses in autoimmunity. Tim-3 is selectively expressed on Th1 and Tc1 cells, which drive tissue inflammation and autoimmunity. Tim-3 regulates their response by inducing apoptotic cell death or dysfunction by binding to its ligands. TIGIT expressed on Tregs induces IL-10 as well as Fgl2, which selectively inhibit pathogenic Th1 and Th17 responses. TIGIT expressing effector and regulatory T cells engage CD155 on APC thereby dampening IL-12 and enhancing IL-10 secretion and thus inhibiting inflammatory responses

Tim-3 therefore specifically inhibit pro-inflammatory immune responses that drive organ-specific autoimmunity. These co-inhibitory receptors hence seem to play a particularly important role in maintaining peripheral T cell tolerance and preventing autoimmunity.

6 Conclusion

Co-inhibitory receptors play a pivotal role in maintaining immune homeostasis and preventing autoimmunity while at the same time permitting effective immune responses to control cancer and eradicate pathogens. The family of co-inhibitory receptors has grown from the potent, global inhibitor of immune responses, CTLA-4, to include co-inhibitory receptors (Tim-3, Lag-3, and TIGIT) that show more specialized functions in regulating T cell responses. These co-inhibitory receptors and their ligands are highly expressed at tissue sites, where they regulate ongoing effector T cell responses and maintain tissue tolerance. As such, Tim-3, Lag-3, and TIGIT are highly expressed in T cells that are stimulated by persistent antigen including pro-inflammatory T cells in autoimmune diseases, tumor-infiltrating lymphocytes in cancer, and exhausted virus-specific T cells in chronic infections. In fact, these T cells co-express multiple co-inhibitory receptors that functionally synergize to dampen effector T cell responses to prevent immunopathology. At the same time, the synergistic effects observed, when several co-inhibitory receptors are targeted together, show that the inhibitory functions of Tim-3, Lag-3, and TIGIT are not identical but that they have nuanced functions even when expressed together on the population of "exhausted" or "dysfunctional" T cells. Further knowledge on how expression and specialized function of these receptors synergize and regulate effector functions of T cells will open up new areas for therapeutic targeting of these co-inhibitory receptors in chronic human diseases.

Acknowledgements This work was supported by the Swiss National Science Foundation (PP00P3_150663/1 to N.J.), the European Research Council (677200 to N.J.), and the National Institutes of Health (P01 AI073748, P01 NS076410, P01 AI039671, and R01 NS045937 to V.K.K.).

References

Anderson AC, Lord GM, Dardalhon V, Lee DH, Sabatos-Peyton CA, Glimcher LH, Kuchroo VK (2010) T-bet, a Th1 transcription factor regulates the expression of Tim-3. Eur J Immunol 40 (3):859–866. doi:10.1002/eji.200939842

Baitsch L, Baumgaertner P, Devevre E, Raghav SK, Legat A, Barba L, Wieckowski S, Bouzourene H, Deplancke B, Romero P, Rufer N, Speiser DE (2011) Exhaustion of tumor-specific CD8(+) T cells in metastases from melanoma patients. J Clin Invest 121 (6):2350–2360. doi:10.1172/JCI46102

Barber DL, Wherry EJ, Masopust D, Zhu B, Allison JP, Sharpe AH, Freeman GJ, Ahmed R (2006) Restoring function in exhausted CD8 T cells during chronic viral infection. Nature 439 (7077):682–687. doi:10.1038/nature04444

Bettini M, Szymczak-Workman AL, Forbes K, Castellaw AH, Selby M, Pan X, Drake CG, Korman AJ, Vignali DA (2011) Cutting edge: accelerated autoimmune diabetes in the absence of LAG-3. J Immunol 187(7):3493–3498. doi:10.4049/jimmunol.1100714

Bi J, Zhang Q, Liang D, Xiong L, Wei H, Sun R, Tian Z (2014) T-cell Ig and ITIM domain regulates natural killer cell activation in murine acute viral hepatitis. Hepatology 59(5):1715–1725. doi:10.1002/hep.26968

Blackburn SD, Shin H, Haining WN, Zou T, Workman CJ, Polley A, Betts MR, Freeman GJ, Vignali DA, Wherry EJ (2009) Coregulation of CD8 + T cell exhaustion by multiple inhibitory receptors during chronic viral infection. Nat Immunol 10(1):29–37. doi:10.1038/ni.1679

Boles KS, Vermi W, Facchetti F, Fuchs A, Wilson TJ, Diacovo TG, Cella M, Colonna M (2009) A novel molecular interaction for the adhesion of follicular CD4 T cells to follicular DC. Eur J Immunol 39(3):695–703. doi:10.1002/eji.200839116

Bottino C, Castriconi R, Pende D, Rivera P, Nanni M, Carnemolla B, Cantoni C, Grassi J, Marcenaro S, Reymond N, Vitale M, Moretta L, Lopez M, Moretta A (2003) Identification of PVR (CD155) and Nectin-2 (CD112) as cell surface ligands for the human DNAM-1 (CD226) activating molecule. J Exp Med 198(4):557–567. doi:10.1084/jem.20030788

Burton BR, Britton GJ, Fang H, Verhagen J, Smithers B, Sabatos-Peyton CA, Carney LJ, Gough J, Strobel S, Wraith DC (2014) Sequential transcriptional changes dictate safe and effective antigen-specific immunotherapy. Nat Commun 5:4741. doi:10.1038/ncomms5741

Callahan MK, Postow MA, Wolchok JD (2016) Targeting T cell co-receptors for cancer therapy. Immunity 44(5):1069–1078. doi:10.1016/j.immuni.2016.04.023

Camisaschi C, Casati C, Rini F, Perego M, De Filippo A, Triebel F, Parmiani G, Belli F, Rivoltini L, Castelli C (2010) LAG-3 expression defines a subset of CD4(+)CD25(high)Foxp3 (+) regulatory T cells that are expanded at tumor sites. Journal of immunology 184(11):6545–6551. doi:10.4049/jimmunol.0903879

Cao E, Zang X, Ramagopal UA, Mukhopadhaya A, Federov A, Federov E, Zencheck WD, Lary JW, Cole JL, Deng H, Xiao H, DiLorenzo TP, Allison JP, Nathenson SG, Almo SC (2007) T cell immunoglobulin mucin-3 crystal structure reveals a novel ligand binding surface. Immunity 26:311–321

Casado JG, Pawelec G, Morgado S, Sanchez-Correa B, Delgado E, Gayoso I, Duran E, Solana R, Tarazona R (2009) Expression of adhesion molecules and ligands for activating and costimulatory receptors involved in cell-mediated cytotoxicity in a large panel of human melanoma cell lines. Cancer Immunol Immunother 58(9):1517–1526. doi:10.1007/s00262-009-0682-y

Chan CW, Kay LS, Khadaroo RG, Chan MW, Lakatoo S, Young KJ, Zhang L, Gorczynski RM, Cattral M, Rotstein O, Levy GA (2003) Soluble fibrinogen-like protein 2/fibroleukin exhibits immunosuppressive properties: suppressing T cell proliferation and inhibiting maturation of bone marrow-derived dendritic cells. J Immunol 170(8):4036–4044

Chan CJ, Martinet L, Gilfillan S, Souza-Fonseca-Guimaraes F, Chow MT, Town L, Ritchie DS, Colonna M, Andrews DM, Smyth MJ (2014) The receptors CD96 and CD226 oppose each other in the regulation of natural killer cell functions. Nat Immunol 15(5):431–438. doi:10.1038/ni.2850

Chauvin JM, Pagliano O, Fourcade J, Sun Z, Wang H, Sander C, Kirkwood JM, Chen TH, Maurer M, Korman AJ, Zarour HM (2015) TIGIT and PD-1 impair tumor antigen-specific CD8(+) T cells in melanoma patients. J Clin Invest 125(5):2046–2058. doi:10.1172/JCI80445

Chen DS, Mellman I (2013) Oncology meets immunology: the cancer-immunity cycle. Immunity 39(1):1–10. doi:10.1016/j.immuni.2013.07.012

Chew GM, Fujita T, Webb GM, Burwitz BJ, Wu HL, Reed JS, Hammond KB, Clayton KL, Ishii N, Abdel-Mohsen M, Liegler T, Mitchell BI, Hecht FM, Ostrowski M, Shikuma CM, Hansen SG, Maurer M, Korman AJ, Deeks SG, Sacha JB, Ndhlovu LC (2016) TIGIT marks exhausted T cells, correlates with disease progression, and serves as a target for immune restoration in HIV and SIV infection. PLoS Pathog 12(1):e1005349. doi:10.1371/journal.ppat.1005349

Chiba S, Baghdadi M, Akiba H, Yoshiyama H, Kinoshita I, Dosaka-Akita H, Fujioka Y, Ohba Y, Gorman JV, Colgan JD, Hirashima M, Uede T, Takaoka A, Yagita H, Jinushi M (2012) Tumor-infiltrating DCs suppress nucleic acid-mediated innate immune responses through interactions between the receptor TIM-3 and the alarmin HMGB1. Nat Immunol 13(9):832–842. doi:10.1038/ni.2376

Couzin-Frankel J (2013) Breakthrough of the year 2013. Cancer immunotherapy. Science 342 (6165):1432–1433. doi:10.1126/science.342.6165.1432

Coyne CB, Kim KS, Bergelson JM (2007) Poliovirus entry into human brain microvascular cells requires receptor-induced activation of SHP-2. EMBO J 26(17):4016–4028. doi:10.1038/sj.emboj.7601831

da Silva IP, Gallois A, Jimenez-Baranda S, Khan S, Anderson AC, Kuchroo VK, Osman I, Bhardwaj N (2014) Reversal of NK-cell exhaustion in advanced melanoma by Tim-3 blockade. Cancer Immunol Res 2(5):410–422. doi:10.1158/2326-6066.CIR-13-0171

Dardalhon V, Schubart AS, Reddy J, Meyers JH, Monney L, Sabatos CA, Ahuja R, Nguyen K, Freeman GJ, Greenfield EA, Sobel RA, Kuchroo VK (2005) CD226 is specifically expressed on the surface of Th1 cells and regulates their expansion and effector functions. J Immunol 175 (3):1558–1565 175/3/1558 [pii]

Dardalhon V, Anderson AC, Karman J, Apetoh L, Chandwaskar R, Lee DH, Cornejo M, Nishi N, Yamauchi A, Quintana FJ, Sobel RA, Hirashima M, Kuchroo VK (2010) Tim-3/galectin-9 pathway: regulation of Th1 immunity through promotion of CD11b + Ly-6G + myeloid cells. J Immunol 185(3):1383–1392. doi:10.4049/jimmunol.0903275

Dejean AS, Beisner DR, Ch'en IL, Kerdiles YM, Babour A, Arden KC, Castrillon DH, DePinho RA, Hedrick SM (2009) Transcription factor Foxo3 controls the magnitude of T cell immune responses by modulating the function of dendritic cells. Nat Immunol 10(5):504–513. doi:10.1038/ni.1729

DeKruyff RH, Bu X, Ballesteros A, Santiago C, Chim YL, Lee HH, Karisola P, Pichavant M, Kaplan GG, Umetsu DT, Freeman GJ, Casasnovas JM (2010) T cell/transmembrane, Ig, and mucin-3 allelic variants differentially recognize phosphatidylserine and mediate phagocytosis of apoptotic cells. J Immunol 184(4):1918–1930. doi:10.4049/jimmunol.0903059

Doering TA, Crawford A, Angelosanto JM, Paley MA, Ziegler CG, Wherry EJ (2012) Network analysis reveals centrally connected genes and pathways involved in CD8 + T cell exhaustion versus memory. Immunity 37(6):1130–1144. doi:10.1016/j.immuni.2012.08.021

Esensten JH, Helou YA, Chopra G, Weiss A, Bluestone JA (2016) CD28 costimulation: from mechanism to therapy. Immunity 44(5):973–988. doi:10.1016/j.immuni.2016.04.020

Fallarino F, Grohmann U, Hwang KW, Orabona C, Vacca C, Bianchi R, Belladonna ML, Fioretti MC, Alegre ML, Puccetti P (2003) Modulation of tryptophan catabolism by regulatory T cells. Nat Immunol 4(12):1206–1212 10.1038/ni1003ni1003[pii]

Fourcade J, Sun Z, Benallaoua M, Guillaume P, Luescher IF, Sander C, Kirkwood JM, Kuchroo V, Zarour HM (2010) Upregulation of Tim-3 and PD-1 expression is associated with tumor antigen-specific CD8 + T cell dysfunction in melanoma patients. J Exp Med 207 (10):2175–2186. doi:10.1084/jem.20100637

Frebel H, Nindl V, Schuepbach RA, Braunschweiler T, Richter K, Vogel J, Wagner CA, Loffing-Cueni D, Kurrer M, Ludewig B, Oxenius A (2012) Programmed death 1 protects from fatal circulatory failure during systemic virus infection of mice. J Exp Med 209(13): 2485–2499. doi:10.1084/jem.20121015

Fromentin R, Bakeman W, Lawani MB, Khoury G, Hartogensis W, DaFonseca S, Killian M, Epling L, Hoh R, Sinclair E, Hecht FM, Bacchetti P, Deeks SG, Lewin SR, Sekaly RP, Chomont N (2016) CD4 + T cells expressing PD-1, TIGIT and LAG-3 contribute to HIV persistence during ART. PLoS Pathog 12(7):e1005761. doi:10.1371/journal.ppat.1005761

Fuchs A, Cella M, Giurisato E, Shaw AS, Colonna M (2004) Cutting edge: CD96 (tactile) promotes NK cell-target cell adhesion by interacting with the poliovirus receptor (CD155). J Immunol 172(7):3994–3998

Fuhrman CA, Yeh WI, Seay HR, Saikumar Lakshmi P, Chopra G, Zhang L, Perry DJ, McClymont SA, Yadav M, Lopez MC, Baker HV, Zhang Y, Li Y, Whitley M, von Schack D, Atkinson MA, Bluestone JA, Brusko TM (2015) Divergent phenotypes of human regulatory T cells expressing the receptors TIGIT and CD226. J Immunol 195(1):145–155. doi:10.4049/jimmunol.1402381

Gagliani N, Magnani CF, Huber S, Gianolini ME, Pala M, Licona-Limon P, Guo B, Herbert DR, Bulfone A, Trentini F, Di Serio C, Bacchetta R, Andreani M, Brockmann L, Gregori S, Flavell RA, Roncarolo MG (2013) Coexpression of CD49b and LAG-3 identifies human and mouse T regulatory type 1 cells. Nat Med 19(6):739–746. doi:10.1038/nm.3179

Gao X, Zhu Y, Li G, Huang H, Zhang G, Wang F, Sun J, Yang Q, Zhang X, Lu B (2012) TIM-3 expression characterizes regulatory T cells in tumor tissues and is associated with lung cancer progression. PLoS ONE 7(2):e30676. doi:10.1371/journal.pone.0030676

Gautron AS, Dominguez-Villar M, de Marcken M, Hafler DA (2014) Enhanced suppressor function of TIM-3 + FoxP3 + regulatory T cells. Eur J Immunol 44(9):2703–2711. doi:10.1002/eji.201344392

Gleason MK, Lenvik TR, McCullar V, Felices M, O'Brien MS, Cooley SA, Verneris MR, Cichocki F, Holman CJ, Panoskaltsis-Mortari A, Niki T, Hirashima M, Blazar BR, Miller JS (2012) Tim-3 is an inducible human natural killer cell receptor that enhances interferon gamma production in response to galectin-9. Blood 119(13):3064–3072. doi:10.1182/blood-2011-06-360321

Golden-Mason L, Palmer BE, Kassam N, Townshend-Bulson L, Livingston S, McMahon BJ, Castelblanco N, Kuchroo V, Gretch DR, Rosen HR (2009) Negative immune regulator Tim-3 is overexpressed on T cells in hepatitis C virus infection and its blockade rescues dysfunctional CD4 + and CD8 + T cells. J Virol 83:9122–9130
Gravitz L (2013) Cancer immunotherapy. Nature 504(7480):S1. doi:10.1038/504S1a
Grohmann U, Orabona C, Fallarino F, Vacca C, Calcinaro F, Falorni A, Candeloro P, Belladonna ML, Bianchi R, Fioretti MC, Puccetti P (2002) CTLA-4-Ig regulates tryptophan catabolism in vivo. Nat Immunol 3(11):1097–1101 10.1038/ni846 ni846 [pii]
Grosso JF, Kelleher CC, Harris TJ, Maris CH, Hipkiss EL, De Marzo A, Anders R, Netto G, Getnet D, Bruno TC, Goldberg MV, Pardoll DM, Drake CG (2007) LAG-3 regulates CD8 + T cell accumulation and effector function in murine self- and tumor-tolerance systems. J Clin Invest 117(11):3383–3392. doi:10.1172/JCI31184
Gupta S, Thornley TB, Gao W, Larocca R, Turka LA, Kuchroo VK, Strom TB (2012) Allograft rejection is restrained by short-lived TIM-3 + PD-1 + Foxp3 + Tregs. J Clin Invest 122 (7):2395–2404. doi:10.1172/JCI45138
Hafalla JC, Claser C, Couper KN, Grau GE, Renia L, de Souza JB, Riley EM (2012) The CTLA-4 and PD-1/PD-L1 inhibitory pathways independently regulate host resistance to plasmodium-induced acute immune pathology. PLoS Pathog 8(2):e1002504. doi:10.1371/journal.ppat.1002504
Hafler JP, Maier LM, Cooper JD, Plagnol V, Hinks A, Simmonds MJ, Stevens HE, Walker NM, Healy B, Howson JM, Maisuria M, Duley S, Coleman G, Gough SC, Worthington J, Kuchroo VK, Wicker LS, Todd JA (2009) CD226 Gly307Ser association with multiple autoimmune diseases. Genes Immun 10(1):5–10. doi:10.1038/gene.2008.82
Hannier S, Tournier M, Bismuth G, Triebel F (1998) CD3/TCR complex-associated lymphocyte activation gene-3 molecules inhibit CD3/TCR signaling. J Immunol 161(8):4058–4065
Hodi FS, O'Day SJ, McDermott DF, Weber RW, Sosman JA, Haanen JB, Gonzalez R, Robert C, Schadendorf D, Hassel JC, Akerley W, van den Eertwegh AJ, Lutzky J, Lorigan P, Vaubel JM, Linette GP, Hogg D, Ottensmeier CH, Lebbe C, Peschel C, Quirt I, Clark JI, Wolchok JD, Weber JS, Tian J, Yellin MJ, Nichol GM, Hoos A, Urba WJ (2010) Improved survival with ipilimumab in patients with metastatic melanoma. N Engl J Med 363(8):711–723. doi:10.1056/NEJMoa1003466
Huang CT, Workman CJ, Flies D, Pan X, Marson AL, Zhou G, Hipkiss EL, Ravi S, Kowalski J, Levitsky HI, Powell JD, Pardoll DM, Drake CG, Vignali DA (2004) Role of LAG-3 in regulatory T cells. Immunity 21(4):503–513. doi:10.1016/j.immuni.2004.08.010
Huang YH, Zhu C, Kondo Y, Anderson AC, Gandhi A, Russell A, Dougan SK, Petersen BS, Melum E, Pertel T, Clayton KL, Raab M, Chen Q, Beauchemin N, Yazaki PJ, Pyzik M, Ostrowski MA, Glickman JN, Rudd CE, Ploegh HL, Franke A, Petsko GA, Kuchroo VK, Blumberg RS (2015) CEACAM1 regulates TIM-3-mediated tolerance and exhaustion. Nature 517(7534):386–390. doi:10.1038/nature13848
Huard B, Prigent P, Tournier M, Bruniquel D, Triebel F (1995) CD4/major histocompatibility complex class II interaction analyzed with CD4- and lymphocyte activation gene-3 (LAG-3)-Ig fusion proteins. Eur J Immunol 25(9):2718–2721. doi:10.1002/eji.1830250949
Inozume T, Yaguchi T, Furuta J, Harada K, Kawakami Y, Shimada S (2016) Melanoma cells control antimelanoma CTL responses via interaction between TIGIT and CD155 in the effector phase. J Invest Dermatol 136(1):255–263. doi:10.1038/JID.2015.404
Jayaraman P, Sada-Ovalle I, Beladi S, Anderson AC, Dardalhon V, Hotta C, Kuchroo VK, Behar SM (2010) Tim3 binding to galectin-9 stimulates antimicrobial immunity. J Exp Med 207(11):2343–2354. doi:10.1084/jem.20100687
Jin HT, Anderson AC, Tan WG, West EE, Ha SJ, Araki K, Freeman GJ, Kuchroo VK, Ahmed R (2010) Cooperation of Tim-3 and PD-1 in CD8 T-cell exhaustion during chronic viral infection. Proc Natl Acad Sci USA 107(33):14733–14738. doi:10.1073/pnas.1009731107

Johnston RJ, Comps-Agrar L, Hackney J, Yu X, Huseni M, Yang Y, Park S, Javinal V, Chiu H, Irving B, Eaton DL, Grogan JL (2014) The immunoreceptor TIGIT regulates antitumor and antiviral CD8(+) T cell effector function. Cancer Cell 26(6):923–937. doi:10.1016/j.ccell.2014.10.018

Joller N, Hafler JP, Brynedal B, Kassam N, Spoerl S, Levin SD, Sharpe AH, Kuchroo VK (2011) Cutting edge: TIGIT has T cell-intrinsic inhibitory functions. J Immunol 186(3):1338–1342. doi:10.4049/jimmunol.1003081

Joller N, Lozano E, Burkett PR, Patel B, Xiao S, Zhu C, Xia J, Tan TG, Sefik E, Yajnik V, Sharpe AH, Quintana FJ, Mathis D, Benoist C, Hafler DA, Kuchroo VK (2014) Treg Cells expressing the coinhibitory molecule TIGIT selectively inhibit proinflammatory Th1 and Th17 cell responses. Immunity 40(4):569–581. doi:10.1016/j.immuni.2014.02.012

Jones RB, Ndhlovu LC, Barbour JD, Sheth PM, Jha AR, Long BR, Wong JC, Satkunarajah M, Schweneker M, Chapman JM, Gyenes G, Vali B, Hyrcza MD, Yue FY, Kovacs C, Sassi A, Loutfy M, Halpenny R, Persad D, Spotts G, Hecht FM, Chun TW, McCune JM, Kaul R, Rini JM, Nixon DF, Ostrowski MA (2008) Tim-3 expression defines a novel population of dysfunctional T cells with highly elevated frequencies in progressive HIV-1 infection. J Exp Med 205(12):2763–2779. doi:10.1084/jem.20081398

Kang CW, Dutta A, Chang LY, Mahalingam J, Lin YC, Chiang JM, Hsu CY, Huang CT, Su WT, Chu YY, Lin CY (2015) Apoptosis of tumor infiltrating effector TIM-3 + CD8 + T cells in colon cancer. Sci Rep 5:15659. doi:10.1038/srep15659

Kasagi S, Kawano S, Kumagai S (2011) PD-1 and autoimmunity. Critical reviews in immunology 31(4):265–295

Kisielow M, Kisielow J, Capoferri-Sollami G, Karjalainen K (2005) Expression of lymphocyte activation gene 3 (LAG-3) on B cells is induced by T cells. Eur J Immunol 35(7):2081–2088. doi:10.1002/eji.200526090

Kouo T, Huang L, Pucsek AB, Cao M, Solt S, Armstrong T, Jaffee E (2015) Galectin-3 shapes antitumor immune responses by suppressing CD8 + T Cells via LAG-3 and inhibiting expansion of plasmacytoid dendritic cells. Cancer Immunol Res 3(4):412–423. doi:10.1158/2326-6066.CIR-14-0150

Kurtulus S, Sakuishi K, Ngiow SF, Joller N, Tan DJ, Teng MW, Smyth MJ, Kuchroo VK, Anderson AC (2015) TIGIT predominantly regulates the immune response via regulatory T cells. J Clin Invest 125 (11):4053–4062. doi:10.1172/JCI81187

Lazar-Molnar E, Chen B, Sweeney KA, Wang EJ, Liu W, Lin J, Porcelli SA, Almo SC, Nathenson SG, Jacobs WR Jr (2010) Programmed death-1 (PD-1)-deficient mice are extraordinarily sensitive to tuberculosis. Proc Natl Acad Sci USA 107(30):13402–13407. doi:10.1073/pnas.1007394107

Lee SY, Goverman JM (2013) The influence of T cell Ig mucin-3 signaling on central nervous system autoimmune disease is determined by the effector function of the pathogenic T cells. J Immunol 190(10):4991–4999. doi:10.4049/jimmunol.1300083

Lee J, Su EW, Zhu C, Hainline S, Phuah J, Moroco JA, Smithgall TE, Kuchroo VK, Kane LP (2011) Phosphotyrosine-dependent coupling of Tim-3 to T-cell receptor signaling pathways. Mol Cell Biol 31(19):3963–3974. doi:10.1128/MCB.05297-11

Levin SD, Taft DW, Brandt CS, Bucher C, Howard ED, Chadwick EM, Johnston J, Hammond A, Bontadelli K, Ardourel D, Hebb L, Wolf A, Bukowski TR, Rixon MW, Kuijper JL, Ostrander CD, West JW, Bilsborough J, Fox B, Gao Z, Xu W, Ramsdell F, Blazar BR, Lewis KE (2011) Vstm3 is a member of the CD28 family and an important modulator of T-cell function. Eur J Immunol 41(4):902–915. doi:10.1002/eji.201041136

Li M, Xia P, Du Y, Liu S, Huang G, Chen J, Zhang H, Hou N, Cheng X, Zhou L, Li P, Yang X, Fan Z (2014) T-cell immunoglobulin and ITIM domain (TIGIT) receptor/poliovirus receptor (PVR) ligand engagement suppresses interferon-gamma production of natural killer cells via beta-arrestin 2-mediated negative signaling. J Biol Chem 289(25):17647–17657. doi:10.1074/jbc.M114.572420

Liang B, Workman C, Lee J, Chew C, Dale BM, Colonna L, Flores M, Li N, Schweighoffer E, Greenberg S, Tybulewicz V, Vignali D, Clynes R (2008) Regulatory T cells inhibit dendritic cells by lymphocyte activation gene-3 engagement of MHC class II. J Immunol 180(9): 5916–5926

Linsley P, Brady W, Urnes M, Grosmaire L, Damle N, Ledbetter J (1991) CTLA-4 is a second receptor for the B cell activation antigen B7. J Exp Med 174(3):561–569

Liu S, Zhang H, Li M, Hu D, Li C, Ge B, Jin B, Fan Z (2013) Recruitment of Grb2 and SHIP1 by the ITT-like motif of TIGIT suppresses granule polarization and cytotoxicity of NK cells. Cell Death Differ 20(3):456–464. doi:10.1038/cdd.2012.141

Lozano E, Dominguez-Villar M, Kuchroo V, Hafler DA (2012) The TIGIT/CD226 axis regulates human T cell function. J Immunol. doi:10.4049/jimmunol.1103627

Lozano E, Joller N, Cao Y, Kuchroo V, Hafler DA (2013) The CD226/CD155 Interaction Regulates the Proinflammatory (Th1/Th17)/Anti-Inflammatory (Th2) Balance in Humans. J Immunol. doi:10.4049/jimmunol.1300945

Matsuzaki J, Gnjatic S, Mhawech-Fauceglia P, Beck A, Miller A, Tsuji T, Eppolito C, Qian F, Lele S, Shrikant P, Old LJ, Odunsi K (2010) Tumor-infiltrating NY-ESO-1-specific CD8 + T cells are negatively regulated by LAG-3 and PD-1 in human ovarian cancer. Proc Natl Acad Sci USA 107(17):7875–7880. doi:10.1073/pnas.1003345107

McMahan RH, Golden-Mason L, Nishimura MI, McMahon BJ, Kemper M, Allen TM, Gretch DR, Rosen HR (2010) Tim-3 expression on PD-1 + HCV-specific human CTLs is associated with viral persistence, and its blockade restores hepatocyte-directed in vitro cytotoxicity. J Clin Invest 120(12):4546–4557. doi:10.1172/JCI43127

Mendelsohn CL, Wimmer E, Racaniello VR (1989) Cellular receptor for poliovirus: molecular cloning, nucleotide sequence, and expression of a new member of the immunoglobulin superfamily. Cell 56(5):855–865 0092-8674(89)90690-9 [pii]

Meyers JH, Sabatos CA, Chakravarti S, Kuchroo VK (2005) The TIM gene family regulates autoimmune and allergic diseases. Trends Mol Med 11(8):362–369. doi:10.1016/j.molmed.2005.06.008

Monney L, Sabatos CA, Gaglia JL, Ryu A, Waldner H, Chernova T, Manning S, Greenfield EA, Coyle AJ, Sobel RA, Freeman GJ, Kuchroo VK (2002) Th1-specific cell surface protein Tim-3 regulates macrophage activation and severity of an autoimmune disease. Nature 415 (6871):536–541. doi:10.1038/415536a415536a

Moskophidis D, Lechner F, Pircher H, Zinkernagel RM (1993) Virus persistence in acutely infected immunocompetent mice by exhaustion of antiviral cytotoxic effector T cells. Nature 362(6422):758–761. doi:10.1038/362758a0

Nakayama M, Akiba H, Takeda K, Kojima Y, Hashiguchi M, Azuma M, Yagita H, Okumura K (2009) Tim-3 mediates phagocytosis of apoptotic cells and cross-presentation. Blood 113 (16):3821–3830. doi:10.1182/blood-2008-10-185884

Ndhlovu LC, Lopez-Verges S, Barbour JD, Jones RB, Jha AR, Long BR, Schoeffler EC, Fujita T, Nixon DF, Lanier LL (2012) Tim-3 marks human natural killer cell maturation and suppresses cell-mediated cytotoxicity. Blood 119(16):3734–3743. doi:10.1182/blood-2011-11-392951

Nebbia G, Peppa D, Schurich A, Khanna P, Singh HD, Cheng Y, Rosenberg W, Dusheiko G, Gilson R, ChinAleong J, Kennedy P, Maini MK (2012) Upregulation of the Tim-3/galectin-9 pathway of T cell exhaustion in chronic hepatitis B virus infection. PloS one 7(10):e47648. doi:10.1371/journal.pone.0047648

Ngiow SF, von Scheidt B, Akiba H, Yagita H, Teng MW, Smyth MJ (2011) Anti-TIM3 antibody promotes T cell IFN-gamma-mediated antitumor immunity and suppresses established tumors. Cancer Res 71 (10):3540–3551. doi:10.1158/0008-5472.CAN-11-0096

Nishikawa H, Sakaguchi S (2010) Regulatory T cells in tumor immunity. Int J Cancer 127(4): 759–767. doi:10.1002/ijc.25429

Nishimura H, Nose M, Hiai H, Minato N, Honjo T (1999) Development of lupus-like autoimmune diseases by disruption of the PD-1 gene encoding an ITIM motif-carrying immunoreceptor. Immunity 11(2):141–151 S1074-7613(00)80089-8 [pii]

Nishimura H, Okazaki T, Tanaka Y, Nakatani K, Hara M, Matsumori A, Sasayama S, Mizoguchi A, Hiai H, Minato N, Honjo T (2001) Autoimmune dilated cardiomyopathy in PD-1 receptor-deficient mice. Science 291(5502):319–322. doi:10.1126/science.291.5502.319291/5502/319

Oikawa T, Kamimura Y, Akiba H, Yagita H, Okumura K, Takahashi H, Zeniya M, Tajiri H, Azuma M (2006) Preferential involvement of Tim-3 in the regulation of hepatic CD8 + T cells in murine acute graft-versus-host disease. J Immunol 177(7):4281–4287

Okazaki T, Okazaki IM, Wang J, Sugiura D, Nakaki F, Yoshida T, Kato Y, Fagarasan S, Muramatsu M, Eto T, Hioki K, Honjo T (2011) PD-1 and LAG-3 inhibitory co-receptors act synergistically to prevent autoimmunity in mice. J Experimental Med 208(2):395–407. doi:10.1084/jem.20100466

Pauken KE, Wherry EJ (2015) Overcoming T cell exhaustion in infection and cancer. Trends Immunol 36(4):265–276. doi:10.1016/j.it.2015.02.008

Qu HQ, Bradfield JP, Grant SF, Hakonarson H, Polychronakos C, Type IDGC (2009) Remapping the type I diabetes association of the CTLA4 locus. Genes and immunity 10(Suppl 1):27–32. doi:10.1038/gene.2009.88

Rangachari M, Zhu C, Sakuishi K, Xiao S, Karman J, Chen A, Angin M, Wakeham A, Greenfield EA, Sobel RA, Okada H, McKinnon PJ, Mak TW, Addo MM, Anderson AC, Kuchroo VK (2012) Bat3 promotes T cell responses and autoimmunity by repressing Tim-3-mediated cell death and exhaustion. Nat Med 18(9):1394–1400. doi:10.1038/nm.2871

Richter K, Agnellini P, Oxenius A (2010) On the role of the inhibitory receptor LAG-3 in acute and chronic LCMV infection. Int Immunol 22(1):13–23. doi:10.1093/intimm/dxp107

Robert C, Schachter J, Long GV, Arance A, Grob JJ, Mortier L, Daud A, Carlino MS, McNeil C, Lotem M, Larkin J, Lorigan P, Neyns B, Blank CU, Hamid O, Mateus C, Shapira-Frommer R, Kosh M, Zhou H, Ibrahim N, Ebbinghaus S, Ribas A, K-investigators (2015) Pembrolizumab versus Ipilimumab in Advanced Melanoma. N Engl J Med 372(26):2521–2532. doi:10.1056/NEJMoa1503093

Sabatos CA, Chakravarti S, Cha E, Schubart A, Sanchez-Fueyo A, Zheng XX, Coyle AJ, Strom TB, Freeman GJ, Kuchroo VK (2003) Interaction of Tim-3 and Tim-3 ligand regulates T helper type 1 responses and induction of peripheral tolerance. Nat Immunol 4(11):1102–1110 10.1038/ni988ni988[pii]

Sada-Ovalle I, Chavez-Galan L, Torre-Bouscoulet L, Nava-Gamino L, Barrera L, Jayaraman P, Torres-Rojas M, Salazar-Lezama MA, Behar SM (2012) The Tim3-galectin 9 pathway induces antibacterial activity in human macrophages infected with mycobacterium tuberculosis. J Immunol 189(12):5896–5902. doi:10.4049/jimmunol.1200990

Sakuishi K, Apetoh L, Sullivan JM, Blazar BR, Kuchroo VK, Anderson AC (2010) Targeting Tim-3 and PD-1 pathways to reverse T cell exhaustion and restore anti-tumor immunity. J Exp Med 207(10):2187–2194. doi:10.1084/jem.20100643

Sakuishi K, Ngiow SF, Sullivan JM, Teng MW, Kuchroo VK, Smyth MJ, Anderson AC (2013) TIM3 + FOXP3 + regulatory T cells are tissue-specific promoters of T-cell dysfunction in cancer. Oncoimmunology 2(4):e23849. doi:10.4161/onci.23849

Sanchez-Fueyo A, Tian J, Picarella D, Domenig C, Zheng XX, Sabatos CA, Manlongat N, Bender O, Kamradt T, Kuchroo VK, Gutierrez-Ramos JC, Coyle AJ, Strom TB (2003) Tim-3 inhibits T helper type 1-mediated auto- and alloimmune responses and promotes immunological tolerance. Nat Immunol 4(11):1093–1101. doi:10.1038/ni987

Santiago C, Ballestros A, Martinez-Munoz L, Mellado M, Kaplan GG, Freeman GJ, Casanovas JM (2007a) Structures of T Cell Immunoglobulin Mucin Protein 4 show a metal-ion-dependent ligand binding site where phosphatidylserine binds. Immunity 27:941–945

Santiago C, Ballestros A, Tami C, Martinez-Munoz L, Kaplan GG, Casanovas JM (2007b) Structures of T Cell immunoglobulin mucin receptors 1 and 2 reveal mechanisms for regulation of immune responses by the TIM receptor family. Immunity 26:299–310

Sarhan D, Cichocki F, Zhang B, Yingst A, Spellman SR, Cooley S, Verneris MR, Blazar BR, Miller JS (2016) Adaptive NK cells with Low TIGIT expression are inherently resistant to myeloid-derived suppressor cells. Cancer Res 76(19):5696–5706. doi:10.1158/0008-5472. CAN-16-0839

Schildberg FA, Klein SR, Freeman GJ, Sharpe AH (2016) Coinhibitory pathways in the B7-CD28 ligand-receptor family. Immunity 44(5):955–972. doi:10.1016/j.immuni.2016.05.002

Sega EI, Leveson-Gower DB, Florek M, Schneidawind D, Luong RH, Negrin RS (2014) Role of lymphocyte activation gene-3 (Lag-3) in conventional and regulatory T cell function in allogeneic transplantation. PLoS ONE 9(1):e86551. doi:10.1371/journal.pone.0086551

Singer M, Wang C, Cong L, Marjanovic ND, Kowalczyk MS, Zhang H, Nyman J, Sakuishi K, Kurtulus S, Gennert D, Xia J, Kwon JY, Nevin J, Herbst RH, Yanai I, Rozenblatt-Rosen O, Kuchroo VK, Regev A, Anderson AC (2016) A Distinct Gene Module for Dysfunction Uncoupled from Activation in Tumor-Infiltrating T Cells. Cell 166 (6):1500–1511 e1509. doi:10.1016/j.cell.2016.08.052

Song YW, Im CH, Park JH, Lee YJ, Lee EY, Lee EB, Park K (2011) T-cell immunoglobulin and mucin domain 3 genetic polymorphisms are associated with rheumatoid arthritis independent of a shared epitope status. Human Immunol 72(8):652–655. doi:10.1016/j.humimm.2011.04.007

Stanietsky N, Simic H, Arapovic J, Toporik A, Levy O, Novik A, Levine Z, Beiman M, Dassa L, Achdout H, Stern-Ginossar N, Tsukerman P, Jonjic S, Mandelboim O (2009) The interaction of TIGIT with PVR and PVRL2 inhibits human NK cell cytotoxicity. Proc Natl Acad Sci USA 106(42):17858–17863. doi:10.1073/pnas.0903474106

Stanietsky N, Rovis TL, Glasner A, Seidel E, Tsukerman P, Yamin R, Enk J, Jonjic S, Mandelboim O (2013) Mouse TIGIT inhibits NK-cell cytotoxicity upon interaction with PVR. Eur J Immunol 43(8):2138–2150. doi:10.1002/eji.201243072

Stengel KF, Harden-Bowles K, Yu X, Rouge L, Yin J, Comps-Agrar L, Wiesmann C, Bazan JF, Eaton DL, Grogan JL (2012) Structure of TIGIT immunoreceptor bound to poliovirus receptor reveals a cell-cell adhesion and signaling mechanism that requires cis-trans receptor clustering. Proc Natl Acad Sci USA. doi:10.1073/pnas.1120606109

Tirosh I, Izar B, Prakadan SM, Wadsworth MH 2nd, Treacy D, Trombetta JJ, Rotem A, Rodman C, Lian C, Murphy G, Fallahi-Sichani M, Dutton-Regester K, Lin JR, Cohen O, Shah P, Lu D, Genshaft AS, Hughes TK, Ziegler CG, Kazer SW, Gaillard A, Kolb KE, Villani AC, Johannessen CM, Andreev AY, Van Allen EM, Bertagnolli M, Sorger PK, Sullivan RJ, Flaherty KT, Frederick DT, Jane-Valbuena J, Yoon CH, Rozenblatt-Rosen O, Shalek AK, Regev A, Garraway LA (2016) Dissecting the multicellular ecosystem of metastatic melanoma by single-cell RNA-seq. Science 352(6282):189–196. doi:10.1126/science.aad0501

Tivol EA, Borriello F, Schweitzer AN, Lynch WP, Bluestone JA, Sharpe AH (1995) CTLA-4 deficient mice exhibit massive lymphoproliferation and multi-organ lymphatic infiltration: a critical negative immunoregulatory role of CTLA-4. Immunity 3:541–547

Topalian SL, Hodi FS, Brahmer JR, Gettinger SN, Smith DC, McDermott DF, Powderly JD, Carvajal RD, Sosman JA, Atkins MB, Leming PD, Spigel DR, Antonia SJ, Horn L, Drake CG, Pardoll DM, Chen L, Sharfman WH, Anders RA, Taube JM, McMiller TL, Xu H, Korman AJ, Jure-Kunkel M, Agrawal S, McDonald D, Kollia GD, Gupta A, Wigginton JM, Sznol M (2012) Safety, activity, and immune correlates of anti-PD-1 antibody in cancer. N Engl J Med 366(26):2443–2454. doi:10.1056/NEJMoa1200690

Triebel F, Jitsukawa S, Baixeras E, Roman-Roman S, Genevee C, Viegas-Pequignot E, Hercend T (1990) LAG-3, a novel lymphocyte activation gene closely related to CD4. J Exp Med 171 (5):1393–1405

Vaccari M, Boasso A, Fenizia C, Fuchs D, Hryniewicz A, Morgan T, Weiss D, Doster MN, Heraud JM, Shearer GM, Franchini G (2012) Fatal pancreatitis in simian immunodeficiency virus SIV(mac251)-infected macaques treated with 2',3'-dideoxyinosine and stavudine following cytotoxic-T-lymphocyte-associated antigen 4 and indoleamine 2,3-dioxygenase blockade. J Virol 86(1):108–113. doi:10.1128/JVI.05609-11

Wang M, Ji B, Wang J, Cheng X, Zhou Q, Zhou J, Cao C, Guo Q (2014) Tim-3 polymorphism downregulates gene expression and is involved in the susceptibility to ankylosing spondylitis. DNA and Cell Biology 33(10):723–728. doi:10.1089/dna.2014.2456

Wang SC, Li YH, Piao HL, Hong XW, Zhang D, Xu YY, Tao Y, Wang Y, Yuan MM, Li DJ, Du MR (2015a) PD-1 and Tim-3 pathways are associated with regulatory CD8 + T-cell function in decidua and maintenance of normal pregnancy. Cell Death Dis 6:e1738. doi:10.1038/cddis.2015.112

Wang F, Hou H, Wu S, Tang Q, Liu W, Huang M, Yin B, Huang J, Mao L, Lu Y, Sun Z (2015b) TIGIT expression levels on human NK cells correlate with functional heterogeneity among healthy individuals. Eur J Immunol 45(10):2886–2897. doi:10.1002/eji.201545480

Waterhouse P, Penninger J, Timms E, Wakeham A, Shahinian A, Lee K, Thompson C, Griesser H, Mak T (1995) Lymphoproliferative disorders with early lethality in mice deficient in CTLA-4. Science 270:985–988

Wherry EJ, Kurachi M (2015) Molecular and cellular insights into T cell exhaustion. Nat Rev Immunol 15(8):486–499. doi:10.1038/nri3862

Wing K, Onishi Y, Prieto-Martin P, Yamaguchi T, Miyara M, Fehervari Z, Nomura T, Sakaguchi S (2008) CTLA-4 control over Foxp3 + regulatory T cell function. Science 322 (5899):271–275. doi:10.1126/science.1160062

Woo SR, Turnis ME, Goldberg MV, Bankoti J, Selby M, Nirschl CJ, Bettini ML, Gravano DM, Vogel P, Liu CL, Tangsombatvisit S, Grosso JF, Netto G, Smeltzer MP, Chaux A, Utz PJ, Workman CJ, Pardoll DM, Korman AJ, Drake CG, Vignali DA (2012) Immune inhibitory molecules LAG-3 and PD-1 synergistically regulate T-cell function to promote tumoral immune escape. Cancer Res 72(4):917–927. doi:10.1158/0008-5472.CAN-11-1620

Workman CJ, Vignali DA (2003) The CD4-related molecule, LAG-3 (CD223), regulates the expansion of activated T cells. Eur J Immunol 33(4):970–979. doi:10.1002/eji.200323382

Workman CJ, Dugger KJ, Vignali DA (2002) Cutting edge: molecular analysis of the negative regulatory function of lymphocyte activation gene-3. J Immunol 169(10):5392–5395

Workman CJ, Cauley LS, Kim IJ, Blackman MA, Woodland DL, Vignali DA (2004) Lymphocyte activation gene-3 (CD223) regulates the size of the expanding T cell population following antigen activation in vivo. J Immunol 172(9):5450–5455

Workman CJ, Wang Y, El Kasmi KC, Pardoll DM, Murray PJ, Drake CG, Vignali DA (2009) LAG-3 regulates plasmacytoid dendritic cell homeostasis. J Immunol 182(4):1885–1891. doi:10.4049/jimmunol.0800185

Wu W, Shi Y, Li S, Zhang Y, Liu Y, Wu Y, Chen Z (2012) Blockade of Tim-3 signaling restores the virus-specific CD8(+) T-cell response in patients with chronic hepatitis B. European J Immunol 42(5):1180–1191. doi:10.1002/eji.201141852

Xu F, Liu J, Liu D, Liu B, Wang M, Hu Z, Du X, Tang L, He F (2014) LSECtin expressed on melanoma cells promotes tumor progression by inhibiting antitumor T-cell responses. Cancer Res 74(13):3418–3428. doi:10.1158/0008-5472.CAN-13-2690

Xu L, Huang Y, Tan L, Yu W, Chen D, Lu C, He J, Wu G, Liu X, Zhang Y (2015) Increased Tim-3 expression in peripheral NK cells predicts a poorer prognosis and Tim-3 blockade improves NK cell-mediated cytotoxicity in human lung adenocarcinoma. Int Immunopharmacol 29(2):635–641. doi:10.1016/j.intimp.2015.09.017

Yan J, Zhang Y, Zhang JP, Liang J, Li L, Zheng L (2013) Tim-3 expression defines regulatory T cells in human tumors. PLoS ONE 8(3):e58006. doi:10.1371/journal.pone.0058006

Yang ZZ, Grote DM, Ziesmer SC, Niki T, Hirashima M, Novak AJ, Witzig TE, Ansell SM (2012) IL-12 upregulates TIM-3 expression and induces T cell exhaustion in patients with follicular B cell non-Hodgkin lymphoma. J Clin Invest 122(4):1271–1282. doi:10.1172/JCI59806

Yang X, Jiang X, Chen G, Xiao Y, Geng S, Kang C, Zhou T, Li Y, Guo X, Xiao H, Hou C, Wang R, Lin Z, Li X, Feng J, Ma Y, Shen B, Li Y, Han G (2013) T cell Ig mucin-3 promotes homeostasis of sepsis by negatively regulating the TLR response. J Immunol 190(5): 2068–2079. doi:10.4049/jimmunol.1202661

Yu X, Harden K, Gonzalez LC, Francesco M, Chiang E, Irving B, Tom I, Ivelja S, Refino CJ, Clark H, Eaton D, Grogan JL (2009) The surface protein TIGIT suppresses T cell activation by promoting the generation of mature immunoregulatory dendritic cells. Nat Immunol 10(1): 48–57. doi:10.1038/ni.1674

Zhang Y, Maksimovic J, Naselli G, Qian J, Chopin M, Blewitt ME, Oshlack A, Harrison LC (2013) Genome-wide DNA methylation analysis identifies hypomethylated genes regulated by FOXP3 in human regulatory T cells. Blood 122(16):2823–2836. doi:10.1182/blood-2013-02-481788

Zhou Q, Munger ME, Veenstra RG, Weigel BJ, Hirashima M, Munn DH, Murphy WJ, Azuma M, Anderson AC, Kuchroo VK, Blazar BR (2011) Coexpression of Tim-3 and PD-1 identifies a CD8 + T-cell exhaustion phenotype in mice with disseminated acute myelogenous leukemia. Blood 117(17):4501–4510. doi:10.1182/blood-2010-10-310425

Zhu C, Anderson AC, Schubart A, Xiong H, Imitola J, Khoury SJ, Zheng XX, Strom TB, Kuchroo VK (2005) The Tim-3 ligand galectin-9 negatively regulates T helper type 1 immunity. Nat Immunol 6(12):1245–1252. doi:10.1038/ni1271

Zhu C, Sakuishi K, Xiao S, Sun Z, Zaghouani S, Gu G, Wang C, Tan DJ, Wu C, Rangachari M, Pertel T, Jin HT, Ahmed R, Anderson AC, Kuchroo VK (2015) An IL-27/NFIL3 signalling axis drives Tim-3 and IL-10 expression and T-cell dysfunction. Nat Commun 6:6072. doi:10.1038/ncomms7072

Zhu Y, Paniccia A, Schulick AC, Chen W, Koenig MR, Byers JT, Yao S, Bevers S, Edil BH (2016) Identification of CD112R as a novel checkpoint for human T cells. J Exp Med 213(2):167–176. doi:10.1084/jem.20150785

Part III
Intracelluar Immune Checkpoint Molecules

SOCS1: Regulator of T Cells in Autoimmunity and Cancer

Subburaj Ilangumaran, Diwakar Bobbala and Sheela Ramanathan

Abstract SOCS1 is a negative feedback regulator of cytokine and growth factor receptor signaling, and plays an indispensable role in attenuating interferon gamma signaling. Studies on SOCS1-deficient mice have established a crucial role for SOCS1 in regulating $CD8^+$ T cell homeostasis. In the thymus, SOCS1 prevents thymocytes that had failed positive selection from surviving and expanding, ensures negative selection and prevents inappropriate developmental skewing toward the CD8 lineage. In the periphery, SOCS1 not only controls production of T cell stimulatory cytokines but also attenuates the sensitivity of $CD8^+$ T cells to synergistic cytokine stimulation and antigen non-specific activation. As cytokine stimulation of $CD8^+$ T lymphocytes increases their sensitivity to low affinity TCR ligands, SOCS1 likely contributes to peripheral T cell tolerance by putting brakes on aberrant T cell activation driven by inflammatory cytokines. In addition, SOCS1 is critical to maintain the stability of T regulatory cells and control their plasticity to become pathogenic Th17 and Th1 cells under the harmful influence of inflammatory cytokines. SOCS1 also regulates T cell activation by dendritic cells via modulating their generation, maturation, antigen presentation, costimulatory signaling, and cytokine production. The above control mechanisms of SOCS1 on T cells, T regulatory cells and dendritic cells collectively contribute to immunological tolerance and prevent autoimmune manifestation. On other hand, silencing SOCS1 in dendritic cells or $CD8^+$ T cells stimulates efficient antitumor immunity. Thus, even though SOCS1 is not a cell surface checkpoint inhibitor, its regulatory functions on T cell responses qualify SOCS1as a "non-classical" checkpoint blocker. SOCS1 also functions as a tumor suppressor in cancer cells by regulating oncogenic signal transduction pathways. The loss of SOCS1 expression observed in many tumors may have an impact on classical checkpoint pathways. The potential to exploit SOCS1 to treat inflammatory/autoimmune diseases and elicit antitumor immunity is discussed.

S. Ilangumaran (✉) · D. Bobbala · S. Ramanathan
Immunology Division, Faculty of Medicine and Health Sciences, Department of Pediatrics, Université de Sherbrooke, 3001 North 12th avenue, Sherbrooke, QC J1H 5N4, Canada
e-mail: Subburaj.Ilangumaran@Usherbrooke.ca

Current Topics in Microbiology and Immunology (2017) 410:159–190
DOI 10.1007/82_2017_63

Published Online: 13 September 2017

Contents

1 Introduction ... 160
2 SOCS1-Dependent Regulation of Cytokine Signaling ... 161
3 Regulation of T Cell Activation and Homeostasis by SOCS1 ... 164
4 Role of SOCS1 in T Cell Development ... 168
5 Regulation of Treg Plasticity and Function by SOCS1 ... 172
6 Influence of SOCS1 on Antigen Presenting Cell Functions ... 175
7 SOCS1 as an Immune Checkpoint Molecule ... 177
8 Targeting SOCS1 in Autoimmunity and Cancer ... 179
References ... 180

1 Introduction

T lymphocytes play a central role in immune protection against pathogens and provide immune surveillance against neoplastic cells from growing into life-threatening tumors. They also play a key role in the pathogenesis of autoimmune diseases. Physiologic and pathogenic activations of T cells require two essential signals. Signal 1 is delivered via the T cell antigen receptor (TCR) following its engagement by antigenic peptides presented by the MHC:peptide complex. However, concomitant signaling via costimulatory receptors (mainly CD28 and many others such as ICOS, CD27, 4-1BB/CD137, OX40/CD134, etc.) is also required to achieve full activation, elicit cell proliferation, and develop effector functions (Baxter and Hodgkin 2002; Chen and Flies 2013). In the absence of the second signal, TCR-stimulated T cells undergo anergy, which can be prevented by IL-2, an important product of costimulatory signaling (Acuto and Michel 2003; Schwartz 2003; Wells et al. 2001). Besides these two essential signals, activated T cells invariably get additional help from inflammatory cytokines (mainly type-I interferons, IL-12 and IL-27 for CD8 T cells, and IL-1 for CD4 T cells), which provide the third signal to amplify T cell responses (Ben-Sasson et al. 2009; Curtsinger and Mescher 2010; Curtsinger et al. 2005b; Haring et al. 2006; Kolumam et al. 2005). Indeed, loss of the cytokine-mediated signal could lead to impaired effector functions and induction of tolerance (Curtsinger et al. 2003, 2005a; Mescher et al. 2007). Several other cytokines including IL-2, IL-4, IL-6, IL-15, IL-17, and IL-21 also contribute to cell proliferation, differentiation, acquisition of distinct effector functions, prevention of exhaustion and generation of memory cells. Thus, cytokine signals profoundly modulate the quality, type, and magnitude of the T cell response toward pathogen-derived antigens, which also hold good for autoantigens and tumor antigens (Cox et al. 2011).

Regulating the amplitude and duration of the T cell response is essential to minimize potential collateral damage to normal tissues while eliminating harmful pathogens. This balance is achieved through several mechanisms acting at each of the three nodes of the T cell activation cascade namely, the TCR, costimulatory receptors and the cytokine receptors. These regulatory mechanisms may operate either in a cell-autonomous manner or be modulated following interaction of activated T cells with ligands expressed on other cells. The signal 1 is negatively regulated by downmodulation of the TCRζ chain via CBL-mediated ubiquitination as well as through protein tyrosine phosphatases and many other negative feedback regulators that impact on the various components of the TCR signaling machinery (Acuto et al. 2008; Baniyash 2004; Naramura et al. 2002; Stanford et al. 2012; Wang et al. 2001). Induction of co-inhibitory receptors such as CTLA-4, PD-1, LAG3, TIM3, etc., not only attenuate costimulatory signals by competing for stimulatory ligands (B7-1/CD80, B7-2/CD86) but also inhibit various signaling pathways downstream of TCR by engaging newly induced co-inhibitory ligands such as PD-L1/CD274 that are expressed on antigen presenting cells (APC) and cancer cells (Chen and Flies 2013; Paterson et al. 2009). The third signal provided by cytokines could be regulated by several feedback negative regulatory mechanisms; for example, the one involving suppressor of cytokine signaling (SOCS) family proteins, of which CISH (CIS-1; cytokine-inducible SH2-containing protein 1) has also been recently implicated in modulating TCR signaling (Palmer et al. 2015; Palmer and Restifo 2009; Yasukawa et al. 2000; Yoshimura 2013; Yoshimura et al. 2007).

The checks and balances imposed by the above control mechanisms tightly regulate the temporal sequence, amplitude, and duration of T cell activation thereby modulating the overall T cell response to ultimately benefit the host. Whereas the loss of these control mechanisms invariably leads to autoimmune pathologies, the tumor microenvironment exploits them to dampen antitumor T cell responses. Indeed, the signal 2 of T cell activation influenced by costimulatory and co-inhibitory signaling receptors, particularly CTLA4 and PD-1, is profoundly modulated in autoimmunity and in cancer, earning these molecules the label "checkpoint inhibitors", even though the other regulatory nodes in the T cell activation cascade can also function as important checkpoints to ensure appropriate and adequate T cell response. Many of the "classical" checkpoint inhibitors that impinge on the signal 2 of T cell activation cascade are reviewed elsewhere in this volume. In this review, we will focus on the role of SOCS1 as an important "non-classical" checkpoint regulator of T cell activation at multiple levels and its implications to autoimmunity and cancer.

2 SOCS1-Dependent Regulation of Cytokine Signaling

Three laboratories independently discovered SOCS1 in 1997 as a negative regulator of cytokine receptor signaling that inhibited the Janus kinase (JAK)-signal transducer and activator of transcription (STAT) pathway (Endo et al. 1997; Naka et al. 1997; Starr et al. 1997). SOCS1 shares structural similarity with the previously known CISH and several other newly identified proteins designated SOCS2 to SOCS7, which together constitute the SOCS family proteins (Alexander 2002; Yoshimura et al. 1995). SOCS proteins contain a central SRC homology (SH2)

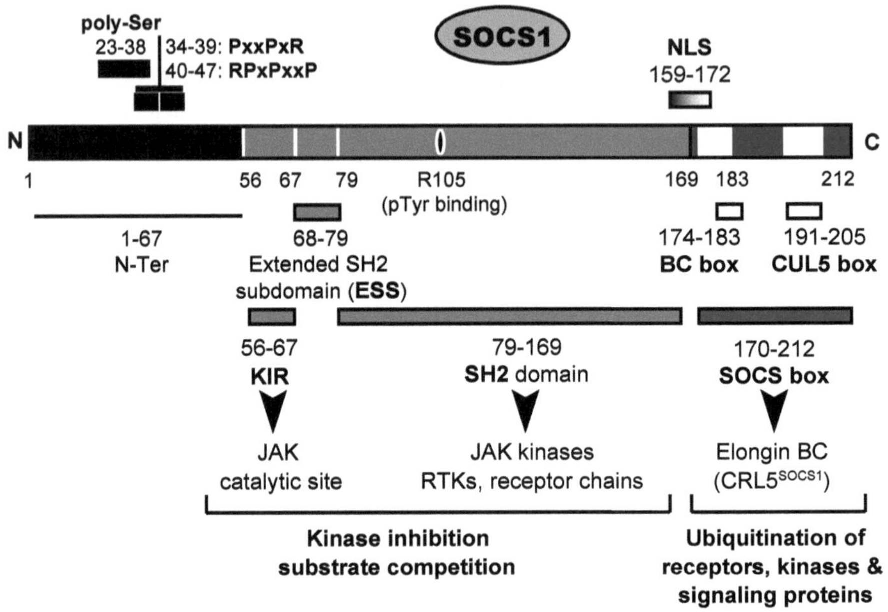

Fig. 1 Structure and functions of SOCS1. SOCS1 contains a central SH2 domain, which mediates interaction with JAK kinases, RTKs and receptor chains. The extended SH2 subdomain (ESS) and the kinase inhibitory region (KIR) are also required to block JAK kinase activity. The N-terminus of SOCS1 contains serine and proline-rich sequences, and the latter may bind SH3 domain-containing proteins. The C-terminal SOCS box includes the BC box and the Cul box, which interact with Elongins B and C, and Cullin 5, respectively. The SOCS box facilitates the assembly of the E3 ubiquitin ligase CRL^{SOCS1}, which promotes ubiquitination of many SOCS1-interacting proteins. SOCS1 harbors a nuclear localization signal (NLS) between the SH2 domain and the SOCS box that is likely involved in its nuclear functions, which are not yet well characterized. The boundaries of the various structural domains and motifs of SOCS1, and their molecular functions are indicated

domain and a conserved structural motif at the C-terminus called the SOCS box (Fig. 1), which is also shared by several proteins outside the SOCS family (Alexander 2002; Kile et al. 2002). The N-terminal region of SOCS proteins is highly variable in length and largely accounts for the difference in their size that range from 198 to 579 amino acids (Fujimoto and Naka 2003). The N-terminal segment may harbor additional structural motifs such as the kinase inhibitory region (KIR) adjacent to the SH2 domain of SOCS1 and SOCS3 (Nicholson et al. 1999; Sasaki et al. 1999; Yasukawa et al. 1999), the N-terminal conserved motif (NTCR) in SOCS4 and SOCS5 (Feng et al. 2012) or a nuclear localization signal (NLS) in SOCS6 and SOCS7 (Hwang et al. 2007; Kremer et al. 2007; Martens et al. 2004). SOCS1 also harbors a functional NLS between the SH2 domain and the SOCS box (Baetz et al. 2008; Strebovsky et al. 2011).

Most SOCS proteins are induced following cytokine stimulation and inhibit further signaling in a negative feedback manner that involves the SH2 domain and the SOCS box (Alexander 2002; Trengove and Ward 2013). The SH2 domain facilitates binding to phosphorylated tyrosine residues on JAK kinases, receptor chains and signaling proteins. The SOCS box, by virtue of its ability to interact with a Cullin5, Elongin B and C, assembles an E3 ubiquitin ligase that facilitates ubiquitination of the SOCS-interacting proteins and their degradation by proteasomes (Babon et al. 2009; Ilangumaran et al. 2004; Okumura et al. 2012). While most SOCS proteins attenuate cytokine signaling by promoting ubiquitination-dependent turnover of receptor chains, and presumably the associated signaling molecules, SOCS1 and SOCS3 mediate their functions mainly by inhibiting JAK kinase activity (Boyle et al. 2007; Yoshimura and Yasukawa 2012; Zhang et al. 2001). At least a part of this selective mode of action might result from reduced affinity of SOCS1 and SOCS3 toward Cullin5 compared to other SOCS family members (Babon et al. 2009).

Structure–function studies on SOCS1 and SOCS3 have revealed the requirement of the SH2 domain, adjacent N-terminal sequence (called extended SH2 domain or ESS) and the KIR to inhibit JAK activity (Nicholson et al. 1999; Sasaki et al. 1999; Yasukawa et al. 1999). It was initially proposed that while the SH2 domain binds the phospho-Tyr residue within the activation loop of JAKs, the KIR sequence occupies the substrate-binding pocket as a pseudosubstrate, thereby inhibiting enzyme activity (Yasukawa et al. 1999). Structural analysis by nuclear magnetic imaging of the tripartite complex formed by SOCS3, cytoplasmic domains of the IL-6 receptor subunit gp130 and the catalytic JH1 domain of JAK2 indicated that the SH2 domain interacts with gp130 while the KIR domain binds JH1 outside the substrate-binding pocket of JAK1, JAK2 and TYK2 (but not JAK3, which lacks the conserved GQM motif) and inhibit the kinase activity in a non-competitive manner (Babon et al. 2012). However, subsequent studies on the crystal structure of SOCS3-JAK2-gp130 ternary complex ($SOCS3_{22-185}$–$JAK2_{JH1}$–$gp130_{750-764}$) showed that the KIR of SOCS3 occupies the substrate-binding groove of JAK, supporting the original model proposed by Yasukawa and colleagues, without affecting ATP binding to the catalytic site (Babon et al. 2014b; Kershaw et al. 2013; Yasukawa et al. 1999). Like SOCS3, SOCS1 binding to cytokine receptors could subsequently lead to inhibition of the JAKs bound to the receptor chains and signal attenuation. Differences in the amino acid sequence surrounding the phosphor-Tyr-binding groove of the SH2 domain may determine the specificity and potency of SOCS1 and SOCS3 to inhibit different cytokines; for example, IFNγ and IL-2 family cytokines by SOCS1 and IL-6 by SOCS3 (Alexander et al. 1999; Babon et al. 2014a; Croker et al. 2003; Marine et al. 1999; Sporri et al. 2001; Wormald et al. 2006). Similarly, the ability of SOCS3 to inhibit STAT3 activation downstream of IL-6 but not IL-10 receptor, and hence attenuate pro-inflammatory but not anti-inflammatory STAT3 signaling, is determined by the ability of SOCS3 to bind IL-6 receptor but not the IL-10 receptor (Yasukawa et al. 2003). SOCS1 binding via SH2 domain to phosphorylated cytokine receptor chains will also result in competition for other signal transducers such as STAT molecules from being recruited to the receptor and become activated (Trengove and Ward 2013).

Even though SOCS molecules were originally discovered as JAK-binding proteins, SOCS1 was also shown to bind several receptor tyrosine kinases (RTK) including c-KIT, FLT3, CSF-1R, and MET, and modulate downstream signaling (Bourette et al. 2001; De Sepulveda et al. 1999; Gui et al. 2015; Kazi et al. 2014). Indeed, it has now become clear that the SOCS proteins can be grouped by their ability to inhibit predominantly the JAK-STAT pathway (CISH, SOCS2) or RTK signaling (SOCS6, SOCS7), or both (SOCS1, SOCS3, SOCS4, and SOCS5) (Kazi et al. 2014; Trengove and Ward 2013). Non-receptor tyrosine kinases other than the JAK family kinases such as TEC, SYK, and FAK have also been reported to interact with and inhibited by SOCS1 (Liu et al. 2003; Matsuda et al. 2000; Ohya et al. 1997).

While SOCS1-dependent regulation of the JAK-STAT pathway in vivo has been well documented using mice and cells lacking SOCS1 (discussed below), most studies on SOCS1-mediated inhibition of RTKs and NRTKS have been carried out in cell line models. The potential implications of the latter in the immune system also remain to be explored. For instance, HGF-induced MET signaling has been implicated in several immune cell functions including T cell recruitment to the heart during transplant rejection (Ilangumaran et al. 2016; Komarowska et al. 2015). As SOCS1 is an indispensable regulator of MET signaling in hepatocytes and several other RTKs, it will be worthwhile to investigate the role of SOCS1 (and other SOCS proteins) in regulating RTK signaling in T cells and other immune cells.

3 Regulation of T Cell Activation and Homeostasis by SOCS1

Early studies on SOCS1-deficient mice reported neonatal fatality from severe liver damage and systemic inflammation that was accompanied by profound thymic atrophy and marked depletion of T lymphocytes in spleen (Naka et al. 1998; Starr et al. 1998). Subsequent studies that demonstrated SOCS1 as an indispensable regulator of IFNγ signaling also suggested a role for SOCS1 in regulating T cell activation (Alexander et al. 1999; Marine et al. 1999). Perinatal death of *Socs1*-null mice can be reversed not only by simultaneous ablation of the *Ifng* gene or neutralization of IFNγ but also by concomitant RAG deficiency that blocks T cell development and causes peripheral T lymphopenia (Alexander et al. 1999; Marine et al. 1999). Selective loss of $TCR\alpha\beta^+$ T cells due to TCRα deficiency also prolonged viability of SOCS1-deficient mice, even though they eventually succumbed to severe colitis (Chinen et al. 2006). Analysis of the T cell compartment in *Socs1*$^{-/-}$*Ifng*$^{-/-}$ mice revealed accumulation of $CD8^+$ T cells and reduced CD4/CD8 ratio (Cornish et al. 2003b; Ilangumaran et al. 2003a). SOCS1-deficient $CD8^+$ T cells display a $CD44^{hi}$ $CD62L^{lo}$ $CD122^{hi}$ $Ly6c^{hi}$ phenotype, a characteristic of memory $CD8^+$ T cells and "memory-like" cells that arise from cytokine-driven homeostatic expansion under conditions of T lymphopenia

(Jameson 2002; Murali-Krishna and Ahmed 2000). CD44 is also upregulated following T cell activation; however, SOCS1-deficient $CD8^+$ T cells do not display other activation markers such as CD25 or CD69 (Ilangumaran et al. 2003a). *Socs1*$^{fl/-}$Lck-Cre mice lacking SOCS1 specifically in T cells did not develop lethal inflammatory disease, but displayed all changes in the peripheral T cell compartment caused by global SOCS1 deficiency, including enlarged lymph nodes, increased frequency of $CD8^+$ T cells and altered CD4/CD8 ratio, and elevated CD44 expression on CD8 T cells, indicating that SOCS1 is a cell-intrinsic regulator of $CD8^+$ T cell activation and homeostasis (Chong et al. 2003).

In peripheral T cells, SOCS1 deficiency does not affect TCR-induced calcium flux (Cornish et al. 2003a) or the proximal signaling events (our unpublished data). On the other hand, SOCS1-deficient $CD8^+$ T cells showed increased STAT5 phosphorylation, expression of the anti-apoptotic protein Bcl-xL and cell proliferation in response to IL-15, which promotes homeostatic expansion of memory $CD8^+$ T cells, or the autocrine T cell growth factor IL-2 (Cornish et al. 2003a; Ilangumaran et al. 2003a). IL-7, which is critical for homeostatic expansion of naïve T cells, induced strong STAT5 phosphorylation in SOCS1-deficient T cells but not cell proliferation (Cornish et al. 2003a; Ilangumaran et al. 2003a). Following adoptive transfer to *Rag1*$^{-/-}$ hosts, SOCS1-deficient $CD8^+$ T cells underwent robust homeostatic expansion that was markedly diminished in the absence of IL-7 or IL-15 in the recipient mice (Cornish et al. 2003b; Davey et al. 2005; Ilangumaran et al. 2003a; Ramanathan et al. 2006; Rodriguez et al. 2013). However, only the loss of IL-15 prevented $CD8^+$ T cell accumulation in SOCS1-deficient mice indicating that SOCS1 is an important regulator of IL-15 signaling in these cells (Ramanathan et al. 2006).

IL-15 deficiency in SOCS1-deficient mice also reversed the high CD44 expression on $CD8^+$ T cells without affecting the $CD122^{hi}$ $Ly6C^{hi}$ phenotype, suggesting that SOCS1 controls not only IL-15 but also other factors that contribute to the activation of $CD8^+$ T cells (Ramanathan et al. 2006). The $CD8^+$ T cell compartment in SOCS1-deficient mice still harbors a significant proportion of $CD44^{lo}$ $CD62L^{hi}$ $CD122^{lo}$ $Ly6c^{lo}$ naïve cells (Ilangumaran et al. 2003a), indicating that deregulated IL-15 signaling alone is not sufficient for the accumulation of memory phenotype $CD8^+$ T cells in SOCS1 knockout mice. Analysis of the phenotype of SOCS1-deficient $CD8^+$ T cells expressing the male antigen-specific transgenic TCR H-Y, which is one of the least reactive transgenic TCRs toward environmental antigens, in non-lymphopenic and lymphatic setting shed light on antigen non-specific activation of SOCS1-deficient T cells (Ramanathan et al. 2006). This study revealed that acquisition of the memory phenotype by SOCS1-deficient $CD8^+$ T lymphocytes did not require stimulation by cognate antigen, but required the presence of other activated $CD8^+$ T cells. As $CD8^+$ T cells from *Socs1*$^{-/-}$*Ifng*$^{-/-}$ mice that proliferate in response to IL-15 in vitro also produce abundant TNFα, it is possible that the effector molecules of activated SOCS1-deficient $CD8^+$ T cells may contribute, directly and indirectly (through activation of other cell types), to stimulate other naïve $CD8^+$ T cells. In support of this possibility, *Socs1*$^{fl/-}$LysM-Cre mice that seem to lack SOCS1 in myeloid cells as well as in T cells developed lethal inflammatory disease after four months of age (Chong et al. 2005). $CD8^+$ T cells from *Socs1*$^{fl/-}$LysM-Cre, but not *Socs1*$^{fl/-}$Lck-Cre, mice were also reported to display $CD44^{hi}$ $CD69^{hi}$ $CD25^{hi}$ $CD62L^{lo}$ activated cell phenotype, accompanied by elevated serum levels of IL-12, suggesting that macrophage and dendritic cell-derived IL-12 could contribute to antigen non-specific activation of SOCS1-deficient T cells. However, these findings still do not adequately explain how the initial activation of naïve $CD8^+$ T cells occurs in SOCS1-deficient mice.

Further investigations on SOCS1-deficient $CD8^+$ T cells led to the finding that they can be activated in an antigen non-specific manner by combinations of cytokines, particularly to IL-15 and IL-21, and that the availability of IL-21 markedly diminished the requirement of IL-15 needed for activation (Gagnon et al. 2007). Strikingly, wildtype $CD8^+$ T cells also proliferated in response to synergistic stimulation by IL-15 and IL-21, albeit to a significantly lower extent compared to SOCS1-deficient cells (Gagnon et al. 2007; Zeng et al. 2005). In fact, naïve $CD44^{lo}$ $CD8^+$ T cells expressing polyclonal TCR or transgenic TCR such as H-Y or P14 (specific to the glycoprotein antigen of lymphocytic choriomeningitis virus) showed robust antigen-independent proliferation in the presence of IL-7 and IL-21, and to a lesser extent to IL-15 and IL-21 (Gagnon et al. 2007). SOCS1 deficiency did not augment the naïve $CD8^+$ T cell response to IL-7 and IL-21, but increased responsiveness to the IL-15 and IL-21 combination. Follow-up studies on wild type $CD8^+$ T cells revealed that IL-6, which activates STAT3 similarly to IL-21, could substitute for IL-21 in inducing antigen-independent proliferation of naïve $CD8^+$ T cells (Gagnon et al. 2008). SOCS1 deficiency prolonged the IL-21-induced STAT3 activation in $CD8^+$ T cells. Concomitant stimulation by IL-6 or IL-21 augmented phosphorylation and DNA-binding activity of STAT5 induced by IL-7 or IL-15, although mechanisms underlying this cytokine synergy in activating naïve $CD8^+$ T cells remain to be fully elucidated (Gagnon et al. 2008). These cytokine-stimulated $CD8^+$ T cells show increased TCR responsiveness to limiting concentrations of antigenic peptides and, more importantly, toward altered peptide ligands with low reactivity toward the TCR, in terms of cell proliferation, IFNγ secretion, granzyme B expression, and cytolytic functions (Gagnon et al. 2008; Ramanathan et al. 2011; Rodriguez et al. 2013). Given that IL-6 and IL-15 are expressed by antigen presenting cells and several other cell types following innate immune stimulation, it is likely that SOCS1 may serve to restrain antigen non-specific activation of $CD8^+$ T cells by cytokines and, in that process, also induce their ability to respond to weakly agonistic TCR ligands such as those derived from autoantigens (Fig. 2). In fact, SOCS1-deficient $CD8^+$ T cells fail to undergo homeostatic expansion in TAP1-deficient hosts that express very low levels of MHC-I molecules, indicating that, besides cytokine stimulation, TCR engagement is needed to stimulate SOCS1-deficient $CD8^+$ T cells (Davey et al. 2005).

The checkpoint function of SOCS1 in modulating TCR reactivity (via controlling cytokine responses) may serve two complementary purposes. First, SOCS1 expression may shape the $CD8^+$ T cell response during infections by focusing the response toward dominant antigens through reducing reactivity toward weak

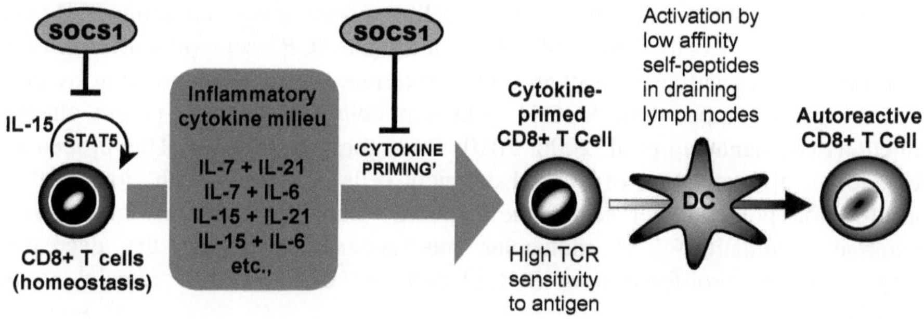

Fig. 2 Possible role of SOCS1 in preventing autoreactive CD8$^+$ T cell activation. SOCS1 regulates homeostasis of CD8$^+$ T cells by regulating STAT5 activation by IL-15. Antigen non-specific activation of naïve CD8$^+$ T cells by inflammatory cytokines leads to increase in antigen sensitivity. By regulating this 'cytokine priming' of CD8$^+$ T lymphocytes, SOCS1 may limit activation of potentially autoreactive cells

antigens. Second, SOCS1 may serve to prevent accidental stimulation of potentially autoreactive cells, whose TCR sensitivity toward autoantigens might lie close to the threshold needed for negative selection in the thymus. While the first possibility has not been yet investigated in detail, there are several indirect and direct evidence that support the second function. Indeed, SOCS1-null mice that express different MHC-I or MHC-II restricted transgenic TCRs survive considerably longer than those bearing a polyclonal TCR: OT-I (specific to ovalbumin, 6 weeks), P14 (4 weeks), Pmel-1 (specific to the melanocyte antigen gp100, 6 weeks) and H-Y (females 8 weeks) and OT-II (5 weeks) (Cornish et al. 2003b; Ramanathan et al. 2006; Rodriguez et al. 2013). OT-I TCR transgenic mice survive a little longer in RAG1 deficient background and die within 10 weeks even under germ-free conditions (Cornish et al. 2003b; Ramanathan et al. 2006; Rodriguez et al. 2013). The in vivo activation of SOCS1-deficient CD8$^+$ T cells was studied following adoptive transfer of SOCS1-deficient TCR transgenic CD8$^+$ T cells into recipients expressing the cognate antigen (Davey et al. 2005; Ramanathan et al. 2010; Rodriguez et al. 2013). This approach was necessary because the transgenic T cells are either deleted by thymic negative selection in mice bearing the cognate antigen (OT-I cells (Davey et al. 2005)), or systemic inflammation caused by SOCS1 deficiency precluded their analysis within the antigen-bearing host (P14 and Pmel-1 cells (Ramanathan et al. 2010; Rodriguez et al. 2013)). These studies showed that SOCS1-deficient TCR transgenic CD8$^+$ T cells displayed potent antigen-specific effector functions, recognized the tissues expressing the cognate antigen more efficiently and caused severe autoimmune manifestation (autoimmune type 1 diabetes by OT-I and P14 cells, and cutaneous inflammation by Pmel-1 cells), indicating that SOCS1 is necessary to control the pathogenicity of potentially autoreactive CD8 T cells (Davey et al. 2005; Ramanathan et al. 2010; Rodriguez et al. 2013).

Intriguingly, SOCS1-deficient $CD8^+$ T cells expressing the transgenic TCR such as P14 and Pmel-1 cells, and $CD8^{lo}$ H-Y transgenic TCR^+ cells (that arise in male H-Y mice in Rag-sufficient background) proliferate poorly to stimulation by cognate antigens compared to SOCS1 sufficient cells but develop potent effector functions (Ramanathan et al. 2006, 2010; Rodriguez et al. 2013). This propensity for effector differentiation of SOCS1-deficient cells is not due to the loss of their proliferation potential per se, as they undergo massive cell division following cytokine stimulation. The molecular mechanisms underlying the decreased antigen-induced proliferation of SOCS1-deficient $CD8^+$ T cells remains to be elucidated.

4 Role of SOCS1 in T Cell Development

The critical role of SOCS1 in regulating the $CD8^+$ T cell compartment already manifests during T cell development and maturation in the thymus, whereas *Socs1*$^{-/-}$ mice show severe thymic atrophy and reduction in thymic cell numbers, these abnormalities are not observed in *Socs1*$^{-/-}$*Ifng*$^{-/-}$ mice (Alexander et al. 1999; Naka et al. 1998). As IFNγ is known to cause thymic atrophy, the macroscopic abnormalities observed in SOCS1-deficient thymus are caused mainly by uncontrolled IFNγ signaling (Alexander et al. 1999; Marine et al. 1999). SOCS1 is highly expressed in thymocytes from wild type as well as IL-7R, JAK3, STAT5, or RAG2 deficient mice, indicating that *Socs1* expression occurs early during T cell development, even prior to the expression of rearranged TCRβ chain and independently of the IL-7-mediated signals (Marine et al. 1999). Hence, SOCS1 expression in thymocytes is considered to be constitutive, although stem cell factor (Kit ligand), which is critical for T lymphocyte development and induces the *Socs1* gene in bone marrow-derived mast cells (De Sepulveda et al. 1999; Waskow et al. 2002), could contribute to *Socs1* gene expression in T cell progenitors. Over expression of SOCS1 prevented fetal liver-derived hematopoietic progenitor cells from progressing beyond the stage-I $CD4^-CD8^-$ double negative cells (DN1; $CD44^+$ $CD25^-$) in fetal thymic organ cultures (FTOC) (Trop et al. 2001). As DN1 thymocytes express c-KIT and undergo IL-7-mediated expansion, this developmental arrest could result from inhibition of both Kit and IL-7 receptor signaling by SOCS1 (De Sepulveda et al. 1999; Trop et al. 2001). DN cells rearrange the genes coding for TCRβ, which complexes with pre-TCRα to form the pre-TCR complex that delivers the signal for progression through the DN4 ($CD44^-$ $CD25^-$) stage and gives rise to $CD4^+CD8^+$ double positive (DP) cells. Intriguingly, engagement of the pre-TCR in DN3 ($CD44^-$ $CD25^+$) thymocytes profoundly diminishes *Socs1* gene expression (Trop et al. 2001), although the underlying mechanism remains unknown. Forced expression of SOCS1 in DN3 cells did not affect differentiation toward DP cells but markedly reduced their number in FTOC (Trop et al. 2001). Consistent with the regulatory role for SOCS1 in early T cell development, SOCS1 transgenic mice show a developmental block from DN2 to DN3 stage and a marked reduction in cellularity (Fujimoto et al. 2000).

Hence, it has been proposed that (i) SOCS1 expression in DN cells serves to limit their expansion by IL-7 and Kit ligand and (ii) downmodulation of SOCS1 by pre-TCR signaling would facilitate the selective expansion of DN3 cells that have completed the β selection process.

Analyses of T cell developmental stages in *Socs1*$^{-/-}$ and *Socs1*$^{fl/-}$Lck-Cre mice did not reveal any obvious abnormality within the DN developmental stages (Chong et al. 2003; Ilangumaran et al. 2003a), indicating that either SOCS1 is dispensable for regulating the early DN developmental stages, or that the SOCS1-deficient cells proceed rapidly through these stages in vivo without allowing accumulation of the expanded DN cells. In agreement with the latter possibility, the SOCS1-deficient thymi generate a greater number of CD8^{+}CD4^{-} single positive (SP) thymocytes. Intrathymic transfer of SOCS1-deficient DN thymocytes into the thymus of SOCS1 sufficient *Rag1*$^{-/-}$ mice or in FTOC cultures results in an increased generation of CD8^{+} SP T cells, indicating that SOCS1 exerts a cell-intrinsic regulatory role in developing thymocytes (Cornish et al. 2003b; Ilangumaran et al. 2003b). Several lines of evidence indicate a critical regulatory role for SOCS1 at the DP and CD8^{+} SP stages. Whereas DN and CD4^{+} SP cells express the *Socs1* gene only after exposure to γ_c cytokines, DP and CD8^{+} SP cells show constitutive *Socs1* gene expression (Ilangumaran et al. 2003b). Based on the expression of a Cre-induced human CD4 as a reporter downstream of the *Socs1* promoter, Chong et al. reported high SOCS1 expression in DP cells compared to all other thymocyte subsets (Chong et al. 2003). SOCS1-deficient DP cells displayed high sensitivity to IL-7 in terms of STAT5 phosphorylation, and purified DP cells cultured in the presence of IL-7 gave rise to abundant CD8^{+} SP cells even in the absence of TCR stimulation, implicating IL-7 in DP to SP differentiation (Chong et al. 2003). Singer and colleagues observed that constitutive SOCS1 expression in pre-selection DP cells (prior to positive selection), which express IL-7Rα, IL-4Rα and γ_c, prevents them from responding to IL-7 or IL-4, thus contributing to the smaller size of DP thymocytes (Yu et al. 2006). This study also showed that TCRαβ signaling, which delivers the positive selection signal, downmodulates *Socs1* expression in DP thymocytes via unknown mechanisms and restores their responsiveness to IL-7 and IL-4, which induce the pro-survival gene *Bcl2*. It has been proposed that SOCS1 may serve to actively suppress cytokine signaling to prevent the rescue of CD4^{+}CD8^{+} double positive (DP) thymocytes that are destined to die by neglect through the induction of pro-survival factors by cytokines (Yu et al. 2006). Increased IL-7 signaling in positively selected, SOCS1-downregulated DP cells lead to induction of the transcription factor Runx3, which promotes CD8 lineage specification (Park et al. 2010).

Whereas purified SOCS1-deficient DP cells fail to proliferate in response to IL-7, IL-2 or IL-15 in vitro, CD8^{+} SP cells undergo robust proliferation following incubation with IL-15 or IL-2 but not IL-7, despite showing prolonged IL-7-induced STAT5 phosphorylation (Ilangumaran et al. 2003b). In agreement with the role of SOCS1 in regulating cytokine-mediated CD8^{+} SP cell expansion, only the CD8^{+} SP subset shows increased in vivo proliferation in SOCS1-deficient thymus, which displays an enlarged medulla wherein the SP cells undergo maturation before emigration to the periphery (Chong et al. 2003; Ilangumaran et al. 2010). Analysis of SOCS1-deficient thymus also lacking IL-7, IL-15 or both implicated these two cytokines in increasing the number of $CD8^+$ SP cells and skewing the CD4:CD8 ratio (Ramanathan et al. 2006). Collectively these studies suggest that SOCS1 regulates cytokine-mediated survival and/or differentiation of the DP cell pool and expansion of $CD8^+$ SP cells (Chong et al. 2003; Ilangumaran et al. 2003b).

Intriguingly, restoration of SOCS1 expression specifically in T cells of mice in which the *Socs1* gene is ablated in all cells (*Socs1*$^{-/-}$Tg) prevents lethality, but thymocytes from these mice show accelerated maturation of DP toward $CD4^+$ and $CD8^+$ SP cells (Hanada et al. 2003). Peripheral T cells in these mice also display $CD44^{hi}$ activated phenotype. These observations suggest that the T cell developmental abnormalities observed in SOCS1-deficient mice do not arise exclusively from T cell-intrinsic defects. Thymus tissues from these mice contained increased frequency of $CD11c^+CD11b^+$ dendritic cells (DC) that displayed MHC-II. As DCs can produce the T cell stimulatory cytokine IL-15 in response to myriad stimuli (Mattei et al. 2001), it is possible that SOCS1 may regulate T cell development also in a T cell-extrinsic manner by controlling IL-15 availability within the thymic microenvironment. This prediction remains to be tested.

Even though most studies on SOCS1 functions in T cells focused on $CD8^+$ T cells, SOCS1 might also regulate the $CD4^+$ T cell compartment, which is more apparent in the thymus than in the periphery. Thymic cellularity of *Socs1*$^{-/-}$*Ifng*$^{-/-}$ mice is comparable to that of *Ifng*$^{-/-}$ controls, yet the former harbored nearly twice the number of $CD4^+$ SP cells and six times more $CD8^+$ SP cells (Ramanathan et al. 2006). Analysis of recent thymic emigrants revealed that increased numbers of $CD4^+$ T cells are exported from SOCS1-deficient thymi compared to $CD8^+$ T cells (Ilangumaran et al. 2010). As $CD4^+$ SP cells in SOCS1-deficient thymi do not undergo increased cycling, their increased generation likely results from accelerated maturation from DP cells (Ilangumaran et al. 2010). In fact, IL-21 (another γ_c cytokine) has been shown to expand the DP cell pool undergoing positive selection in an in vitro T cell development model, and increase the number of DP, $CD4^+$ SP, and $CD8^+$ SP cells in vivo following systemic administration (Rafei et al. 2013). IL-21 has also been reported to promote T cell survival (Ostiguy et al. 2007). As SOCS1 attenuates IL-21-induced STAT5 activation in $CD8^+$ T cells (Gagnon et al. 2007), it is not unlikely that SOCS1 may regulate IL-21-mediated DP cell expansion, and hence the loss of SOCS1 would increase $CD4^+$ SP cell generation and export. Notwithstanding the increased generation and export of $CD4^+$ SP cells in SOCS1-deficient mice, pronounced cytokine-driven expansion of $CD8^+$ T cells outnumber $CD4^+$ T cells in the periphery.

Investigations into the role of SOCS1 in thymic selection have reported varying findings. Using the staphylococcal enterotoxin B (SEB) model of negative selection that specifically deletes TCRV$\beta8^+$ thymocytes in FTOC, Chong et al. reported that SOCS1 deficiency did not affect deletion of either $CD4^+$ or $CD8^+$ T cells bearing TCRVβ8 (Chong et al. 2003). Similarly, negative selection of $CD8^+$ SP cells expressing the male antigen-specific H-Y transgenic TCR in male mice was not affected by SOCS1 deficiency (Ramanathan et al. 2010). On the other hand, SOCS1 was reported to regulate both positive and negative selection of $CD4^+$ T cells expressing the MHC-II -restricted transgenic TCR AND, which is specific to the pigeon cytochrome C (PCC) and the I-E^k molecule (Catlett and Hedrick 2005). Loss of SOCS1 in *Rag1*$^{-/-}$TcrAND mice resulted in the generation of CD8 SP cells bearing the $V\alpha 11^+$ transgenic TCR, accompanied by a reduction in the number of CD4 SP cells. SOCS1 deficiency enabled the generation of these aberrant $CD8^+$ SP cells even in the absence of MHC-I molecules, but not in the absence of the non-selecting $H2^d$, suggesting that these cells were selected by MHC-II but diverted toward the inappropriate lineage in the absence of SOCS1 (Fig. 3). Upon adoptive transfer to mice bearing $H2^{b/s}$, in which TcrAND T cells are positively selected by

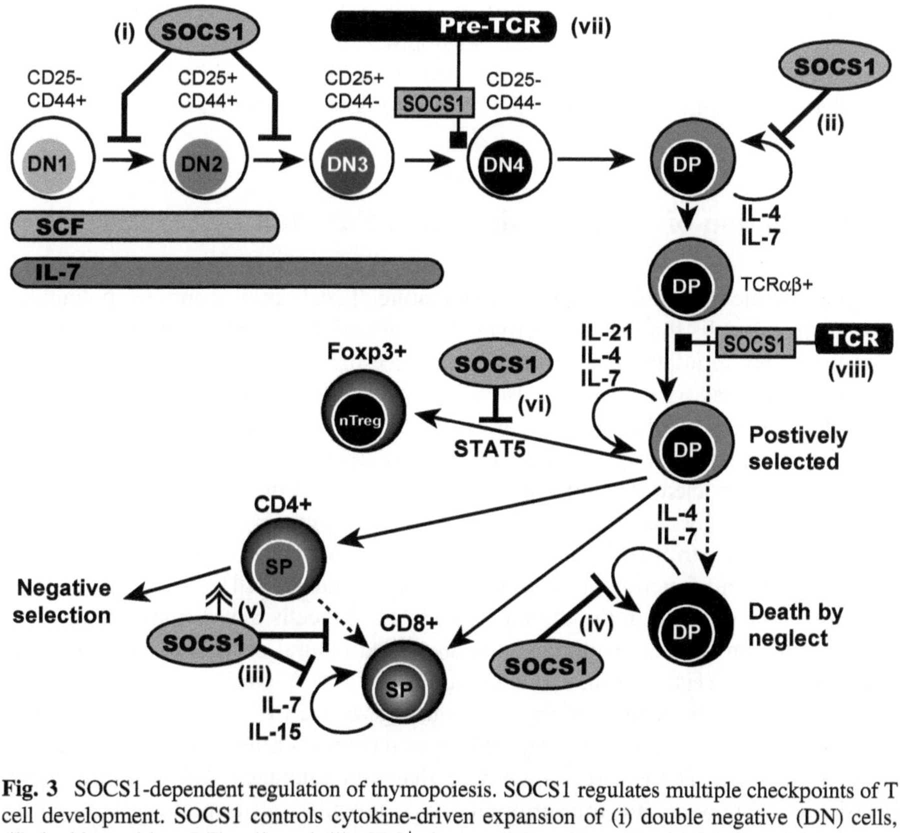

Fig. 3 SOCS1-dependent regulation of thymopoiesis. SOCS1 regulates multiple checkpoints of T cell development. SOCS1 controls cytokine-driven expansion of (i) double negative (DN) cells, (ii) double positive (DP) cells and (iii) $CD8^+$ single positive (SP) cells. SOCS1 is also implicated in preventing the cytokine-mediated rescue of DP cells that had failed positive selection (iv), and in promoting negative selection of $CD4^+$ SP cells by preventing their aberrant skewing toward $CD8^+$ SP cells (v). SOCS1 also controls generation of natural T regulatory cells (vi). Intriguingly, signaling via pre-TCR at the DN3 stage and TCRαβ at the DP stage downmodulate SOCS1 expression (vii, viii), presumably to allow selective expansion of signaled cells at these critical developmental stages

$H2^b$ and then negatively selected by $H\text{-}2^s$, SOCS1-deficient TcrAND $CD4^+$ and $CD8^+$ SP cells failed to undergo negative selection, suggesting that SOCS1 may serve to control generation of potentially autoreactive T cells. The defective positive and negative selection of TcrAND T cells caused by the lack of SOCS1 was reversed by IFNγ deficiency, demonstrating that SOCS1 shields developing thymocytes from the deleterious effects of systemic inflammation (Catlett and Hedrick 2005).

Collectively, SOCS1 impacts on four T cell developmental checkpoints (Fig. 3): (i) at the DN1-DN2 stage, to control signals delivered by SCF and IL-7; (ii) at the DN3 stage, to block the expansion of thymocytes that have not succeeded the β selection process; (iii) in DP thymocytes, to prevent cells that had failed the positive selection from undergoing cytokine-driven survival and expansion; (iv) in post-selection DP thymocytes, to ensure negative selection and prevent inappropriate developmental skewing toward the CD8 SP lineage. Even though SOCS1 deficiency would deregulate all these developmental stages, only the final outcome —accumulation $CD8^+$ SP cells—is discernible in SOCS1-deficient mice, presumably due to dynamic transition of the immature developmental stages.

5 Regulation of Treg Plasticity and Function by SOCS1

Despite thymic negative selection, the mature T cell pool contains potentially autoreactive T cells that can get activated in the periphery and become pathogenic. These cells are regulated by peripheral tolerance mechanisms, in which regulatory T cells (Treg) play a key role. Tregs inhibit activated T cells via direct contact and through secreted cytokines such as TGFβ and IL-10. The development of a multi-organ inflammatory disease in SOCS1-deficient mice, characterized by abundant mononuclear cell infiltration, could result not only from aberrant generation of potentially autoreactive T cells in the thymus and their activation by cytokines in the periphery, but also from the loss of SOCS1-dependent control of Treg stability and functions (Takahashi and Yoshimura 2014) (Fig. 4).

Most Tregs develop in the thymus from $CD4^+$ SP cells that display high avidity TCR interaction with autoantigens, and are called natural or thymus-derived Tregs (nTreg or tTreg) (Hsieh et al. 2012; Ohkura et al. 2013). Under certain conditions, naïve $CD4^+$ T cells in the periphery can also give rise to Tregs, which are designated as induced or peripheral Tregs (iTreg or pTreg) (Schmitt and Williams 2013). Both nTregs and iTregs are required to maintain tolerance and immune homeostasis. The Treg-specific transcription factor Foxp3 is an important regulator of Treg development and functions. However, Foxp3 expression is not static even in terminally differentiated Tregs. Following adoptive transfer to lymphopenic hosts or under inflammatory conditions, Tregs may lose Foxp3 expression and regulatory functions, become the so-called "exFoxp3" cells or "lapsed Tregs", acquire Th1 or Th17 effector-memory phenotype and exacerbate an inflammatory response (Sakaguchi et al. 2013).

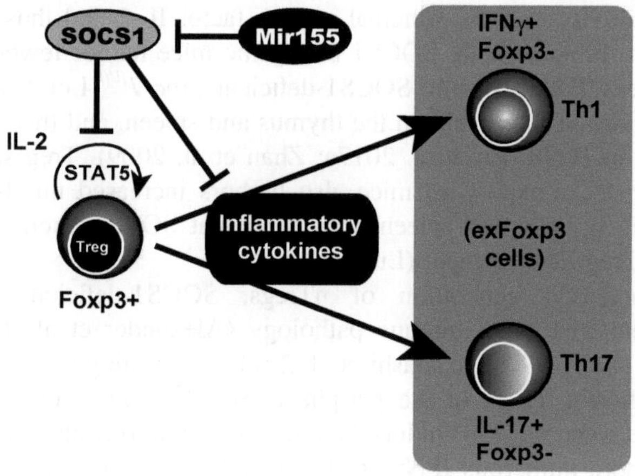

Fig. 4 Regulation of T regulatory cells by SOCS1. SOCS1 controls homeostasis of nTregs by regulating IL-2-mediated STAT5 signaling. SOCS1 also plays a crucial role in regulating Treg plasticity by controlling their responsiveness to inflammatory cytokines that leads to loss of Foxp3 expression (exFoxp3 cells) and differentiation toward Th1 and Th17 cells. SOCS1 also controls the Th1 differentiation pathway by inhibiting IFNγ signaling. By regulating SOCS1 expression, miR-155 exerts control over Treg homeostasis

The phenotypic and functional changes associated with such "Treg plasticity" arise from modulation of the Foxp3 promoter activity in uncommitted Foxp3$^+$ cells, which constitute a small proportion of Foxp3$^+$ cells, and their selective expansion (Hori 2014; Komatsu et al. 2009; Takahashi and Yoshimura 2014). In addition to the TCR signal-responsive Foxp3 promoter, three highly conserved non-coding DNA sequences (CNS 1-3) regulate Foxp3 expression and Treg plasticity. The TGFβ-responsive CNS1 has Smad2/3 binding sites for induced by TGFβ, while CNS2 binds TCR-induced CREB and IL-2-induced STAT5, and CNS3 has binding sites for c-Rel activated by TCR/CD28 co-stimulation. The CNS2 also harbors CpG islands that are hypomethylated in committed Tregs, which is further enhanced by IL-2 signaling (Chen et al. 2011). These Treg-specific demethylated regions (TSDR) become strongly methylated in exFoxp3 cells (reviewed in Takahashi and Yoshimura 2014). The stability of nTreg is regulated by stable expression of Foxp3, and sustained by Smad2/3, STAT5, and NF-κB signaling pathways. Takahashi and colleagues have shown that IFNγ-induced STAT1 can destabilize Foxp3 expression, leading to conversion of nTregs into Th1 cells, and that SOCS1 plays an essential role in preserving Treg functions by regulating the IFNγ signaling pathway (Takahashi et al. 2011).

Investigations into the Foxp3-regulated genes in Tregs led Rudensky and colleagues to identify miR155 and its target SOCS1 as key regulators of Treg homeostasis (Lu et al. 2017a). Loss of miR155 resulted in fewer Treg cells in the thymus and affected their homeostasis in the periphery. Mechanistically, loss of miR155 causes marked increase in SOCS1 expression in Tregs, leading to decreased sensitivity to their principal growth factor IL-2 and thus reducing their homeostatic fit. In agreement, SOCS1 transgenic mice harbor fewer nTregs in the thymus, whereas T cell-specific SOCS1-deficient (*Socs1*$^{fl/fl}$ Lck-Cre) mice harbor significantly more of these cells in the thymus and spleen, and this increase occurs independently of IFNγ (Lu et al. 2017a; Zhan et al. 2009). Treg-specific SOCS1 knockout (*Socs1*$^{fl/fl}$ Foxp3-Cre) mice also harbors increased numbers of Foxp3$^+$ CD4 T cells in thymus and spleen, indicating that SOCS1 exerts a cell-intrinsic regulation of Treg development (Lu et al. 2010).

Despite increased generation of nTregs, SOCS1-deficient mice develop IFNγ-dependent Th1 type immune pathology (Alexander et al. 1999; Lu et al. 2010; Marine et al. 1999; Takahashi et al. 2011), suggesting defective functioning of SOCS1-deficient Tregs in the periphery. *Socs1*$^{+/-}$ mice also develop severe dextran sulfate sodium (DSS)-induced colitis compared to control mice, with fewer Foxp3$^+$ and more numerous IFNγ$^+$ CD4$^+$ T cells in the intestinal lamina propria (Horino et al. 2008). In addition, SOCS1-deficient CD4$^+$ T cells display higher sensitivity to IFNγ-mediated blockade of Treg differentiation induced by TCR/CD28 stimulation in the presence of TGFβ (Horino et al. 2008). On the other hand, T cell-specific SOCS1 knockout mice showed resistance to experimental autoimmune encephalomyelitis (EAE), a Th17-mediated disease, due to the requirement of SOCS1 to preserve Th17 cell differentiation by preventing Th1 skewing (Tanaka et al. 2008). All these observations suggested a complex regulation of Th cell differentiation and Treg functions by SOCS1 in the periphery. Yoshimura and colleagues elucidated the underlying mechanisms by addressing the role of SOCS1 in regulating nTreg plasticity (Takahashi et al. 2011). In the classical in vivo functional assay, Tregs from *Socs1*$^{fl/fl}$ Lck-Cre mice failed to suppress the induction of colitis by wildtype CD4$^+$ T cells in *Rag1*$^{-/-}$ mice, accompanied by faster loss of Foxp3 and other markers of suppressive activity (CD25 and CTLA4) compared to wildtype Tregs (Takahashi et al. 2011). In this setting, both the Foxp3-maintaining and the Foxp3-losing fractions of SOCS1-deficient Tregs produced higher amounts of IFNγ, despite maintaining IL-10 production. SOCS1-deficient Tregs showed rapid methylation of the CNS2 region of the Foxp3 locus in vivo and in vitro, which was prevented by ablation of the *Ifng* gene, indicating that SOCS1 is essential to thwart conversion of Tregs to IFNγ producing exFoxp3 cells. However, *Socs1*$^{-/-}$*Ifng*$^{-/-}$ Tregs, which did not become exFoxp3 cells in lymphopenic mice, also failed to inhibit colitis induction. The inability of *Socs1*$^{-/-}$*Ifng*$^{-/-}$ Tregs to maintain their regulatory function was explained by their propensity of differentiating toward pathogenic Th17 cells in the absence of IFNγ. In agreement with the role of SOCS1 modulating both Treg and Th17 differentiation, transfection of naïve CD4$^+$ T cells with pre-miR155 promoted, whereas anti-miR155 inhibited, the development of Tregs and Th17 cells under the respective polarizing conditions, accompanied by modulation of SOCS1 and its targets STAT5 and STAT3 (Yao et al. 2012). Increased activation of STAT1 and STAT3 in SOCS1-deficient Tregs, the former by IFNγ and the latter presumably by inflammatory cytokines, is believed to underlie their defective immune regulatory function (Takahashi and Yoshimura 2014). The molecular mechanisms by which

STAT1 and STAT3 destabilize Foxp3 expression and disrupt Treg stability are unclear. Although cytokine-induced STAT molecules are the principal inducers of *Socs1,* they synergize with TCR-stimulated NF-κB to upregulate SOCS1 expression in Tregs, as the deficiency of Ubc13, a ubiquitin conjugating enzyme involved in activating IKK upstream of NF-κB, resulted in a similar loss of Treg function as caused by SOCS1 deficiency (Chang et al. 2012). Hence, SOCS1 plays a crucial role in protecting Tregs from converting to effector T cells under the harmful influence of inflammatory cytokines (Fig. 4). In addition to becoming pathogenic Th1 and Th17 cells themselves under inflammatory conditions, exFoxp3 cells can provide help to activate naïve CD8 T cells (Bailey-Bucktrout et al. 2013; Komatsu et al. 2014; Sharma et al. 2010), raising such a possibility in the activation of potentially autoreactive $CD8^+$ T cells in SOCS1-deficient mice.

6 Influence of SOCS1 on Antigen Presenting Cell Functions

Early investigations into the lethal phenotype of SOCS1-deficient mice revealed that SOCS1 is an important regulator of macrophage activation by innate immune stimuli such as LPS and CpG DNA, and by IFNγ. In response to these stimuli, SOCS1-deficient macrophages produce abundant quantities of inflammatory cytokines such as TNFα and IL-6, as well as IL-12, a key cytokine in promoting the Th1 immune response (Kinjyo et al. 2002; Nakagawa et al. 2002). SOCS1-deficient macrophages show increased ability to control *M. tuberculosis* via increased IFNγ production, which in turn relies on strong MyD88-dependent TLR2 activation and enhanced IL-12 production (Carow et al. 2011). The heightened activation of SOCS1-deficient macrophages by LPS via the toll like receptor 4 (TLR4), results from increased activation of NF-κB and STAT1 (Kinjyo et al. 2002; Nakagawa et al. 2002). SOCS1 blocks LPS-induced macrophage activation by promoting ubiquitination and proteasomal degradation of the key signaling adaptor MAL/TIRAP in the My88-dependent pathway, and possibly by interfering with the key TLR4 signaling adaptor TRAF6 further downstream, thereby blocking NF-κB activation (Kinjyo et al. 2002; Mansell et al. 2006). SOCS1 may also directly inhibit NF-κB signaling by promoting ubiquitination and degradation of the p65RelA component as shown in other cell types (Maine et al. 2007; Ryo et al. 2003). The same mechanisms likely underlie SOCS1-dependent regulation LPS signaling in dendritic cells (DC), which also produce copious amount of IL-12 and IFNγ in the absence of SOCS1 (see below).

SOCS1 exerts control over differentiation of DCs and their functions in stimulating T cell responses. Splenic DCs from SOCS1-deficient mice show an increased frequency of $CD8\alpha^+$ conventional DCs, which are critical for cross-priming $CD8^+$ T cells in vivo (Hanada et al. 2005; Heath and Carbone 2009; Tsukada et al. 2005). This increment could arise from increased generation of these cells due to enhanced

Flt3 and GM-CSF signaling caused by SOCS1 deficiency, as FLT3 is required for the development of CD8α^+ DCs and GM-CSF skews DC development away from plasmacytoid DCs (Wu and Liu 2007). SOCS1 inhibits Flt3-induced mitogenic responses and blocks GM-CSF signaling by promoting ubiquitination and proteasomal degradation of the β_c chain of the GM-CSF receptor (Bunda et al. 2013; De Sepulveda et al. 1999). Expression of the *Socs1* gene increases during culture with GM-CSF and IL-4, and is further augmented by LPS (Jackson et al. 2004; Shen et al. 2004). On the other hand, LPS-induced SOCS1 has been shown to block GM-CSF-mediated DC differentiation from human CD14$^+$ monocytes and murine bone marrow precursors (Bartz et al. 2006). In addition to modulating DC generation, SOCS1 may control IL-15-dependent maturation and survival of DCs (Dubois et al. 2005; Hanada et al. 2005). Intriguingly, CD8α^+ DCs enriched from spleen show much lower levels of SOCS1 expression than CD4$^+$ or CD4$^-$CD8$^-$ DC subsets, which might underlie the intrinsically stronger ability of CD8α+ DCs to stimulate Th1 responses, and its further augmentation by SOCS1 deficiency (Hanada et al. 2005).

In response to TLR stimulation by LPS or CpG DNA, SOCS1-deficient DCs produce abundant quantities of IL-12 as well as IFNγ, which could contribute to the Th1-type immunopathology observed in SOCS1-deficient mice (Hanada et al. 2005; Tsukada et al. 2005). In addition to the Th1 cytokines, SOCS1-deficient DCs produce abundant quantities of the TNF family growth factors for B cells BAFF/BLyS and APRIL, which likely contribute to the elevated levels of autoantibodies in SOCS1-deficient mice (Hanada et al. 2003). Even though SOCS1-deficient DCs do not efficiently support CD4$^+$ T cell activation in vitro due to increased induction of indoleamine-2,3-dioxygenase (IDO) by autocrine IFNγ, they may efficiently promote T activation in vivo wherein the use of IFNγ by other cells would diminish its availability to T cells (Hanada et al. 2005; Tsukada et al. 2005). This notion is supported by strong proliferation of allogeneic T cells cultured with bone marrow-derived DCs (BMDC) from *Socs1*$^{-/-}$*Ifng*$^{-/-}$ mice, and the heightened ability of SOCS1-deficient and *Socs1*-silenced DCs to stimulate IFNγ-producing Th1 cells and CD8$^+$ T cells in vivo (Evel-Kabler et al. 2006; Hanada et al. 2003, 2005; Shen et al. 2004). In mice bearing a polyclonal TCR repertoire, immunization with *Socs1*-silenced or SOCS1-deficient BMDCs that were pulsed with antigenic peptides led to efficient activation antigen-specific CD8$^+$ T cells, which displayed increased proliferation, IFNγ and IL-2 production and effector functions, and caused regression of tumors expressing the cognate antigen (Evel-Kabler et al. 2006; Hanada et al. 2005; Shen et al. 2004). The ability of *Socs1*-silenced DCs to activate antigen-specific CD8$^+$ T cells in vivo was further augmented by innate immune stimuli such as LPS that stimulate DC maturation (Evel-Kabler et al. 2006; Shen et al. 2004).

The ability of SOCS1 to control T cell activation at the level of DCs may operate via limiting and modulating their generation, maturation, antigen presentation, expression of costimulatory signals, and production of cytokines. Of these, increased production of IL-12 and IFNγ by SOCS1-deficient DCs has been well documented. The propensity for SOCS1-deficient DCs to produce copious amounts of IFNγ is attributed to increased STAT1 signaling and expression of the Eomes transcription factor (Hanada et al. 2005). Using DCs derived from mice lacking the IL-12 subunit p35, or the IL-12Rβ, Chen and colleagues have shown that SOCS1 regulates the quantity and duration of LPS-induced IL-12 production in DCs by controlling autocrine IL-12 signaling (Evel-Kabler et al. 2006). Splenic DCs isolated from SOCS1-deficient mice express elevated levels of costimulatory molecules CD80, CD86 and CD40, and MHC class-II, although their expression levels were not altered in *Socs1*-silenced BMDCs at steady state (Evel-Kabler et al. 2006; Hanada et al. 2003). However, *Socs1*-silenced BMDCs exposed to *Candida albicans* show increased expression of costimulatory molecules and MHC-II (Shi et al. 2015). A recent study has shown that constitutive SOCS1 expression in DCs downmodulates MHC and costimulatory molecules, and that these DCs elicit T cell hyporesponsiveness and facilitate the survival of allogeneic pancreatic islets (Lu et al. 2017b). SOCS1 likely regulates the expression of MHC and costimulatory molecules by regulating both TLR and autocrine IFNγ signaling. Few studies have addressed the influence of SOCS1 on antigen processing and presentation. As SOCS1-deficient BMDCs show increased expression of several IFNγ-responsive genes at the basal level (Hanada et al. 2005), and IFNγ is a key modulator of MHC-I antigen processing pathway, it is quite likely that SOCS1 may regulate antigen processing and presentation functions of DCs.

7 SOCS1 as an Immune Checkpoint Molecule

SOCS1 is critical to control aberrant activation of potentially autoreactive T cells and preserve self-tolerance. This has been clearly demonstrated using SOCS1-deficient mice harboring TCR transgenic $CD8^+$ T cells (Davey et al. 2005; Ramanathan et al. 2010). In mice expressing the cognate transgenic antigen in pancreatic islets, SOCS1-deficient TCR transgenic CD8+ T cells cause autoimmune type-1 diabetes whereas SOCS1-sufficient cells do not. Similarly, SOCS1-deficient mice harboring Pmel-1 TCR transgenic $CD8^+$ T cells develop reactivity toward the endogenous melanocyte antigen gp100 expressed in normal keratinocytes of *Rag1*$^{-/-}$ mice, resulting in widespread cutaneous inflammatory lesions (Rodriguez et al. 2013). Hence, even though the activated phenotype of polyclonal $CD8^+$ T cells in SOCS1-deficient cells might arise from antigen non-specific activation by cytokines (Ilangumaran et al. 2003a; Metcalf et al. 2002; Starr et al. 1998), this activated T cell pool may contain potentially autoreactive T cells. In a mouse model of acute inflammatory arthritis induced by methylated bovine serum albumin (mBSA) BSA, SOCS1-deficient mice develop severe disease accompanied by not only increased myeloid cell infiltration but also an accumulation of activated, mBSA-specific $CD4^+$ T cells in the draining lymph nodes (Egan et al. 2003). The systemic lupus erythematosus (SLE)-like syndrome that develops in SOCS1-deficient mice does not occur in the absence of $CD4^+$ T cells, suggesting that the general inflammatory conditions caused by SOCS1 deficiency stimulate autoreactive $CD4^+$ T cells that provide help to B lymphocytes (Fujimoto et al. 2004; Hanada et al. 2003). Similarly, decreased activation of Th1 and Th17 cells in mice lacking miR155 (and thus expressing more SOCS1), following immunization with the myelin oligodendrocyte glycoprotein (MOG) peptide MOG_{35-55}, could arise not only from impaired Th cell differentiation, but also from impaired antigen-specific T cell activation (Murugaiyan et al. 2011; O'Connell et al. 2010). Moreover, adoptively transferred wild type $CD4^+$ T cells undergo efficient activation in *Mir155*$^{-/-}$ mice and cause disease, suggesting that the cytokine products of SOCS1-deficient APC are sufficient to activate autoreactive $CD4^+$ T cells (O'Connell et al. 2010). Thus, SOCS1 deficiency either in T cells or loss of SOCS1-dependent control of APCs could promote activation of autoreactive T cells.

As discussed in the previous section, loss of SOCS1 expression in DCs or antitumor T cells would boost antitumor CTL response. However, tumor cells often lose SOCS1 expression and this may dampen antitumor immunity. In fact, SOCS1 has been extensively studied for its role as tumor suppressor in primary human cancers and mouse models. Following the seminal finding that the *SOCS1* gene promoter is repressed by CpG methylation in hepatocellular carcinoma specimens (Yoshikawa et al. 2001), numerous reported have shown similar findings in other cancers including acute and chronic myeloid leukemia, breast cancer, glioblastoma, ovarian cancer, etc., (reviewed in (Inagaki-Ohara et al. 2013; Sasi et al. 2014; Trengove and Ward, 2013)). SOCS1 expression is also downmodulated by microRNAs including miR155, miR19a, and miR30d in breast cancer, multiple myeloma and prostate cancer (Jiang et al. 2010; Kobayashi et al. 2012; Pichiorri et al. 2008). In melanoma and prostate cancer, SOCS1 protein expression correlates with disease severity and metastasis (Chevrier et al. 2017; Huang et al. 2008). Observations in SOCS1-deficient mice also support the tumor suppressor function of SOCS1: *Socs1*$^{-/-}$*Ifng*$^{-/-}$ mice that harbor activated T cells are more susceptible to radiation-induced T cell leukemia (Metcalf et al. 2002). On the other hand, mice lacking SOCS1 in all tissues except T cells spontaneously develop colorectal carcinoma, which is dependent on IFNγ signaling in the colonic epithelium (Hanada et al. 2006). Both *Socs1*$^{-/-}$*Ifng*$^{-/-}$ mice and hepatocyte-specific SOCS1-deficient mice show high susceptibility to chemically induced hepatocellular carcinoma (Yeganeh et al. 2016; Yoshida et al. 2004). While hematological malignancies are thwarted by SOCS1 mainly by putting brakes on JAK kinases, in epithelial cancers SOCS1 blocks oncogenic receptor tyrosine kinase signaling and diverse signaling molecules that are beyond the scope of this discussion. Clearly, increased tissue inflammation in SOCS1-deficient tissues contributes to neoplastic transformation and cancer growth (Inagaki-Ohara et al. 2013). However, it is possible that SOCS1 deficiency in cancer cells may also indirectly hinder antitumor immune response through "adaptive immune resistance", a process by which cancer cells inhibit tumor antigen-specific T cells by exploiting IFNγ produced by immune cells (Ribas 2015). Hence, even though loss of SOCS1-dependent control of IFNγ signaling would theoretically enhance antigen presentation in cancer cells, their increased IFNγ sensitivity can boost the expression of PD-L1 and IDO to hamper antitumor T cell response (McGray et al. 2014; Spranger et al. 2013). Thus, loss of SOCS1 in cancer cells might exert an indirect checkpoint blockade on antitumor immunity. It will be worth investigating how SOCS1 deficiency in tumor cells impacts on antitumor immune responses.

8 Targeting SOCS1 in Autoimmunity and Cancer

Relieving the checkpoint functions of SOCS1 to boost antitumor immunity would require its repression in APCs or antitumor T cells, whereas its functional reconstitution in inflammatory cells would be needed to promote its checkpoint functions in case of auto-inflammatory diseases. As an intracellular signal regulator, SOCS1 differs from the cell surface receptor–ligand pairs of classical immune checkpoint regulators (Mahoney et al. 2015). SOCS1 also lacks enzymatic activity, and thus is not an ideal "druggable" target. Hence, approaches other than antibodies or small molecule modulators are needed to modulate SOCS1 expression in target cells.

Johnson and colleagues developed a cell-penetrating derivative of the KIR region of SOCS1 that functions as SOCS1 mimetic in inhibiting cytokine responses (Ahmed et al. 2015). This SOCS1 mimetic has been shown to inhibit EAE by suppressing MOG peptide-specific T cell activation and IL-17A production (Jager et al. 2011). The SOCS1 KIR peptide also partially rescues the functional defects in Ubc13 deficient Tregs, which fail to upregulate SOCS1 due to impaired NF-κB signaling, in vitro and in vivo (Chang et al. 2012). Similar SOCS1 mimetic peptides have been developed with the idea of using them in topical applications, especially for cutaneous inflammatory conditions such as psoriasis (Madonna et al. 2013). A cell-penetrating version of the full-length SOCS1 molecule (CP-SOCS1) has also been reported that inhibits pro-inflammatory cytokine production in cell lines (DiGiandomenico et al. 2009). Delivery through liposomes may be another option, as SOCS1 released from alveolar macrophages has been shown to inhibit IFNγ signaling in airway epithelial cells (Bourdonnay et al. 2015). These SOCS1 delivery approaches may be particularly useful in controlling inflammatory manifestations of unknown etiology and hinder activation of potentially autoreactive cells.

SOCS1 can be exploited to boost antitumor T cell responses at least in two ways (Fig. 5). As documented in animal models, silencing SOCS1 in DCs could be exploited to stimulate antitumor T cell response in the setting of DC-based cancer vaccines to improve their immunogenic potential (Anguille et al. 2014; Evel-Kabler et al. 2006; Palucka and Banchereau 2012; Shen et al. 2004). An alternative approach could use *SOCS1* gene silencing in T cells in the setting of adoptive cell therapy to select for poly-specific antitumor $CD8^+$ T cells from tumor-infiltrating lymphocytes (TIL) (Ji et al. 2015; Palmer and Restifo 2009; Rosenberg and Restifo 2015). As SOCS1-deficient $CD8^+$ T cells show increased cytokine responsiveness and cytokine pre-stimulation increases antigen sensitivity (Ramanathan et al. 2010, 2011), lowering SOCS1 expression in TILs would facilitate the cytokine-mediated expansion of $CD8^+$ T cells and also would allow selection of those bearing low

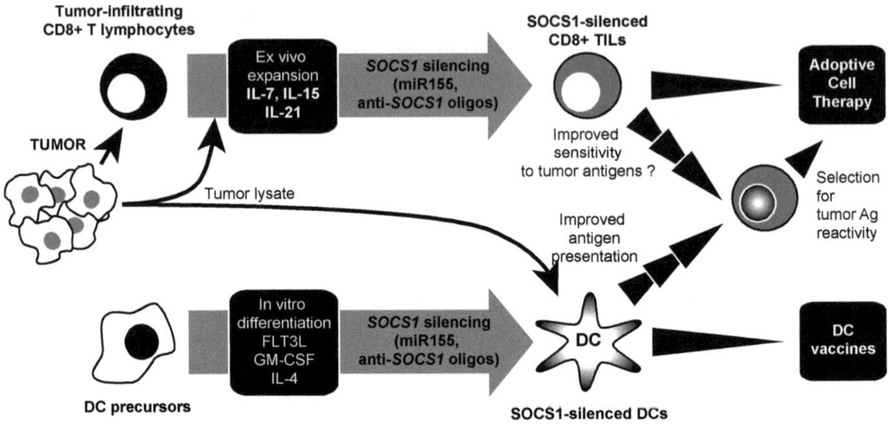

Fig. 5 Targeting SOCS1 for cancer immunotherapy. SOCS1 silencing in $CD8^+$ T cells that are expanded from tumor-infiltrating lymphocytes (TILs) could be useful to generate tumor antigen reactive CTLs for adoptive cell therapy. Alternatively, SOCS1 silencing in dendritic cells (DC) would improve their antigen presenting potential that can be exploited for antitumor DC vaccines or for ex vivo selection and expansion of antitumor CTLS for adoptive cell therapy

avidity TCR toward unknown tumor neoantigens. On the other hand, tackling the possible adaptive immune resistance in SOCS1-deficient tumors would require restoration of SOCS1 function. While the SOCS1 mimetic KIR peptide inhibits cancer cells in vitro, *SOCS1* gene therapy has shown promise in suppressing tumor growth (Flowers et al. 2005; Liu et al. 2013; Natatsuka et al. 2015; Souma et al. 2012). However, as the feasibility of delivering SOCS1 to all cells of primary cancer could be uncertain, combinatorial checkpoint therapy using anti-PD-1 or anti-PDL1 antibody and IDO inhibitors (Mahoney et al. 2015) would be a more practical approach to deal with the adaptive immune resistance in SOCS1-deficient tumors.

Acknowledgments SI and SR acknowledge funding support from Natural Sciences and Engineering Research Council of Canada (NSERC) Discovery Grant.

References

Acuto O, Di Bartolo V, Michel F (2008) Tailoring T-cell receptor signals by proximal negative feedback mechanisms. Nat Rev Immunol 8:699–712

Acuto O, Michel F (2003) CD28-mediated co-stimulation: a quantitative support for TCR signalling. Nat Rev Immunol 3:939–951

Ahmed CM, Larkin J 3rd, Johnson HM (2015) SOCS1 mimetics and antagonists: a complementary approach to positive and negative regulation of immune function. Front Immunol 6:183

Alexander WS (2002) Suppressors of cytokine signalling (SOCS) in the immune system. Nat Rev Immunol 2:1–7

Alexander WS, Starr R, Fenner JE, Scott CL, Handman E, Sprigg NS, Corbin JE, Cornish AL, Darwiche R, Owczarek CM et al (1999) SOCS1 is a critical inhibitor of interferon gamma signaling and prevents the potentially fatal neonatal actions of this cytokine. Cell 98:597–608

Anguille S, Smits EL, Lion E, van Tendeloo VF, Berneman ZN (2014) Clinical use of dendritic cells for cancer therapy. Lancet Oncol 15:e257–e267

Babon JJ, Sabo JK, Zhang JG, Nicola NA, Norton RS (2009) The SOCS box encodes a hierarchy of affinities for Cullin5: implications for ubiquitin ligase formation and cytokine signalling suppression. J Mol Biol 387:162–174

Babon JJ, Kershaw NJ, Murphy JM, Varghese LN, Laktyushin A, Young SN, Lucet IS, Norton RS, Nicola NA (2012) Suppression of cytokine signaling by SOCS3: characterization of the mode of inhibition and the basis of its specificity. Immunity 36:239–250

Babon JJ, Lucet IS, Murphy JM, Nicola NA, Varghese LN (2014a) The molecular regulation of Janus kinase (JAK) activation. Biochem J 462:1–13

Babon JJ, Varghese LN, Nicola NA (2014b) Inhibition of IL-6 family cytokines by SOCS3. Semin Immunol 26:13–19

Baetz A, Koelsche C, Strebovsky J, Heeg K, Dalpke AH (2008) Identification of a nuclear localization signal in suppressor of cytokine signaling 1. Faseb J 22:4296–4305

Bailey-Bucktrout SL, Martinez-Llordella M, Zhou X, Anthony B, Rosenthal W, Luche H, Fehling HJ, Bluestone JA (2013) Self-antigen-driven activation induces instability of regulatory T cells during an inflammatory autoimmune response. Immunity 39:949–962

Baniyash M (2004) TCR zeta-chain downregulation: curtailing an excessive inflammatory immune response. Nat Rev Immunol 4:675–687

Bartz H, Avalos NM, Baetz A, Heeg K, Dalpke AH (2006) Involvement of suppressors of cytokine signaling in toll-like receptor-mediated block of dendritic cell differentiation. Blood 108:4102–4108

Baxter AG, Hodgkin PD (2002) Activation rules: the two-signal theories of immune activation. Nat Rev Immunol 2:439–446

Ben-Sasson SZ, Hu-Li J, Quiel J, Cauchetaux S, Ratner M, Shapira I, Dinarello CA, Paul WE (2009) IL-1 acts directly on CD4 T cells to enhance their antigen-driven expansion and differentiation. Proc Natl Acad Sci USA 106:7119–7124

Bourdonnay E, Zaslona Z, Penke LR, Speth JM, Schneider DJ, Przybranowski S, Swanson JA, Mancuso P, Freeman CM, Curtis JL, Peters-Golden M (2015) Transcellular delivery of vesicular SOCS proteins from macrophages to epithelial cells blunts inflammatory signaling. J Exp Med 212:729–742

Bourette RP, De Sepulveda P, Arnaud S, Dubreuil P, Rottapel R, Mouchiroud G (2001) Suppressor of cytokine signaling 1 interacts with the macrophage colony-stimulating factor receptor and negatively regulates its proliferation signal. J Biol Chem 276:22133–22139

Boyle K, Egan P, Rakar S, Willson TA, Wicks IP, Metcalf D, Hilton DJ, Nicola NA, Alexander WS, Roberts AW, Robb L (2007) The SOCS box of suppressor of cytokine signaling-3 contributes to the control of G-CSF responsiveness in vivo. Blood 110:1466–1474

Bunda S, Kommaraju K, Heir P, Ohh M (2013) SOCS-1 mediates ubiquitylation and degradation of GM-CSF receptor. PLoS ONE 8:e76370

Carow B, Ye X, Gavier-Widen D, Bhuju S, Oehlmann W, Singh M, Skold M, Ignatowicz L, Yoshimura A, Wigzell H, Rottenberg ME (2011) Silencing suppressor of cytokine signaling-1 (SOCS1) in macrophages improves Mycobacterium tuberculosis control in an interferon-gamma (IFN-gamma)-dependent manner. J Biol Chem 286:26873–26887

Catlett IM, Hedrick SM (2005) Suppressor of cytokine signaling 1 is required for the differentiation of CD4+ T cells. Nat Immunol 6:715–721

Chang JH, Xiao Y, Hu H, Jin J, Yu J, Zhou X, Wu X, Johnson HM, Akira S, Pasparakis M et al (2012) Ubc13 maintains the suppressive function of regulatory T cells and prevents their conversion into effector-like T cells. Nat Immunol 13:481–490

Chen L, Flies DB (2013) Molecular mechanisms of T cell co-stimulation and co-inhibition. Nat Rev Immunol 13:227–242
Chen Q, Kim YC, Laurence A, Punkosdy GA, Shevach EM (2011) IL-2 controls the stability of Foxp3 expression in TGF-beta-induced Foxp3+ T cells in vivo. J Immunol 186:6329–6337
Chevrier M, Bobbala D, Villalobos-Hernandez A, Khan MG, Ramanathan S, Saucier C, Ferbeyre G, Geha S, Ilangumaran S (2017) Expression of SOCS1 and the downstream targets of its putative tumor suppressor functions in prostate cancer. BMC Cancer 17:157
Chinen T, Kobayashi T, Ogata H, Takaesu G, Takaki H, Hashimoto M, Yagita H, Nawata H, Yoshimura A (2006) Suppressor of cytokine signaling-1 regulates inflammatory bowel disease in which both IFNgamma and IL-4 are involved. Gastroenterology 130:373–388
Chong MM, Cornish AL, Darwiche R, Stanley EG, Purton JF, Godfrey DI, Hilton DJ, Starr R, Alexander WS, Kay TW (2003) Suppressor of cytokine signaling-1 is a critical regulator of interleukin-7-dependent CD8+ T cell differentiation. Immunity 18:475–487
Chong MM, Metcalf D, Jamieson E, Alexander WS, Kay TW (2005) Suppressor of cytokine signaling-1 in T cells and macrophages is critical for preventing lethal inflammation. Blood 106:1668–1675
Cornish AL, Chong MM, Davey GM, Darwiche R, Nicola NA, Hilton DJ, Kay TW, Starr R, Alexander WS (2003a) Suppressor of cytokine signaling-1 regulates signaling in response to interleukin-2 and other gamma c-dependent cytokines in peripheral T cells. J Biol Chem 278:22755–22761
Cornish AL, Davey GM, Metcalf D, Purton JF, Corbin JE, Greenhalgh CJ, Darwiche R, Wu L, Nicola NA, Godfrey DI et al (2003b) Suppressor of cytokine signaling-1 has IFN-gamma-independent actions in T cell homeostasis. J Immunol 170:878–886
Cox MA, Harrington LE, Zajac AJ (2011) Cytokines and the inception of CD8 T cell responses. Trends Immunol 32:180–186
Croker BA, Krebs DL, Zhang JG, Wormald S, Willson TA, Stanley EG, Robb L, Greenhalgh CJ, Forster I, Clausen BE et al (2003) SOCS3 negatively regulates IL-6 signaling in vivo. Nat Immunol 4:540–545
Curtsinger JM, Lins DC, Mescher MF (2003) Signal 3 determines tolerance versus full activation of naive CD8 T cells: dissociating proliferation and development of effector function. J Exp Med 197:1141–1151
Curtsinger JM, Lins DC, Johnson CM, Mescher MF (2005a) Signal 3 tolerant CD8 T cells degranulate in response to antigen but lack granzyme B to mediate cytolysis. J Immunol 175:4392–4399
Curtsinger JM, Valenzuela JO, Agarwal P, Lins D, Mescher MF (2005b) Type I IFNs provide a third signal to CD8 T cells to stimulate clonal expansion and differentiation. J Immunol 174:4465–4469
Curtsinger JM, Mescher MF (2010) Inflammatory cytokines as a third signal for T cell activation. Curr Opin Immunol 22:333–340
Davey GM, Starr R, Cornish AL, Burghardt JT, Alexander WS, Carbone FR, Surh CD, Heath WR (2005) SOCS-1 regulates IL-15-driven homeostatic proliferation of antigen-naive CD8 T cells, limiting their autoimmune potential. J Exp Med 202:1099–1108
De Sepulveda P, Okkenhaug K, Rose JL, Hawley RG, Dubreuil P, Rottapel R (1999) Socs1 binds to multiple signalling proteins and suppresses steel factor- dependent proliferation. EMBO J 18:904–915
DiGiandomenico A, Wylezinski LS, Hawiger J (2009) Intracellular delivery of a cell-penetrating SOCS1 that targets IFN-gamma signaling. Sci Signal 2:ra37
Dubois SP, Waldmann TA, Muller JR (2005) Survival adjustment of mature dendritic cells by IL-15. Proc Natl Acad Sci USA 102:8662–8667
Egan PJ, Lawlor KE, Alexander WS, Wicks IP (2003) Suppressor of cytokine signaling-1 regulates acute inflammatory arthritis and T cell activation. J Clin Invest 111:915–924
Endo TA, Masuhara M, Yokouchi M, Suzuki R, Sakamoto H, Mitsui K, Matsumoto A, Tanimura S, Ohtsubo M, Misawa H et al (1997) A new protein containing an SH2 domain that inhibits JAK kinases. Nature 387:921–924

Evel-Kabler K, Song XT, Aldrich M, Huang XF, Chen SY (2006) SOCS1 restricts dendritic cells' ability to break self tolerance and induce antitumor immunity by regulating IL-12 production and signaling. J Clin Invest 116:90–100

Feng ZP, Chandrashekaran IR, Low A, Speed TP, Nicholson SE, Norton RS (2012) The N-terminal domains of SOCS proteins: a conserved region in the disordered N-termini of SOCS4 and 5. Proteins 80:946–957

Flowers LO, Subramaniam PS, Johnson HM (2005) A SOCS-1 peptide mimetic inhibits both constitutive and IL-6 induced activation of STAT3 in prostate cancer cells. Oncogene 24:2114–2120

Fujimoto M, Naka T (2003) Regulation of cytokine signaling by SOCS family molecules. Trends Immunol 24:659–666

Fujimoto M, Naka T, Nakagawa R, Kawazoe Y, Morita Y, Tateishi A, Okumura K, Narazaki M, Kishimoto T (2000) Defective thymocyte development and perturbed homeostasis of T cells in STAT-induced STAT inhibitor-1/suppressors of cytokine signaling-1 transgenic mice. J Immunol 165:1799–1806

Fujimoto M, Tsutsui H, Xinshou O, Tokumoto M, Watanabe D, Shima Y, Yoshimoto T, Hirakata H, Kawase I, Nakanishi K et al (2004) Inadequate induction of suppressor of cytokine signaling-1 causes systemic autoimmune diseases. Int Immunol 16:303–314

Gagnon J, Ramanathan S, Leblanc C, Cloutier A, McDonald PP, Ilangumaran S (2008) IL-6, in synergy with IL-7 or IL-15, stimulates TCR-independent proliferation and functional differentiation of CD8+ T lymphocytes. J Immunol 180:7958–7968

Gagnon J, Ramanathan S, Leblanc C, Ilangumaran S (2007) Regulation of IL-21 signaling by suppressor of cytokine signaling-1 (SOCS1) in CD8(+) T lymphocytes. Cell Signal 19:806–816

Gui Y, Yeganeh M, Donates YC, Tobelaim WS, Chababi W, Mayhue M, Yoshimura A, Ramanathan S, Saucier C, Ilangumaran S (2015) Regulation of MET receptor tyrosine kinase signaling by suppressor of cytokine signaling 1 in hepatocellular carcinoma. Oncogene 34:5718–5728

Hanada T, Yoshida H, Kato S, Tanaka K, Masutani K, Tsukada J, Nomura Y, Mimata H, Kubo M, Yoshimura A (2003) Suppressor of cytokine signaling-1 is essential for suppressing dendritic cell activation and systemic autoimmunity. Immunity 19:437–450

Hanada T, Tanaka K, Matsumura Y, Yamauchi M, Nishinakamura H, Aburatani H, Mashima R, Kubo M, Kobayashi T, Yoshimura A (2005) Induction of hyper Th1 cell-type immune responses by dendritic cells lacking the suppressor of cytokine signaling-1 gene. J Immunol 174:4325–4332

Hanada T, Kobayashi T, Chinen T, Saeki K, Takaki H, Koga K, Minoda Y, Sanada T, Yoshioka T, Mimata H et al (2006) IFNgamma-dependent, spontaneous development of colorectal carcinomas in SOCS1-deficient mice. J Exp Med 203:1391–1397

Haring JS, Badovinac VP, Harty JT (2006) Inflaming the CD8+ T cell response. Immunity 25:19–29

Heath WR, Carbone FR (2009) Dendritic cell subsets in primary and secondary T cell responses at body surfaces. Nat Immunol 10:1237–1244

Hori S (2014) Lineage stability and phenotypic plasticity of Foxp3(+) regulatory T cells. Immunol Rev 259:159–172

Horino J, Fujimoto M, Terabe F, Serada S, Takahashi T, Soma Y, Tanaka K, Chinen T, Yoshimura A, Nomura S et al (2008) Suppressor of cytokine signaling-1 ameliorates dextran sulfate sodium-induced colitis in mice. Int Immunol 20:753–762

Hsieh CS, Lee HM, Lio CW (2012) Selection of regulatory T cells in the thymus. Nat Rev Immunol 12:157–167

Huang FJ, Steeg PS, Price JE, Chiu WT, Chou PC, Xie K, Sawaya R, Huang S (2008) Molecular basis for the critical role of suppressor of cytokine signaling-1 in melanoma brain metastasis. Can Res 68:9634–9642

Hwang MN, Min CH, Kim HS, Lee H, Yoon KA, Park SY, Lee ES, Yoon S (2007) The nuclear localization of SOCS6 requires the N-terminal region and negatively regulates Stat3 protein levels. Biochem Biophys Res Commun 360:333–338

Ilangumaran S, Gagnon J, Leblanc C, Poussier P, Ramanathan S (2010) Increased generation of CD8 single positive cells in SOCS1-deficient thymus does not proportionately increase their export. Immunol Lett 132:12–17

Ilangumaran S, Ramanathan S, La Rose J, Poussier P, Rottapel R (2003a) Suppressor of cytokine signaling 1 regulates IL-15 receptor signaling in CD8 + CD44high memory T lymphocytes. J Immunol 171:2435–2445

Ilangumaran S, Ramanathan S, Ning T, La Rose J, Reinhart B, Poussier P, Rottapel R (2003b) Suppressor of cytokine signaling 1 attenuates IL-15 receptor signaling in CD8+ thymocytes. Blood 102:4115–4122

Ilangumaran S, Ramanathan S, Rottapel R (2004) Regulation of the immune system by SOCS family adaptor proteins. Semin Immunol 16:351–365

Ilangumaran S, Villalobos-Hernandez A, Bobbala D, Ramanathan S (2016) The hepatocyte growth factor (HGF)-MET receptor tyrosine kinase signaling pathway: Diverse roles in modulating immune cell functions. Cytokine 82:125–139

Inagaki-Ohara K, Kondo T, Ito M, Yoshimura A (2013) SOCS, inflammation, and cancer. Jak-Stat 2:e24053

Jackson SH, Yu CR, Mahdi RM, Ebong S, Egwuagu CE (2004) Dendritic cell maturation requires STAT1 and is under feedback regulation by suppressors of cytokine signaling. J Immunol 172:2307–2315

Jager LD, Dabelic R, Waiboci LW, Lau K, Haider MS, Ahmed CM, Larkin J 3rd, David S, Johnson HM (2011) The kinase inhibitory region of SOCS-1 is sufficient to inhibit T-helper 17 and other immune functions in experimental allergic encephalomyelitis. J Neuroimmunol 232:108–118

Jameson SC (2002) Maintaining the norm: T-cell homeostasis. Nat Rev Immunol 2:547–556

Ji Y, Wrzesinski C, Yu Z, Hu J, Gautam S, Hawk NV, Telford WG, Palmer DC, Franco Z, Sukumar M et al (2015) miR-155 augments CD8+ T-cell antitumor activity in lymphoreplete hosts by enhancing responsiveness to homeostatic gammac cytokines. Proc Natl Acad Sci USA 112:476–481

Jiang S, Zhang HW, Lu MH, He XH, Li Y, Gu H, Liu MF, Wang ED (2010) MicroRNA-155 functions as an OncomiR in breast cancer by targeting the suppressor of cytokine signaling 1 gene. Can Res 70:3119–3127

Kazi JU, Kabir NN, Flores-Morales A, Ronnstrand L (2014) SOCS proteins in regulation of receptor tyrosine kinase signaling. Cell Mol L Sci: CMLS 71:3297–3310

Kershaw NJ, Murphy JM, Liau NP, Varghese LN, Laktyushin A, Whitlock EL, Lucet IS, Nicola NA, Babon JJ (2013) SOCS3 binds specific receptor-JAK complexes to control cytokine signaling by direct kinase inhibition. Nat Struct Mol Biol 20:469–476

Kile BT, Schulman BA, Alexander WS, Nicola NA, Martin HM, Hilton DJ (2002) The SOCS box: a tale of destruction and degradation. Trends Biochem Sci 27:235–241

Kinjyo I, Hanada T, Inagaki-Ohara K, Mori H, Aki D, Ohishi M, Yoshida H, Kubo M, Yoshimura A (2002) SOCS1/JAB is a negative regulator of LPS-induced macrophage activation. Immunity 17:583–591

Kobayashi N, Uemura H, Nagahama K, Okudela K, Furuya M, Ino Y, Ito Y, Hirano H, Inayama Y, Aoki I et al (2012) Identification of miR-30d as a novel prognostic maker of prostate cancer. Oncotarget 3:1455–1471

Kolumam GA, Thomas S, Thompson LJ, Sprent J, Murali-Krishna K (2005) Type I interferons act directly on CD8 T cells to allow clonal expansion and memory formation in response to viral infection. J Exp Med 202:637–650

Komarowska I, Coe D, Wang G, Haas R, Mauro C, Kishore M, Cooper D, Nadkarni S, Fu H, Steinbruchel DA et al (2015) Hepatocyte growth factor receptor c-Met instructs T Cell cardiotropism and promotes T Cell migration to the heart via autocrine chemokine release. Immunity 42:1087–1099

Komatsu N, Mariotti-Ferrandiz ME, Wang Y, Malissen B, Waldmann H, Hori S (2009) Heterogeneity of natural Foxp3+ T cells: a committed regulatory T-cell lineage and an uncommitted minor population retaining plasticity. Proc Natl Acad Sci USA 106:1903–1908

Komatsu N, Okamoto K, Sawa S, Nakashima T, Oh-hora M, Kodama T, Tanaka S, Bluestone JA, Takayanagi H (2014) Pathogenic conversion of Foxp3+ T cells into TH17 cells in autoimmune arthritis. Nat Med 20:62–68

Kremer BE, Adang LA, Macara IG (2007) Septins regulate actin organization and cell-cycle arrest through nuclear accumulation of NCK mediated by SOCS7. Cell 130:837–850

Liu E, Cote JF, Vuori K (2003) Negative regulation of FAK signaling by SOCS proteins. EMBO J 22:5036–5046

Liu L, Li W, Wei X, Cui Q, Lou W, Wang G, Hu X, Qian C (2013) Potent antitumor activity of oncolytic adenovirus-mediated SOCS1 for hepatocellular carcinoma. Gene Ther 20:84–92

Lu LF, Boldin MP, Chaudhry A, Lin LL, Taganov KD, Hanada T, Yoshimura A, Baltimore D, Rudensky AY (2010) Function of miR-146a in controlling Treg cell-mediated regulation of Th1 responses. Cell 142:914–929

Lu X, Chen M, Xue Z, Zhang X, Xu J, Wu L, Deng R, Ma Y (2017a) Dendritic cells that highly express SOCS1 induce T-cell hypo-responsiveness and prolong islet allograft survival. Cell Immunol

Lu X, Chen M, Xue Z, Zhang X, Xu J, Wu L, Deng R, Ma Y (2017b) Dendritic cells that highly express SOCS1 induce T-cell hypo-responsiveness and prolong islet allograft survival. Cell Immunol 314:36–41

Madonna S, Scarponi C, Doti N, Carbone T, Cavani A, Scognamiglio PL, Marasco D, Albanesi C (2013) Therapeutical potential of a peptide mimicking the SOCS1 kinase inhibitory region in skin immune responses. Eur J Immunol 43:1883–1895

Mahoney KM, Rennert PD, Freeman GJ (2015) Combination cancer immunotherapy and new immunomodulatory targets. Nat Rev Drug Discov 14:561–584

Maine GN, Mao X, Komarck CM, Burstein E (2007) COMMD1 promotes the ubiquitination of NF-kappaB subunits through a cullin-containing ubiquitin ligase. EMBO J 26:436–447

Mansell A, Smith R, Doyle SL, Gray P, Fenner JE, Crack PJ, Nicholson SE, Hilton DJ, O'Neill LA, Hertzog PJ (2006) Suppressor of cytokine signaling 1 negatively regulates Toll-like receptor signaling by mediating Mal degradation. Nat Immunol 7:148–155

Marine JC, Topham DJ, McKay C, Wang D, Parganas E, Stravopodis D, Yoshimura A, Ihle JN (1999) SOCS1 deficiency causes a lymphocyte-dependent perinatal lethality. Cell 98:609–616

Martens N, Wery M, Wang P, Braet F, Gertler A, Hooghe R, Vandenhaute J, Hooghe-Peters EL (2004) The suppressor of cytokine signaling (SOCS)-7 interacts with the actin cytoskeleton through vinexin. Exp Cell Res 298:239–248

Matsuda T, Yamamoto T, Kishi H, Yoshimura A, Muraguchi A (2000) SOCS-1 can suppress CD3zeta- and Syk-mediated NF-AT activation in a non- lymphoid cell line. FEBS Lett 472:235–240

Mattei F, Schiavoni G, Belardelli F, Tough DF (2001) IL-15 is expressed by dendritic cells in response to type I IFN, double-stranded RNA, or lipopolysaccharide and promotes dendritic cell activation. J Immunol 167:1179–1187

McGray AJ, Hallett R, Bernard D, Swift SL, Zhu Z, Teoderascu F, Vanseggelen H, Hassell JA, Hurwitz AA, Wan Y, Bramson JL (2014) Immunotherapy-induced CD8+ T cells instigate immune suppression in the tumor. Mol Ther: J Am Soc Gene Ther 22:206–218

Mescher MF, Agarwal P, Casey KA, Hammerbeck CD, Xiao Z, Curtsinger JM (2007) Molecular basis for checkpoints in the CD8 T cell response: tolerance versus activation. Semin Immunol 19:153–161

Metcalf D, Mifsud S, Di Rago L, Nicola NA, Hilton DJ, Alexander WS (2002) Polycystic kidneys and chronic inflammatory lesions are the delayed consequences of loss of the suppressor of cytokine signaling-1 (SOCS-1). Proc Natl Acad Sci USA 99:943–948

Murali-Krishna K, Ahmed R (2000) Cutting edge: naive T cells masquerading as memory cells. J Immunol 165:1733–1737

Murugaiyan G, Beynon V, Mittal A, Joller N, Weiner HL (2011) Silencing microRNA-155 ameliorates experimental autoimmune encephalomyelitis. J Immunol 187:2213–2221
Naka T, Matsumoto T, Narazaki M, Fujimoto M, Morita Y, Ohsawa Y, Saito H, Nagasawa T, Uchiyama Y, Kishimoto Ts (1998) Accelerated apoptosis of lymphocytes by augmented induction of Bax in SSI-1 (STAT-induced STAT inhibitor-1) deficient mice. Proc Natl Acad Sci USA 95:15577–15582
Naka T, Narazaki M, Hirata M, Matsumoto T, Minamoto S, Aono A, Nishimoto N, Kajita T, Taga T, Yoshizaki K et al (1997) Structure and function of a new STAT-induced STAT inhibitor. Nature 387:924–929
Nakagawa R, Naka T, Tsutsui H, Fujimoto M, Kimura A, Abe T, Seki E, Sato S, Takeuchi O, Takeda K et al (2002) SOCS-1 participates in negative regulation of LPS responses. Immunity 17:677–687
Naramura M, Jang IK, Kole H, Huang F, Haines D, Gu H (2002) c-Cbl and Cbl-b regulate T cell responsiveness by promoting ligand-induced TCR down-modulation. Nat Immunol 3:1192–1199
Natatsuka R, Takahashi T, Serada S, Fujimoto M, Ookawara T, Nishida T, Hara H, Nishigaki T, Harada E, Murakami T et al (2015) Gene therapy with SOCS1 for gastric cancer induces G2/M arrest and has an antitumour effect on peritoneal carcinomatosis. Br J Cancer 113:433–442
Nicholson SE, Willson TA, Farley A, Starr R, Zhang JG, Baca M, Alexander WS, Metcalf D, Hilton DJ, Nicola NA (1999) Mutational analyses of the SOCS proteins suggest a dual domain requirement but distinct mechanisms for inhibition of LIF and IL-6 signal transduction. EMBO J 18:375–385
O'Connell RM, Kahn D, Gibson WS, Round JL, Scholz RL, Chaudhuri AA, Kahn ME, Rao DS, Baltimore D (2010) MicroRNA-155 promotes autoimmune inflammation by enhancing inflammatory T cell development. Immunity 33:607–619
Ohkura N, Kitagawa Y, Sakaguchi S (2013) Development and maintenance of regulatory T cells. Immunity 38:414–423
Ohya K, Kajigaya S, Yamashita Y, Miyazato A, Hatake K, Miura Y, Ikeda U, Shimada K, Ozawa K, Mano H (1997) SOCS-1/JAB/SSI-1 can bind to and suppress Tec protein-tyrosine kinase. J Biol Chem 272:27178–27182
Okumura F, Matsuzaki M, Nakatsukasa K, Kamura T (2012) The role of elongin BC-containing ubiquitin ligases. Front Oncol 2:10
Ostiguy V, Allard EL, Marquis M, Leignadier J, Labrecque N (2007) IL-21 promotes T lymphocyte survival by activating the phosphatidylinositol-3 kinase signaling cascade. J Leukoc Biol 82:645–656
Palmer DC, Guittard GC, Franco Z, Crompton JG, Eil RL, Patel SJ, Ji Y, Van Panhuys N, Klebanoff CA, Sukumar M et al (2015) Cish actively silences TCR signaling in CD8 + T cells to maintain tumor tolerance. J Exp Med 212:2095–2113
Palmer DC, Restifo NP (2009) Suppressors of cytokine signaling (SOCS) in T cell differentiation, maturation, and function. Trends Immunol 30:592–602
Palucka K, Banchereau J (2012) Cancer immunotherapy via dendritic cells. Nat Rev Cancer 12:265–277
Park JH, Adoro S, Guinter T, Erman B, Alag AS, Catalfamo M, Kimura MY, Cui Y, Lucas PJ, Gress RE et al (2010) Signaling by intrathymic cytokines, not T cell antigen receptors, specifies CD8 lineage choice and promotes the differentiation of cytotoxic-lineage T cells. Nat Immunol 11:257–264
Paterson AM, Vanguri VK, Sharpe AH (2009) SnapShot: B7/CD28 costimulation. Cell 137(974–974):e971
Pichiorri F, Suh SS, Ladetto M, Kuehl M, Palumbo T, Drandi D, Taccioli C, Zanesi N, Alder H, Hagan JP et al (2008) MicroRNAs regulate critical genes associated with multiple myeloma pathogenesis. Proc Natl Acad Sci USA 105:12885–12890
Rafei M, Rouette A, Brochu S, Vanegas JR, Perreault C (2013) Differential effects of gammac cytokines on postselection differentiation of CD8 thymocytes. Blood 121:107–117

Ramanathan S, Dubois S, Chen XL, Leblanc C, Ohashi PS, Ilangumaran S (2011) Exposure to IL-15 and IL-21 enables autoreactive CD8 T cells to respond to weak antigens and cause disease in a mouse model of autoimmune diabetes. J Immunol 186:5131–5141

Ramanathan S, Dubois S, Gagnon J, Leblanc C, Mariathasan S, Ferbeyre G, Rottapel R, Ohashi PS, Ilangumaran S (2010) Regulation of cytokine-driven functional differentiation of CD8 T cells by suppressor of cytokine signaling 1 controls autoimmunity and preserves their proliferative capacity toward foreign antigens. J Immunol 185:357–366

Ramanathan S, Gagnon J, Leblanc C, Rottapel R, Ilangumaran S (2006) Suppressor of cytokine signaling 1 stringently regulates distinct functions of IL-7 and IL-15 in vivo during T lymphocyte development and homeostasis. J Immunol 176:4029–4041

Ribas A (2015) Adaptive immune resistance: how cancer protects from immune attack. Cancer Discov 5:915–919

Rodriguez GM, D'Urbano D, Bobbala D, Chen XL, Yeganeh M, Ramanathan S, Ilangumaran S (2013) SOCS1 prevents potentially skin-reactive cytotoxic T lymphocytes from gaining the ability to cause inflammatory lesions. J Invest Dermatol 133:2013–2022

Rosenberg SA, Restifo NP (2015) Adoptive cell transfer as personalized immunotherapy for human cancer. Science 348:62–68

Ryo A, Suizu F, Yoshida Y, Perrem K, Liou YC, Wulf G, Rottapel R, Yamaoka S, Lu KP (2003) Regulation of NF-kappaB signaling by Pin1-dependent prolyl isomerization and ubiquitin-mediated proteolysis of p65/RelA. Mol Cell 12:1413–1426

Sakaguchi S, Vignali DA, Rudensky AY, Niec RE, Waldmann H (2013) The plasticity and stability of regulatory T cells. Nat Rev Immunol 13:461–467

Sasaki A, Yasukawa H, Suzuki A, Kamizono S, Syoda T, Kinjyo I, Sasaki M, Johnston JA, Yoshimura A (1999) Cytokine-inducible SH2 protein-3 (CIS3/SOCS3) inhibits Janus tyrosine kinase by binding through the N-terminal kinase inhibitory region as well as SH2 domain. Genes Cells 4:339–351

Sasi W, Sharma AK, Mokbel K (2014) The role of suppressors of cytokine signalling in human neoplasms. Mol Biol Int 2014:630797

Schmitt EG, Williams CB (2013) Generation and function of induced regulatory T cells. Front Immunol 4:152

Schwartz RH (2003) T cell anergy. Annu Rev Immunol 21:305–334

Sharma MD, Hou DY, Baban B, Koni PA, He Y, Chandler PR, Blazar BR, Mellor AL, Munn DH (2010) Reprogrammed foxp3(+) regulatory T cells provide essential help to support cross-presentation and CD8(+) T cell priming in naive mice. Immunity 33:942–954

Shen L, Evel-Kabler K, Strube R, Chen SY (2004) Silencing of SOCS1 enhances antigen presentation by dendritic cells and antigen-specific anti-tumor immunity. Nat Biotechnol 22:1546–1553

Shi D, Li D, Yin Q, Qiu Y, Yan H, Shen Y, Lu G, Liu W (2015) Silenced suppressor of cytokine signaling 1 (SOCS1) enhances the maturation and antifungal immunity of dendritic cells in response to Candida albicans in vitro. Immunol Res 61:206–218

Souma Y, Nishida T, Serada S, Iwahori K, Takahashi T, Fujimoto M, Ripley B, Nakajima K, Miyazaki Y, Mori M et al (2012) Antiproliferative effect of SOCS-1 through the suppression of STAT3 and p38 MAPK activation in gastric cancer cells. Int J Cancer 131:1287–1296

Sporri B, Kovanen PE, Sasaki A, Yoshimura A, Leonard WJ (2001) JAB/SOCS1/SSI-1 is an interleukin-2-induced inhibitor of IL-2 signaling. Blood 97:221–226

Spranger S, Spaapen RM, Zha Y, Williams J, Meng Y, Ha TT, Gajewski TF (2013) Up-regulation of PD-L1, IDO, and T(regs) in the melanoma tumor microenvironment is driven by CD8(+) T cells. Sci Transl Med 5:200ra116

Stanford SM, Rapini N, Bottini N (2012) Regulation of TCR signalling by tyrosine phosphatases: from immune homeostasis to autoimmunity. Immunology 137:1–19

Starr R, Metcalf D, Elefanty AG, Brysha M, Willson TA, Nicola NA, Hilton DJ, Alexander WS (1998) Liver degeneration and lymphoid deficiencies in mice lacking suppressor of cytokine signaling-1. Proc Natl Acad Sci USA 95:14395–14399

Starr R, Willson TA, Viney EM, Murray LJ, Rayner JR, Jenkins BJ, Gonda TJ, Alexander WS, Metcalf D, Nicola NA, Hilton DJ (1997) A family of cytokine-inducible inhibitors of signalling. Nature 387:917–921
Strebovsky J, Walker P, Lang R, Dalpke AH (2011) Suppressor of cytokine signaling 1 (SOCS1) limits NFkappaB signaling by decreasing p65 stability within the cell nucleus. FASEB J 25:863–874
Takahashi R, Nishimoto S, Muto G, Sekiya T, Tamiya T, Kimura A, Morita R, Asakawa M, Chinen T, Yoshimura A (2011) SOCS1 is essential for regulatory T cell functions by preventing loss of Foxp3 expression as well as IFN-{gamma} and IL-17A production. J Exp Med 208:2055–2067
Takahashi R, Yoshimura A (2014) SOCS1 and regulation of regulatory T cells plasticity. J Immunol Res 2014:943149
Tanaka K, Ichiyama K, Hashimoto M, Yoshida H, Takimoto T, Takaesu G, Torisu T, Hanada T, Yasukawa H, Fukuyama S et al (2008) Loss of suppressor of cytokine signaling 1 in helper T cells leads to defective Th17 differentiation by enhancing antagonistic effects of IFN-gamma on STAT3 and Smads. J Immunol 180:3746–3756
Trengove MC, Ward AC (2013) SOCS proteins in development and disease. Am J Clin Exp Immunol 2:1–29
Trop S, De Sepulveda P, Zuniga-Pflucker JC, Rottapel R (2001) Overexpression of suppressor of cytokine signaling-1 impairs pre-T-cell receptor-induced proliferation but not differentiation of immature thymocytes. Blood 97:2269–2277
Tsukada J, Ozaki A, Hanada T, Chinen T, Abe R, Yoshimura A, Kubo M (2005) The role of suppressor of cytokine signaling 1 as a negative regulator for aberrant expansion of CD8alpha+ dendritic cell subset. Int Immunol 17:1167–1178
Wang HY, Altman Y, Fang D, Elly C, Dai Y, Shao Y, Liu YC (2001) Cbl promotes ubiquitination of the T cell receptor zeta through an adaptor function of Zap-70. J Biol Chem 276:26004–26011
Waskow C, Paul S, Haller C, Gassmann M, Rodewald HR (2002) Viable c-Kit(W/W) mutants reveal pivotal role for c-kit in the maintenance of lymphopoiesis. Immunity 17:277–288
Wells AD, Walsh MC, Bluestone JA, Turka LA (2001) Signaling through CD28 and CTLA-4 controls two distinct forms of T cell anergy. J Clin Invest 108:895–903
Wormald S, Zhang JG, Krebs DL, Mielke LA, Silver J, Alexander WS, Speed TP, Nicola NA, Hilton DJ (2006) The comparative roles of suppressor of cytokine signaling-1 and -3 in the inhibition and desensitization of cytokine signaling. J Biol Chem 281:11135–11143
Wu L, Liu YJ (2007) Development of dendritic-cell lineages. Immunity 26:741–750
Yao R, Ma YL, Liang W, Li HH, Ma ZJ, Yu X, Liao YH (2012) MicroRNA-155 modulates Treg and Th17 cells differentiation and Th17 cell function by targeting SOCS1. PLoS ONE 7: e46082
Yasukawa H, Misawa H, Sakamoto H, Masuhara M, Sasaki A, Wakioka T, Ohtsuka S, Imaizumi T, Matsuda T, Ihle JN, Yoshimura A (1999) The JAK-binding protein JAB inhibits janus tyrosine kinase activity through binding in the activation loop. EMBO J 18:1309–1320
Yasukawa H, Ohishi M, Mori H, Murakami M, Chinen T, Aki D, Hanada T, Takeda K, Akira S, Hoshijima M et al (2003) IL-6 induces an anti-inflammatory response in the absence of SOCS3 in macrophages. Nat Immunol 4:551–556
Yasukawa H, Sasaki A, Yoshimura A (2000) Negative regulation of cytokine signaling pathways. Annu Rev Immunol 18:143–164
Yeganeh M, Gui Y, Kandhi R, Bobbala D, Tobelaim WS, Saucier C, Yoshimura A, Ferbeyre G, Ramanathan S, Ilangumaran S (2016) Suppressor of cytokine signaling 1-dependent regulation of the expression and oncogenic functions of p 21 in the liver. Oncog PMID:26725321
Yoshida T, Ogata H, Kamio M, Joo A, Shiraishi H, Tokunaga Y, Sata M, Nagai H, Yoshimura A (2004) SOCS1 is a suppressor of liver fibrosis and hepatitis-induced carcinogenesis. J Exp Med 199:1701–1707
Yoshikawa H, Matsubara K, Qian GS, Jackson P, Groopman JD, Manning JE, Harris CC, Herman JG (2001) SOCS-1, a negative regulator of the JAK/STAT pathway, is silenced by

methylation in human hepatocellular carcinoma and shows growth- suppression activity. Nat Genet 28:29–35

Yoshimura A (2013) CIS: the late-blooming eldest son. Nat Immunol 14:692–694

Yoshimura A, Naka T, Kubo M (2007) SOCS proteins, cytokine signalling and immune regulation. Nat Rev Immunol 7:454–465

Yoshimura A, Ohkubo T, Kiguchi T, Jenkins NA, Gilbert DJ, Copeland NG, Hara T, Miyajima A (1995) A novel cytokine-inducible gene CIS encodes an SH2-containing protein that binds to tyrosine-phosphorylated interleukin 3 and erythropoietin receptors. EMBO J 14:2816–2826

Yoshimura A, Yasukawa H (2012) JAK's SOCS: a mechanism of inhibition. Immunity 36:157–159

Yu Q, Park JH, Doan LL, Erman B, Feigenbaum L, Singer A (2006) Cytokine signal transduction is suppressed in preselection double-positive thymocytes and restored by positive selection. J Exp Med 203:165–175

Zeng R, Spolski R, Finkelstein SE, Oh S, Kovanen PE, Hinrichs CS, Pise-Masison CA, Radonovich MF, Brady JN, Restifo NP et al (2005) Synergy of IL-21 and IL-15 in regulating CD8 + T cell expansion and function. J Exp Med 201:139–148

Zhan Y, Davey GM, Graham KL, Kiu H, Dudek NL, Kay T, Lew AM (2009) SOCS1 negatively regulates the production of Foxp3 + CD4 + T cells in the thymus. Immunol Cell Biol 87:473–480

Zhang JG, Metcalf D, Rakar S, Asimakis M, Greenhalgh CJ, Willson TA, Starr R, Nicholson SE, Carter W, Alexander WS et al (2001) The SOCS box of suppressor of cytokine signaling-1 is important for inhibition of cytokine action in vivo. Proc Natl Acad Sci USA 98:13261–13265

Mining the Complex Family of Protein Tyrosine Phosphatases for Checkpoint Regulators in Immunity

Claudia Penafuerte, Luis Alberto Perez-Quintero, Valerie Vinette, Teri Hatzihristidis and Michel L. Tremblay

Abstract The family of protein tyrosine phosphatases (PTPs) includes 107 genes in humans that are diverse in their structures and expression profiles. The majority are present in immune cells and play various roles in either inhibiting or promoting the duration and amplitude of signaling cascades. Several PTPs, including TC-PTP (PTPN2) and SHP-1 (PTPN6), have been recognized as being crucial for maintaining proper immune response and self-tolerance, and have gained recognition as true immune system checkpoint modulators. This chapter details the most recent literature on PTPs and immunity by examining their known functions in regulating signaling from either established checkpoint inhibitors or by their intrinsic properties, as modulators of the immune response. Notably, we review PTP regulatory properties in macrophages, antigen-presenting dendritic cells, and T cells. Overall, we present the PTP gene family as a remarkable source of novel checkpoint inhibitors wherein lies a great number of new targets for immunotherapies.

Contents

1 Protein Tyrosine Phosphatases and the Mammalian Immune System 192
1.1 The Protein Tyrosine Phosphatase Gene Family 193
1.2 Protein Tyrosine Phosphatases in Immunity 193
1.3 Regulation of Protein Tyrosine Phosphatases 194

C. Penafuerte, L.A. Perez-Quintero, and V. Vinette have equally contributed to the writing of this chapter.

C. Penafuerte · L.A. Perez-Quintero · V. Vinette · T. Hatzihristidis · M.L. Tremblay (✉)
Rosalind and Morris Goodman Cancer Research Centre and Department of Biochemistry, McGill University, Montreal, QC, Canada
e-mail: michel.tremblay@mcgill.ca

C. Penafuerte
Biotechnology Research Institute-NRC, 6100 Royalmount, Montreal H4P 2R2, QC, Canada

Current Topics in Microbiology and Immunology (2017) 410:191–214
DOI 10.1007/82_2017_68

Published Online: 20 September 2017

2 The Concept of Immune Checkpoints and PTPs ... 195
3 Protein Tyrosine Phosphatases as Immune Checkpoint Regulators in Monocytes and Macrophages ... 195
3.1 Background ... 195
3.2 PTP Expression in Macrophages ... 196
3.3 PTP Regulation of JAK/STAT Signaling and the Pro-inflammatory Response ... 196
3.4 Regulation of Macrophage Activation and Polarization by PTPs ... 197
4 PTPs as Negative Immune Checkpoint Regulators of Dendritic Cell Function ... 198
4.1 Background ... 198
4.2 PTP Expression in DCs ... 199
4.3 PTP Regulation of TLR and Cytokine Receptor Signals in DCs ... 199
4.4 PTPs as Regulators of DC Maturation and T Cell Differentiation ... 201
4.5 PTP Regulation of Cell Migration and Vesicle Transport ... 202
5 PTPs as Components of Inhibitory Signaling in T Cells ... 202
5.1 PTPs as Transducers of Checkpoint Signals in T Cells ... 202
5.2 PTPs in CTLA-4 Signaling ... 204
5.3 PTPs in PD-1 Signaling ... 204
5.4 Other Inhibitory Molecules of the CD28 Family ... 205
5.5 Other ITIM-Containing Receptors ... 206
5.6 ITSM Signaling ... 206
5.7 SAP Family of Adaptors as "Natural" Checkpoint Inhibitors ... 206
5.8 JAK/STAT Regulation by PTP1B and TC-PTP ... 207
6 Conclusion ... 208
References ... 209

1 Protein Tyrosine Phosphatases and the Mammalian Immune System

Tyrosine phosphorylation has been central to our understanding of cellular signaling. The discovery of protein tyrosine kinases (PTK) adding phosphate moieties on tyrosine residues over 40 years ago, the subsequent discovery of the existence of protein domains able to recognize and interact specifically with phosphorylated tyrosine residues, and finally the discovery of the protein tyrosine phosphatase (PTP) completed the recognition of a highly dynamic system regulating signaling in mammalian cells (Lim and Pawson 2010). These three molecular components of tyrosine signaling permit the qualitative and quantitative regulation of numerous cellular pathways implicated in almost all cellular physiological events, from mitosis and differentiation, cell migration and cell–cell interactions to cell death. Evolution has conveyed great diversity to both PTK and PTP gene families. With 90 PTK genes and 107 genes, the positive and negative interactions of these family members provide diverse mechanisms and pathways that can be modulated to control intrinsic signaling cascades. The transmembrane proteins serving as entry and exit point into cells and their signaling pathways are also conducted by diverse gene families, including membrane-associated members of the PTK and PTP families, but also by several complex families such as cytokine and chemokine receptors, G coupled receptors, nuclear receptors, and other transmembrane proteins.

1.1 The Protein Tyrosine Phosphatase Gene Family

Although the PTKs are extremely well known, the PTPs have received less attention in part because of their more difficult enzymology and substrate identification. Initially identified in 1988 by N. Tonks' laboratory (Charbonneau et al. 1988), the PTP family gene tally was reported to consist of 107 members (Alonso et al. 2004). More recently, additional enzymes seemingly dephosphorylating phosphotyrosine residues were added to reach 125 genes (Alonso and Pulido 2016). The family includes various subgroups based on their target specificity (unique to p-tyr, or dual PTPs that recognize p-ser/p-thr/p-tyr, or even enzymes that act on phospholipids, phospho-sugars, and other molecules). Typical members of this superfamily of structurally related enzymes are defined by a highly conserved catalytic domain of approximately 200–240 amino acids, which includes the active site signature sequence [HCXXGXXR] encompassing an invariant cysteine (Cys) residue that is critical for PTP activity (Tonks 2006) (for more details on the gene family, see Alonso et al. (2004) and http://ptp.cshl.edu). A short video depicting the mechanism of enzymatic activity was published by Feldhammer et al. (2013).

For many years, PTPs were seen as housekeeping enzymes whose general function was to counter that of tyrosine kinases on their given substrates. However, over the past decades, our laboratory and others have presented ample evidence that these enzymes can act in a very exact manner (Tonks 2013). Among others, (i) They can dephosphorylate not only specific tyrosine-phosphorylated substrates, but even target precise phosphotyrosine sites embedded in a string of p-tyr modifications on the same targeted protein (Stuible et al. 2008). (ii) PTPs do not simply downregulate other protein activities, but can also influence specific signaling pathways positively or negatively (Dubé et al. 2005; Elchebly et al. 1999). (iii) They work in coordination with other proteins (adaptor proteins, kinases, as well as other serine/threonine and tyrosine phosphatases) to influence particular pathways and cell behavior (Xu and Qu 2008; St-Pierre and Tremblay 2012; Pike et al. 2014). Therefore, this family represents a diverse and important gene family, where its members are key factors in cell and animal homeostasis. It is also clear that among many functions, they play crucial roles as "checkpoint" regulators in either allowing or disallowing tyrosine kinase-dependent signaling in multiple pathways.

1.2 Protein Tyrosine Phosphatases in Immunity

The important roles associated with tyrosine kinases and tyrosine phosphorylation in immune cell signaling is paralleled by the presence of tyrosine phosphatases in these as well. Indeed, CD45, known as the leucocyte common antigen, was identified early on in the PTP field to contain tyrosine phosphatase domains (Charbonneau et al. 1988; Tonks et al. 1984) and was found to be highly expressed in T and B cells (Ledbetter et al. 1988). A first comprehensive account of PTP gene expression in immune cells was reported by Arimura and Yagi (Arimura and Yagi 2010), who demonstrated that out of the 107 enzymes, between 58 and 76 PTPs were expressed in the most common immune cell lineages, and an additional 12 were expressed in specific lineages. This analysis may be modified based on the threshold of expression of these different PTP RNAs and suggests that approximately 86 different PTPs are present in the immune system with the remaining 21 being either specific to non-immune tissue or present at very low levels in hematopoietic cells.

1.3 Regulation of Protein Tyrosine Phosphatases

At first glance, this impressive tally of PTP expression in immune cells creates a complex landscape of tyrosine phosphorylation modulation. However, other regulatory steps also influence PTP activities, such as post-transcriptional modifications that occur on PTP gene transcripts. For example, most of the PTP gene family transcripts are differentially spliced, giving rise to different forms of PTP proteins. Like the two PTPN2 (encodes for TC-PTP) mRNAs that are localized either to the nucleus or the endoplasmic reticulum (Bussieres-Marmen et al. 2014), these alternatively spliced mRNAs result in important variations in their final enzyme structures and activities (Julien et al. 2011). Furthermore, in immune cells and other tissues, several receptor PTPs undergo alternative splicing events that generate variable extracellular domains, as was demonstrated for CD45, and also leads to different forms of catalytic domains of the receptor PTPs (Zikherman and Weiss 2008).

More recently, studies on the PTP4A1, PTP4A2, and PTP4A3 enzymes demonstrated another level of regulation, as they are exquisitely dependent on protein synthesis for their regulation. Namely, PTP4A2 translation is linked to intracellular Mg^{2+} and glucose levels. Mg^{2+} concentration inversely influences PTP4A2 mRNA available to be translated. Similar to some bacterial magnesium transporters such as MgtA (Cromie and Groisman, 2010), PTP4A2 may possess riboswitch-like motif(s) in its 5’ noncoding RNA that responds to low levels of this specific cation to promote its translation and maintain physiological levels in highly metabolically active cells (Hardy et al. 2015; Gungabeesoon et al. 2016; Uetani et al. 2017).

Finally, PTP members have been reported to present a great number of post-translational modifications. Most are phosphorylation events on tyrosine, serine, and threonine residues. For instance, the SHP-1 and SHP-2 PTPs are themselves regulated by intramolecular interactions of their SH2 domain to specific tyrosine residues that can either promote or impede their enzymatic activity as well as their protein–protein interactions (Pandey et al. 2017). Moreover, other modifications such as glycosylation, sumoylation, and ubiquitination have also been reported and studied in the context of tyrosine phosphatase activity (Tonks 2013).

2 The Concept of Immune Checkpoints and PTPs

To control the vast array of signaling pathways that serve to monitor, induce, and repress immune responses, a series of immune-specific genes have been identified. Working as antigen-presenting complexes, co-stimulatory and co-inhibitory molecules, they induce key inflexion points in the immune system function. Herein, we review the activities of several protein tyrosine phosphatases in different immune cell lineages, which if targeted by small molecules or biologicals will lead to directed positive or negative changes in the immune response. Already, many PTPs have been recognized to possess immune modulatory functions and can thus be included inside the more common immune checkpoint regulators currently used in cancer immunotherapy. Their usefulness as checkpoint targets also manifests through their capacity to exert their action in a dose-dependent manner. Hence, they offer outstanding opportunities as additional tools in the control of the innate immune response—a future for cancer immunotherapy that will be of great value in the management of many diseases.

3 Protein Tyrosine Phosphatases as Immune Checkpoint Regulators in Monocytes and Macrophages

3.1 Background

Monocytes are a type of nucleated white blood cell which are precursors of several immune cells, predominantly dendritic cells, and tissue macrophages, which serve as the first line of defense of innate immunity (Murphy et al. 2008; Murphy and Weaver 2017). As such, modulating the expression of genes that regulate their activity and function would be an optimal method of controlling the immune response. Following an insult as is the case during tumor cell development or the onset of inflammation, circulating monocytes can differentiate into either inflammatory monocytes or macrophages. Inflammatory monocytes then produce various pro-inflammatory cytokines, which lead to the recruitment of other immune cells to the site of inflammation (Murphy and Weaver 2017). On the other hand, macrophages can exist either as resident macrophages in most tissues or evolve from differentiated or mature monocytes. Key to the innate immune response, macrophages are the predominant type of phagocytic cell. They secrete a plethora of cytokines and chemokines such as IL-1β, IL-6, IL-12, TNF-α, and CXCL8 to attract additional immune cells and initiate a local pro-inflammatory cascade (Murphy and Weaver 2017).

Macrophage functions are regulated by several classes of enzymes, notably PTKs and PTPs. Several PTPs, including PTP1B, SHP-1, SHP-2, and PTPN22, negatively regulate macrophage differentiation, activation, and polarization due to their dephosphorylation of PTKs and other signaling substrates. These various functions make them ideal targets for macrophage-based cancer immunotherapy.

3.2 PTP Expression in Macrophages

As stated above, the immune cell PTP transcriptome has been well characterized, with over 64 PTP-encoding genes being expressed in all immune lineages (Arimura and Yagi 2010). On the other hand, 12 PTPs were found to be expressed in peritoneal macrophages but not in all other immune cell lineages: *Ptprm, Ptpro, Ptpn21, Dusp3, Dusp13B, Dusp23, Ssh3, Mtmr7, Eya4, Dusp9, Dusp26, Epm2a*, with *Dusp3* being the most highly expressed PTP gene in macrophages. In addition, 6 PTPs are commonly expressed between peritoneal macrophages and bone marrow-derived immature DCs: *Ptpro, Dusp3, Ssh3, Dusp26, Epm2a,* and *Mtmr7* (Arimura and Yagi 2010). Interestingly, peritoneal macrophage stimulation with pro-inflammatory lipopolysaccharide (LPS) leads to an alteration of PTP expression. For example, *Ptprj* and *Mtmr7* are increased upon LPS stimulation, whereas *Dusp7* and *Cdc25b* are decreased (Arimura and Yagi 2010), indicating a role for pro-inflammatory stimuli in the expression of PTPs in macrophages.

3.3 PTP Regulation of JAK/STAT Signaling and the Pro-inflammatory Response

Multiple studies have identified PTPs as important negative checkpoint regulators of signaling pathways that are essential for the transcriptional regulation and activity of macrophages. Most importantly to date, the family of janus kinase (JAK) and signal transducers and activators of transcription (STAT) proteins were shown to act as critical negative regulators of immune responses, resulting in the modulation of key cytokine signaling pathways. For instance, CD45 is a negative regulator of cytokine signaling in peritoneal macrophages upon IFN-α stimulation through regulation of the TYK2-STAT1/STAT3 axis (Irie-Sasake 2000). SHP-1 is similarly an IFN-α-induced negative regulator of JAK/STAT signaling. SHP1-deficient mice present hyperphosphorylation of JAK1, but not of TYK2 (David et al. 1995). Pike et al. have also demonstrated that PTP1B is a negative regulator of the IL-10-induced transcriptional program in macrophages through the JAK1-TYK2-STAT3 signaling pathway (Pike et al. 2014). Inhibition of PTP1B in peritoneal macrophages leads to STAT3 hyperactivation, which then shuttles to the nucleus and controls cytokine signaling by upregulating the expression of a plethora of anti-inflammatory genes, such as *Etv3, Bcl3, Zfp36, Sbno2,* and *Nfil3*. Furthermore, IL-10 stimulation of PTP1B-deficient macrophages causes an increase of IL-4rα at the cell surface as well as inhibition of LPS-induced macrophage activation (Pike et al. 2014), both of which are important for macrophage activation and polarization toward the anti-inflammatory and tumor-promoting M2 macrophage type. This highlights the crucial role for PTP1B as a negative checkpoint regulator of IL-10-STAT3 cytokine signaling in macrophages and their induction of anti-inflammatory signals. Other groups demonstrated that the absence of PTP1B in myeloid cells provides protection against LPS-induced inflammation due to the systemic increase in IL-10 and STAT3 hyperphosphorylation (Grant et al. 2014). Interestingly, such an induction in IL-10-STAT3 signaling causes increased expression of the suppressor of cytokine signaling 3 (SOCS3), which acts as a negative feedback regulator of the pro-inflammatory IL-6-STAT3 signaling pathway (Pike et al. 2014; Croker et al. 2003). As such, SOCS3 is proposed to be a major factor in reducing inflammation through regulation of pro-inflammatory cytokine signaling, particularly in activated macrophages (Lang et al. 2003). By inhibiting PTP1B, SOCS3 may be hyperactivated upon IL-10 stimulation and negatively regulate pro-inflammatory cytokine signaling, acting as a critical dampening checkpoint system for the pro-inflammatory response in macrophages.

3.4 Regulation of Macrophage Activation and Polarization by PTPs

Macrophages can be activated and polarized into subsets that trigger different immune responses to tackle various infectious agents as well as tumor cells. Several PTPs have been demonstrated to negatively regulate macrophage activation, such as the protein phosphatase SHP-2, which negatively regulates IL-4-induced macrophage activation through the regulation of the JAK1-JAK3-STAT6 signaling pathway (Barron and Wynn 2011; Tao et al. 2014). On the other hand, PTPN22 negatively regulates IFN-γ-induced macrophage activation through the regulation of JAK1/2 and STAT1 signaling pathways, leading to the subsequent induction of IFNg activation sites (Chang et al. 2013; Kozicky and Sly 2015). PTPN22 reciprocally acts as a positive regulator of pro-inflammatory M2 macrophage activation upon IL-4 and IL-13 stimulation in a STAT6-dependent manner (Chang et al. 2013). Similarly, SHP-1 and PTP1B negatively regulate LPS-induced macrophage activation through the regulation of the NF-κB signaling pathway (Kozicky and Sly 2015).

Considering that macrophages are the predominant immune cell type in most cancers and wound sites, understanding that their polarization status is important for tumor progression has led to important breakthroughs for macrophage-based cancer immunotherapy (Mills et al. 2016). More specifically, studies have shown that converting M2 or tumor-promoting macrophages to the pro-inflammatory and anti-tumor M1 macrophage subset leads to tumor regression (Mills et al. 1992; Ruffell and Coussens 2015; Beatty et al. 2011). Further studies have demonstrated that STAT6 is a predominant regulator of macrophage polarization toward the tumor-promoting M2 subset, while NF-κB and interferon regulatory factors (IRFs) are the main transcriptional regulators of the pro-inflammatory M1 macrophages (Covarrubias et al. 2015; Davies et al. 2013; Murray and Wynn 2011). In addition, tumor-associated macrophages (TAMs) have a different signature than either their pro- or anti-inflammatory counterparts, yet they express molecular factors like PD-L1 and PD-L2 that induce inhibitory immune checkpoint receptors to regulate processes like T cell activation (Mantovani et al. 2017). For this reason, TAMs are often targets of checkpoint blockage immunotherapies (Mantovani et al. 2017), emphasizing the importance of regulating the inflammatory status of macrophages and re-educating TAMs and M2 macrophages to the antitumor M1 macrophage type as the first step toward effective cancer immunotherapies. Consequently, it is essential to identify factors that negatively regulate macrophage activation and polarization to improve the efficacy of macrophage-based cancer immunotherapy.

In conclusion, the inhibition of several PTPs in macrophages leads to an upregulation of the anti-inflammatory gene transcriptome and signaling pathways, negatively regulates pro-inflammatory cytokine signaling as well as modulates macrophage activation and polarization. Hence, as they are exquisite checkpoint regulators, targeting these PTPs to promote a pro-inflammatory response and to reprogram these macrophages towards a pro-inflammatory M1 subset will be essential to develop efficient macrophage-based immunotherapies.

4 PTPs as Negative Immune Checkpoint Regulators of Dendritic Cell Function

4.1 Background

Identification of novel tumor-associated antigens (TAAs) has been a critical development in tumor-specific immunotherapies for the treatment of patients with solid cancers. Therapeutic strategies based on dendritic cells (DCs) loaded with either recombinant TAA, tumor cell lysate, or transfected with viral vectors expressing TAA peptides, function through the activation of tumor-specific T cells and consequently the generation of an antitumor response (CTL- and LAK-mediated cytotoxicity) (van Beek et al. 2014). An increasing number of preclinical and clinical studies using DC-based therapies as a single treatment or combined with other therapeutic approaches reinforce the importance of DCs as preferential inducers of antitumor immunity with low toxicity. However, in the past two decades, DC-based therapies have faced several obstacles in the treatment of advanced cancer patients, characterized by a low response rate (below 15%) and only a slight increase in survival (Constantino et al. 2017; Garg et al. 2017; Anguille et al. 2014).

DC-based immunotherapies are compromised by immunosuppressive tumor microenvironments (Frey and Monu 2008). Cancer patients display a significant increase of immature myeloid DCs which consistently induce T cell tolerance or anergy, affecting the development of an effective antitumor response while favoring tumor growth and metastasis (Gabrilovich 2004). Thus, DC immunotherapy effectiveness depends on several important factors, including the nature of the antigens used to prime patient DCs as well as DC maturation and activation status. (Nicolette et al. 2007). The use of mature instead of immature DCs overcomes potential complications arising from the ability of immature DCs to induce regulatory T cells, which would lead to an immunosuppressive effect rather than the priming of the antitumor response (Dhodapkar et al. 2001). Therefore, it is mandatory to identify negative checkpoint regulators of DC maturation and function in order to boost the efficacy of cancer immunotherapies.

4.2 PTP Expression in DCs

A number of PTPs exert negative effects by dephosphorylating and inactivating receptor-associated tyrosine kinases essential for DC activation and maturation (Pawson 2004; An et al. 2006; Ramachandran et al. 2011).

The expression profile of PTPs can vary depending on the DC subset (e.g., plasmacytoid versus myeloid DCs) and maturation state. Bone marrow-derived immature DCs share the expression of 64 PTP genes with the other immune cell lineages and share 6 PTP genes (*Ptpro, Dusp3, Ssh3, Dusp26, Epm2a,* and *Mtmr7*) with peritoneal macrophages. *Ptpro* is the most highly expressed PTP-encoding gene in bone marrow-derived immature DCs. The expression profile of PTP genes varies over time upon maturation stimulus (Amura and Yagi 2010). Most importantly, the quality of DC stimulus that initially triggers the activation of a specific set of receptors highly influences the PTP expression profile. For instance, PTP gamma expression is induced by LPS, CD40L, or TNFα maturation stimuli in a subset of monocyte-derived DCs (DC-SIGN$^+$) and is co-localized with MHC class II molecules on the cell surface (Lissandrini et al. 2006).

4.3 PTP Regulation of TLR and Cytokine Receptor Signals in DCs

Protein tyrosine phosphorylation affects several crucial components of toll-like receptor (TLR) signaling pathways (An et al. 2006; Ramachandran et al. 2011; Akira and Takeda 2004). TLRs are members of a larger family that includes the interleukin-1 receptors (IL-1Rs). The cytoplasmic tails of TLRs and IL-1Rs contain a conserved region of approximately 200 amino acids that is known as the Toll/IL-1R (TIR) domain. TIR domains comprise three conserved boxes, which are crucial for signal transduction. Ligand engagement by TLRs promote conformational changes required for the recruitment and activation of a number of adaptor molecules and kinases, including MyD88 (myeloid differentiation primary-response protein 88), IRAK1 and 4 (IL-1R-associated kinases 1 and 4), TRAF6 (tumor-necrosis-factor-receptor-associated factor 6), TAK1 (transforming growth factor β, TGF-β-activated kinase), TAB1, and TAB2 (TAK1 and 2-binding protein). These events lead to the activation of mitogen-activated protein kinases (MAPKs) and NF-κB signaling pathways and the subsequent expression of pro-inflammatory cytokines and DC maturation (Akira and Takeda 2004) (Fig. 1).

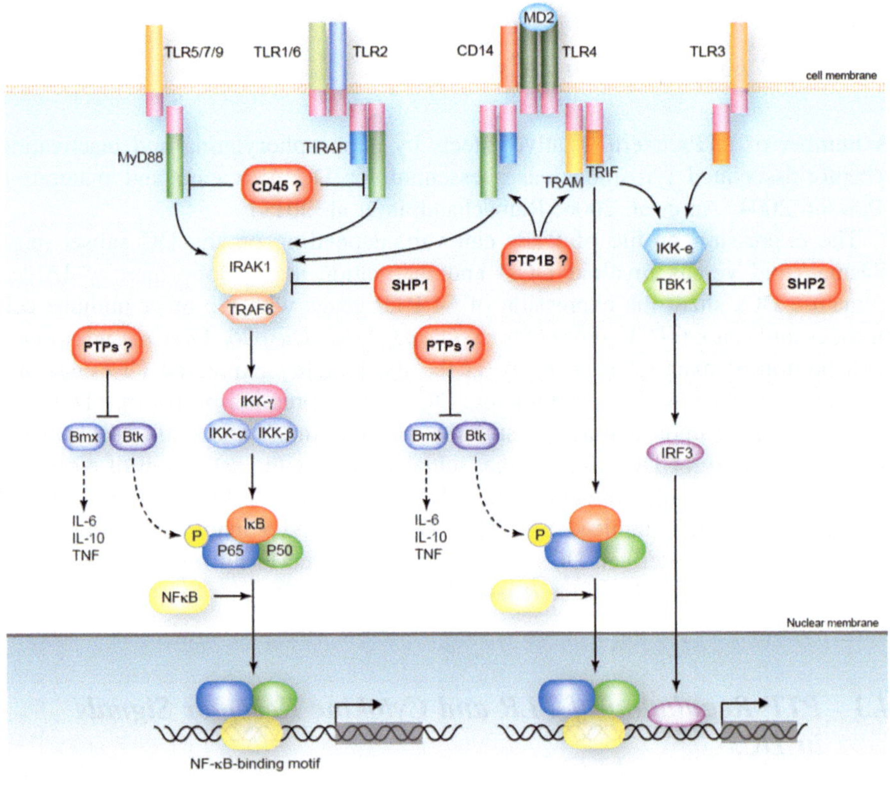

Fig. 1 PTP-mediated regulation of TLR signaling pathway. After ligand binding, TLRs undergo the conformational changes that induce the recruitment of downstream signaling molecules and transcription factors implicated in the regulation of DC functions. Several PTPs such as SHP-1, SHP-2, CD45 and PTP1B have been shown to act as negative regulators by dephosphorylating tyrosine residues on adaptor proteins and protein tyrosine kinases downstream TLR signaling. SHP-1 inhibits IRAK1, limiting the activation of NF-κB signaling pathways downstream of TLR4, whereas SHP-2 inhibits TRIF-dependent activation of TBK1 signals and their target gene expression. In turn, PTP1B might inhibit TLR signaling via suppression of both MyD88-dependent and TRIF-dependent (MyD88-independent) pathways in macrophages and DCs and therefore suppress the production of proimflammatory cytokines. (Adapted from Akira and Takeda2004). Drawed by Noriko Uetani.

Furthermore, TLR engagement induces the expression of TAM PTKs, a family of receptor tyrosine kinases (containing TYRO3, AXL, and MERTK) that are immune checkpoint proteins expressed and involved in the regulation of antigen-presenting cell (APC) activation. These receptors recognize two cognate ligands, protein S (PROS1) and growth-arrest-specific 6 (GAS6). Both ligands contain gamma carboxylated glutamic acid (Gla) residues and are able to bind to phosphatidylserine (PtdSer). As such, these ligands link PtdSer-exposing apoptotic cells to TAM receptor expressing cells (Stitt et al. 1995; Anderson et al. 2003).

TAM signaling inhibits intratumoral DC activation by usurping the pro-inflammatory IFNAR-STAT1 signaling pathways to induce SOCS1 and SOCS3 expression. In turn, SOCS proteins promote the degradation of adaptor molecules implicated in the regulation of TLR, type I IFN, and JAK-STAT signaling pathways (Rothlin et al. 2007). In physiological conditions, cognate ligands taken up by APCs, such as DCs, trigger TAM inhibitory signals to promote T cell tolerance to self and avoid damage to host tissue. However, in the context of cancer, the persistent expression of these negative checkpoint molecules provokes functional exhaustion and impairs both T cell activation and antitumor immunity (Wherry and Kurachi 2015). Therefore, the identification of PTPs implicated in the regulation of TAM receptor signaling in DCs would uncover therapeutic targets to improve the efficacy of DC-based vaccines for cancer immunotherapy.

4.4 PTPs as Regulators of DC Maturation and T Cell Differentiation

Recent studies have identified several PTPs as important negative checkpoint regulators of signaling pathways fundamental for DC maturation and function (Watson et al. 2016). In particular, the JAK-STAT pathways are essential regulators of the immune response and cytokine signaling. Specifically, JAK-STAT pathways control DC differentiation, maturation, activation, and DC-dependent induction of T_H1-cell differentiation (Lugo-Villarino et al. 2003; Li and Watowich 2013; Arima et al. 2010). Early phases of DC differentiation, as well as the commitment of common lymphoid progenitors (CLPs) and common myeloid progenitors (CMPs) to the DC lineage, strictly depend on STAT3 activation as a regulatory checkpoint (Laouar et al. 2003). DC differentiation inversely correlates with STAT3 activation, as mature DCs show low levels of STAT3 activation. STAT3 hyperactivation results in the inhibition of DC maturation and activation in response to diverse stimuli (Kitamura et al. 2005; Hoentjen et al. 2005). On the other hand, STAT1 and STAT4 phosphorylations increase in mature and fully functional DCs with potent antigen presentation capacity for T cell activation and cytokine production. Two tyrosine phosphatase-specific enzymes PTP1B and TC-PTP restrict DC activation by acting as important negative checkpoint regulators of the JAK/STAT signaling pathway. Indeed, our group has reported that downregulation of these two PTPs induces a more immunogenic DC phenotype via STAT1 and STAT4 hyperactivation, leading to robust antigen-specific T cell activation and antitumor responses (Penafuerte et al. 2017).

Other PTPs, such as SHP-1 and the phosphatases that regulate Akt activation (PTEN and SHP-1) have also been identified as negative checkpoint regulators of DC activation. SHP-1 is recruited to the immunoreceptor tyrosine-based inhibitory motif (ITIM) of the inhibitory receptor FcγRII leading to the inhibition of MAPK activation, which consequently impairs DC responsiveness to immune complexes. Both PTEN and SHP-1 attenuate PI3K/Akt signaling, which affects the antigen uptake and presentation capacity of DCs. Hence, the inhibition of these PTP activities enhances DC-mediated immunity and potentiates their antitumor properties (Carmi et al. 2016).

4.5 PTP Regulation of Cell Migration and Vesicle Transport

The actin-filament assembly and disassembly required for vesicle transport and cell migration is tightly regulated by a complex network of mechano-sensing proteins, including focal adhesion kinase (FAK), Src, extracellular signal-regulated kinase (ERK), p130CAS, paxillin, and myosin light-chain kinase (MLCK) (Faure-Andre et al. 2008; Bar-Sagi and Hall 2000). Protein tyrosine phosphorylation and dephosphorylation events promote the formation and disassembly of focal adhesions (sites of contact between the extracellular matrix and the cytoskeleton) to allow cell motility. In particular, the protein tyrosine phosphatases SHP-2, PTP1B, PTPRF (LAR), and PTP-PEST have been implicated in the regulation of cell spreading, migration, and focal adhesion formation in several cell types (Yu et al. 1998; Arregui et al. 1998; Angers-Lousteau et al. 1999).

In conclusion, the inhibition of PTP activity enhances pro-inflammatory cytokine production, chemokine receptor signaling, DC survival, and antigen-specific T cell stimulation. Still, the existing knowledge on the role of PTPs in the regulation of DC functions is limited. Specifically, more efforts are needed to systematically explore the phosphatome expression in DCs as well as the detailed function of PTPs in DC maturation, antigen presentation, and regulation of negative checkpoints.

5 PTPs as Components of Inhibitory Signaling in T Cells

5.1 PTPs as Transducers of Checkpoint Signals in T Cells

CD4 and CD8 T cells are essential players in the adaptive immune response, being effectors and predominant coordinators of immune system decisions. The interactions between CD4 T cells and DCs not only decide if a response occurs, but also the type of response initiated, either humoral, cytotoxic, or both. Given the right cues, CD8 T cells will differentiate into cytotoxic T cells (CTLs), essential effectors of the adaptive cytotoxic immune response. Blocking of T cell interactions with inhibitory ligands present on either DCs or target cells is the mechanistic basis for the already approved checkpoint inhibitor therapies involving CTLA-4 and PD-1.

Due to their modulatory nature, immune checkpoints are signals based mainly on the inhibition of activating receptors. By this definition, PTPs are presumed to be an integral part of checkpoint function. Although an extensive amount of literature is available on CTLA-4 and PD-1, other inhibitory members of the CD28 family or other receptor families recruiting PTPs are considered good candidates for future therapeutic interventions. Here, we will describe the up-to-date knowledge of the PTPs involved in these pathways (Fig. 2).

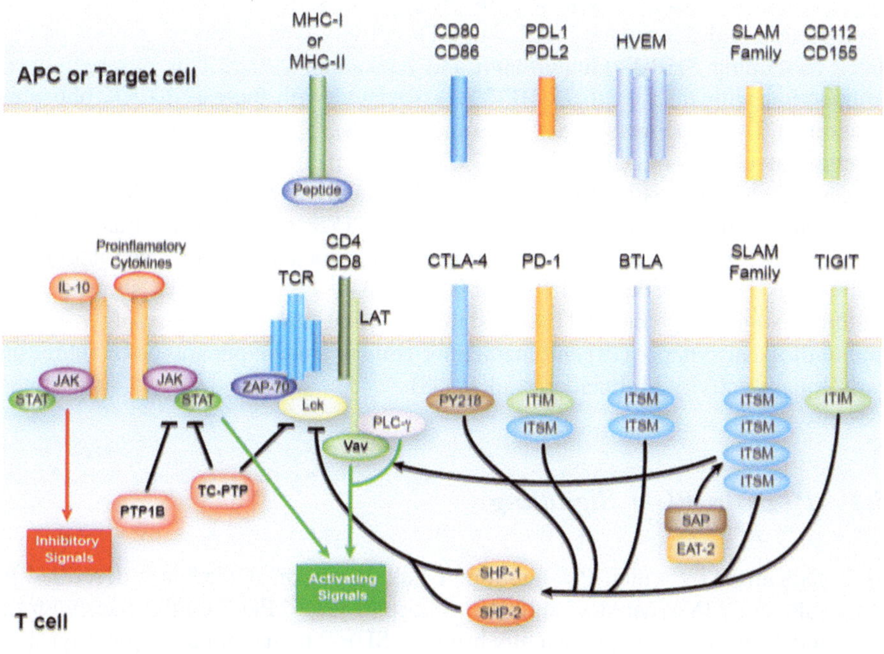

Fig. 2 Protein tyrosine phosphatases and inhibitory checkpoint signals in T cells. Checkpoint receptors as CTLA-4, PD-1, BTLA, SLAM family of receptors and TIGIT recruit the phosphatases SHP-1 and/or SHP-2 to regulate the TCR phosphotyrosine dependent activating signals (*right side*). The TCR-activated kinase Lck is also a target of the phosphatase TC-PTP. The combined action of the phosphatases PTP-1B and TC-PTP dephosphorylating pro-inflammatory JAK and STAT signals promotes a balance shift towards the prevalence of anti-inflammatory signals from cytokines such as IL-10 (*left side*). The adaptor proteins SAP and EAT-2 can reverse the inhibitory signals of SLAM family receptors acting as "natural" checkpoint inhibitors. Drawed by Noriko Uetani.

5.2 PTPs in CTLA-4 Signaling

CTLA-4 engagement by its ligands B7.1 and B7.2 counterbalances activating CD28 signals. However, whether the mechanism involves the recruitment of PTPs is not yet clear. Early work on CTLA-4 showed reduced tyrosine phosphorylation of essential components of the TCR signaling machinery (CD3ζ and LAT) when CTLA-4 is co-engaged with the TCR by antibody crosslinking (Marengère et al. 1996). In the same study, SHP-2 was identified in immunoprecipitates of CTLA-4 in primary T cells and cell lines.

Later studies have supported the role of PTPs in the inhibitory activity of CTLA-4 (Lee et al. 1998; Guntermann and Alexander 2002), yet it is not clear whether the phosphatase responsible is SHP-1 or SHP-2. The short cytoplasmic domain of CTLA-4 contains two potential sites for tyrosine phosphorylation, motifs Y201VKM and Y218FIP. Only the tyrosine at position 218 was found to be able to interact with the SH2 domain of both SHP-1 and SHP-2 (Marengère et al. 1996; Guntermann and Alexander 2002), but PTP activity was observed only in precipitates containing SHP-1 (Guntermann and Alexander 2002). This observation, in conjunction with the fact that SHP-2 is associated with increased ERK phosphorylation in T cells (Frearson and Alexander 1998), opposes available data on the function of CTLA-4 (Calvo et al. 1997; Guntermann and Alexander 2002) and suggests that SHP-1 is responsible for the PTP activity associated with CTLA-4.

Co-engagement of CTLA-4 with TCR and CD28 reduces Ca^{2+} flux and IL-2 secretion. Schneider et al. showed that the contact time and area of T cell interaction with APCs is also reduced. In addition, CTLA-4 stimulated cells showed more motility, suggesting the CTLA-4 signals inhibit the formation of a mature immunological synapse (IS) and maintain T cells in a naive phenotype by inhibiting early kinase activity (Schneider et al. 2008).

5.3 PTPs in PD-1 Signaling

The mechanism by which PD-1 inhibit T-cell activation is elegantly described by Yokosuka et al. (Yokosuka et al. 2012). Phosphorylated PD-1 co-localizes with the activated TCR machinery, transiently bringing SHP-2 to the immunological synapse. Its activity interferes with the phosphorylation of essential components of the TCR machinery such as VAV and PLCγ1. As a result, the actin cytoskeleton changes coordinated by phosphorylated VAV do not take effect, inhibiting the gathering of TCR complexes into bigger clusters required for signal amplification. Similarly, inhibition of PLC-γ1 activity is responsible for the reduction of downstream Ca^{2+}, PKC-θ, and MAPK signaling as well as inhibiting cytokine secretion and activation.

The intra-cytoplasmic domain of PD-1 differs from that of the main receptors of the CD28 family as it contains two well-defined motifs known to recruit inhibitory signals, an ITIM in the tyrosine at position 223 and an ITSM at tyrosine position

248 (Bakdash et al. 2013). SHP-1 and SHP-2 recruitments by ITIM motifs are a well-known mechanism for the modulation of phosphotyrosine signaling. Studies conducted regarding the association of these PTPs with PD-1 tyrosine motifs found that although SHP-2 interacts with the PD-1 cytoplasmic domain, this interaction was dependent on phosphorylation at position 248, but independent of the conservation of the ITIM at position 223 (Okazaki et al. 2001). Later, it was demonstrated that a similar mechanism can be observed in T cells (Latchman et al. 2001). SHP-1 has also been found to precipitate with the intracellular domain of PD-1 at both phosphotyrosine motifs. However, the association at either site does not seem to be relevant to the phosphatase activity or the biological function of PD-1 ligation (Sathish et al. 2001). Although the tyrosine at position Y248 fulfills the consensus sequence of an ITSM, recruitment of SHP-2 is likely the inhibitory mechanism of PD-1, as other ITSM downstream effectors, such as the adaptor SAP or the inositol phosphatase SHIP, do not to interact (Chemnitz et al. 2004).

5.4 *Other Inhibitory Molecules of the CD28 Family*

The therapeutic success obtained with blockers of the CTLA-4 and PD-1 pathways promoted the search for new potential checkpoint signals. Several other members of the CD28-B7 family have been described to promote inhibitory signals, and, of those, BTLA can intrinsically modulate the activation of T cells.

BTLA is a receptor homologous to CTLA-4 and PD-1, sharing similar features like ITIM and ITSM motifs, as well as being known to recruit inhibitory phosphatases, suggesting a regulatory role for BTLA (Watanabe et al. 2003). Indeed, the phosphorylated tyrosine residues can bind SHP-2 and subsequent engagement of BTLA reduces the secretion of IL-2 in activated T cells (Watanabe et al. 2003). Like PD-1, BTLA co-localizes with the TCR in the immunological synapse (Wu et al. 2007) and has been found to inhibit intratumoral T lymphocytes (ITLs) (Fourcade et al. 2012). However, BTLA expression in ITLs has been proposed as a marker of good prognosis (Paulos and June 2010). Transcriptional profiling of activated T cells demonstrated that the genetic footprint after BTLA crosslinking resembled more that of the co-activation receptor ICOS than the profiles obtained by stimulating either CTLA-4 or PD-1 (Wakamatsu et al. 2013). More recently, it was shown that $CD8^{+}BTLA^{+}$ ITLs proliferate more and produce more IL-2 than their $CD8^{+}BTLA^{-}$ counterparts (Haymaker et al. 2015). Increased levels of phosphorylated Akt were found in the $BTLA^{+}$ TILs, suggesting an activating role for BTLA through the recruitment of Grb2 by a third phosphotyrosine motif. The exact mechanism by which BTLA favors inhibitory or activating signals is still unknown, limiting its potential use in a clinical setting.

5.5 Other ITIM-Containing Receptors

TIGIT is a novel inhibitory receptor of the PVR family. Together with the activating receptor DNAM, the inhibitory receptor CD96 and the ligands CD112 and CD155, TIGIT forms an analogous system to the one comprised CD28/CTLA-4/B7.1/B7.2 molecules (Liu et al. 2017). By antagonizing DNAM signals, TIGIT becomes an excellent candidate for checkpoint inhibition in NK and CD8 T cells. TIGIT recruits inhibitory signals through a single ITIM motif, which in NK cells is known to be recognized by the SH2 domain of β-arrestin, thus recruiting SHIP1 and reducing the degradation of phosphorylated IκBα (Li et al. 2014). In T cells, TIGIT co-ligation with CD3 and CD28 inhibits proliferation and IFN-γ secretion, opposing DNAM signaling (Lozano et al. 2012). Whether SHP-1 or SHP-2 are involved is not yet known.

5.6 ITSM Signaling

Recruitment of SHP-2 by the ITSM motif of PD-1 is essential for its inhibitory function. Several other immune receptors families contain this tyrosine motif, including CD31, SIRP-α, siglecs, and the SLAM family of receptors (Shlapatska et al. 2001). ITSM functions are complex and can generate both activating and inhibitory signals. Of importance as modulators of T cell activation is the SLAM family of receptors (Veillette et al. 2009). These receptors and a small family of adaptors which mainly associate with them, the slam-associated protein (SAP) family, are noticeable since mutations are associated with lymphoproliferative and autoimmune diseases (Veillette et al. 2013; Cunninghame Graham et al. 2008). This family is very rich in intracellular ITSM motifs, and most of them are known to recruit inhibitory molecules including SHP-1 and SHP-2.

5.7 SAP Family of Adaptors as "Natural" Checkpoint Inhibitors

The vast recruitment of inhibitory molecules by the SLAM family ITSM give them an important role in immune system regulation. However, ITSM phosphotyrosine motifs have the potential to recruit activating signals as well. In the SLAM family of receptors, this functional switch is controlled by the expression of the SAP family adaptors SAP and EAT-2 (Veillette et al. 2009). Although in most SLAM receptors adaptor binding will turn them into co-activating molecules, the case of SLAM where SAP association helps the recruitment of SHIP adds to signal complexity (Latour et al. 2001). From extensive in vitro and mouse studies, it is hypothesized that expression of these adaptors is important in the control of hematopoietic malignancies by CD8 T cells and NK cells (Pérez-Quintero et al. 2014; Dong et al. 2012). Ectopic expression of these adaptors is a potential strategy to delay tolerance for malignancies of hematopoietic origin.

5.8 *JAK/STAT Regulation by PTP1B and TC-PTP*

Cytokines are the most important soluble factors modulating the immune response. Cytokine receptors belong to the tyrosine kinase family and signals generally require the activation of the JAK family of tyrosine kinases and the JAK-dependent tyrosine phosphorylation of the downstream STAT transcription factors.

The tyrosine phosphatases PTP1B and TC-PTP are highly homologous (70% homology) non-receptor PTPs known to target the JAK-STAT pathway (Pike and Tremblay 2016). By means of a substrate-trapping mutant form of PTP1B, the kinases JAK2 and TYK2 were initially identified as targets of these enzymes. Dephosphorylation of these kinases and downstream transcription factors STAT1 and STAT3 negatively regulate IFN-α and IFN-γ signaling (Myers et al. 2001). Later studies found that PTP1B can directly bind the cytoplasmic domain of the IL-4 receptor, regulating downstream signaling by reducing STAT6 phosphorylation (Lu et al. 2008). In general, PTP1B-deficient mice show signs of inflammation and are prone to autoimmune disease (Berdnikovs et al. 2012). However, the cell-specific consequences of PTP1B deficiency vary. Deficient B cells proliferate more and older mice show a larger production of auto-antibodies (Medgyesi et al. 2014), while deficient macrophages are more sensitive to IL-10 inhibition (Pike et al. 2014). The T cell intrinsic role of PTP1B is not known, and because of its extensive effects on hematopoiesis and other components of the immune system interacting with T cells, the development of a cell-specific mutant is of high relevance.

TC-PTP deficiency leads to premature death in mice due to severe immune defects (You-Ten et al. 1997). Similarly to PTP1B, TC-PTP is a negative regulator of inflammation (Wiede et al. 2017). In T cells, TC-PTP regulates cell activation and proliferation by targeting the dephosphorylation of tyrosine 418 of the kinase LCK which, when phosphorylated, enhances its kinase activity (Wiede et al. 2011). Besides acting directly on TCR signals, TC-PTP targets JAK1 and JAK3 in a murine T cell line, leading to a reduced expression of STAT5. In the same study, it was found that TC-PTP regulates the STAT1 phosphorylation (Heinonen et al. 2009). When deleted in an inducible fashion in adult mice to overcome its lethal effects, mice developed a T cell-dependent autoimmune disease, with increased numbers of T follicular helper (Tfh) cells and germinal center (GC) B cells (Wiede et al. 2017). However, whether the defect observed is intrinsic to T cells is still unknown. Because of its double role in regulating TCR and cytokine signals, the intrinsic effects of TC-PTP inhibition in T cells are of clinical value and have the potential to enhance T cell activity.

Even considering the widespread distribution and intracellular location, pharmacological inhibition of PTP1B and TC-PTP is a potential mechanism for checkpoint inhibition. This may be particularly useful in cancer immunotherapies involving the ex vivo manipulation of cellular products like the chimeric antigen receptor-T (CAR-T) cell-based therapies. One of the main difficulties with CAR-T therapies is the variability and low responsiveness of CAR-T cells resulting from the multiple differentiation states of the T cells isolated for production (Golubovskaya and Wu 2016). By targeting multiple activating pathways, inhibition of TC-PTP and PTP1B can restore T cell differentiation to cell types associated with better CAR-T responses, such as central memory T cells.

6 Conclusion

The scope of regulation provided by the multiple members of the PTP family brings forth a great number of opportunities to modulate them at immune system checkpoints. Among more than 80 different members of this family that are expressed in the immune system, we have described some of the most critical ones that, by their action, can significantly influence the function and efficiency of the immune system. This potential is unfortunately tempered by the lack of small molecule PTP inhibitors currently in the pharmacopeia. Indeed, in spite of much effort and investment in this field, and although several have been moving through clinical trials in different disease applications, not a single PTP inhibitor is currently approved for the clinic. Yet, novel approaches seeking allosteric inhibitors (Chen et al. 2016), antibody modulation, and novel small molecule inhibitors indicate that we must be optimistic that the pharmacology of PTPs will soon be resolved.

We and others have also taken the position that, because of their sequence similarity functional redundancies and poor inhibitor availability, we could better address their targeting as checkpoint inhibitors through a cell therapy approach. As we showed in the case of dendritic cell cancer vaccines (Penafuerte et al. 2017), ex vivo inhibition of PTP1B and TC-PTP provides a remarkable improvement in the efficacy of antigen presentation machinery and simultaneously in the availability of the MHC complex and co-stimulatory molecules. Cell-based therapeutics are increasingly being use into novel clinical regimens. Combine with our growing understanding of the immune system, they open diverse and original approaches in treating cancers and other human diseases.

The future usage of PTP modulators to improve immune responses is a treasure chest that remains to be opened. Therefore, there is an important need for better characterization of PTP function since we have just begun to study a handful of PTPs that are expressed in immune cells. The recent advances on identifying PTP inhibitors as novel immune checkpoints, will no doubt offer promising therapeutic applications in restoring and improving functional innate resistance to cancer.

Acknowledgements We thank N. Uetani for expert graphical assistance. In addition, we recognize the support of the Canadian Institute of Health Research (grant MOP-62887) and the Aclon-Richard and Edith Strauss Foundation to M.L.T., V.V. is a recipient of the Charlotte and Leo Karassik Family Foundation Fellowship, T.H. is a recipient of a Fonds de Recherche du Quebec—Santé studentships. M.L.T. is a holder of the Jeanne and Jean-Louis Lévesque Chair in Cancer Research.

References

Akira S, Takeda K (2004) Toll-like receptor signalling. Nat Rev Immunol 4:499–511. doi:10.1007/s10753-017-0604-7

Alonso A, Pulido R (2016) The extended human PTPome: a growing tyrosine phosphatase family. FEBS J April 283(8):1404–1429

Alonso A, Sasin J, Bottini N, Friedberg I, Friedberg I, Osterman A, Godzik A, Hunter T, Dixon J, Mustelin T (2004) Protein tyrosine phosphatases in the human genome. Cell June 11 117 (6):699–711. Review

An H, Zhao W, Hou J, Zhang Y, Xie Y, Zheng Y et al (2006) SHP-2 phosphatase negatively regulates the TRIF adaptor protein-dependent type I interferon and proinflammatory cytokine production. Immunity 25:919–928. doi:10.1016/j.immuni.2006.10.014

Anderson HA, Maylock CA, Williams JA, Paweletz CP, Shu H, Shacter E (2003) Serum-derived protein S binds to phosphatidylserine and stimulates the phagocytosis of apoptotic cells. Nat Immunol 4:87–91. doi:10.1038/ni871

Angers-Loustau A, Cote JF, Charest A, Dowbenko D, Spencer S, Lasky LA et al (1999) Protein tyrosine phosphatase-PEST regulates focal adhesion disassembly, migration, and cytokinesis in fibroblasts. J Cell Biol 144:1019–1031

Anguille S, Smits EL, Lion E, van Tendeloo VF, Berneman ZN (2014) Clinical use of dendritic cells for cancer therapy. Lancet Oncol 15:e257–e267

Arimura Y, Yagi J (2010) Comprehensive expression profiles of genes for protein tyrosine phosphatases in immune cells. Sci Signal 3(137):1–11

Arima K, Watanabe N, Hanabuchi S, Chang M, Sun SC, Liu YJ (2010) Distinct signal codes generate dendritic cell functional plasticity. Sci Signal 3:ra4

Arregui CO, Balsamo J, Lilien J (1998) Impaired integrin-mediated adhesion and signaling in fibroblasts expressing a dominant-negative mutant PTP1B. J Cell Biol 143:861–873

Bakdash G, Sittig S, van Dijk T, Figdor C, Jolanda de Vries I (2013) The nature of activatory and tolerogenic dendritic cell-derived signal II. Front Immunol 4:53

Barron L, Wynn TA (2011) Macrophage activation governs schistosomiasis-induced inflammation and fibrosis. Eur J Immunol 41(9):2509–2514

Bar-Sagi D, Hall A (2000) Ras and Rho GTPases: a family reunion. Cell 103:227–238

Beatty GL, Chiorean EG, Fishman MP, Saboury B, Teitelbaum UR, Sun W et al (2011) CD40 agonists alter tumor stroma and show efficacy against pancreatic carcinoma in mice and humans. Science 331:612–616

Berdnikovs S, Pavlov VI, Abdala-Valencia H, McCary CA, Klumpp DJ, Tremblay ML, Cook-Mills JM (2012) PTP1B deficiency exacerbates inflammation and accelerates leukocyte trafficking in vivo. J Immunol 188:874–884

Bussières-Marmen S, Hutchins AP, Schirbel A, Rebert N, Tiganis T, Fiocchi C, Miranda-Saavedra D, Tremblay ML (2014) Characterization of PTPN2 and its use as a biomarker. Methods January 15 65(2):239–246

Calvo CR, Amsen D, Kruisbeek AM (1997) Cytotoxic T lymphocyte antigen 4 (CTLA-4) interferes with extracellular signal-regulated kinase (ERK) and Jun NH2 terminal kinase (JNK) activation, but does not affect phosphorylation of T cell receptor ζ and ZAP70. J Exp Med 186:1645–1653

Carmi Y, Prestwood TR, Spitzer MH, Linde IL, Chabon J, Reticker-Flynn NE et al (2016) Akt and SHP-1 are DC-intrinsic checkpoints for tumor immunity. JCI insight 1:e89020

Chang HH, Miaw SC, Tseng W, Sun YW, Liu CC, Tsao HW et al (2013) PTPN22 modulates macrophage polarization and susceptibility to dextran sulfate sodium-induced colitis. J Immunol 191:2134–2143

Charbonneau H, Tonks NK, Walsh KA, Fischer EH (1988) The leukocyte common antigen (CD45): A putative receptor-linked protein tyrosine phosphatase. Proc Natl Acad Sci USA 85:7182–7186

Chemnitz JM, Parry RV, Nichols KE, June CH, Riley JL (2004) SHP-1 and SHP-2 associate with immunoreceptor tyrosine-based switch motif of programmed death 1 upon primary human T cell stimulation, but only receptor ligation prevents T cell activation. J Immunol 173:945–954

Chen YN, LaMarche MJ, Chan HM, Fekkes P, Garcia-Fortanet J, Acker MG, Antonakos B et al (2016) Allosteric inhibition of SHP2 phosphatase inhibits cancers driven by receptor tyrosine kinases. Nature July 7 535(7610):148–152. Epub 2016 Jun 29

Constantino J, Gomes C, Falcao A, Neves BM, Cruz MT (2017) Dendritic cell-based immunotherapy: a basic review and recent advances. Immunol Res. doi:10.1007/s12026-017-8931-1

Covarrubias AJ, Aksoylar HI, Horng T (2015) Control of macrophage metabolism and activation by mTOR and Akt signaling. Semin Immunol 27:286–296

Croker BA, Krebs DL, Zhang JG, Wormald S, Willson TA, Stanley EG et al (2003) SOCS3 negatively regulates IL-6 signaling in vivo. Nat Immunol 4(6):540–545

Cromie MJ1, Groisman EA (2010) Promoter and riboswitch control of the Mg2 + transporter MgtA from Salmonella enterica. J Bacteriol January 192(2):604–607

Cunninghame Graham DS, Vyse TJ, Fortin PR, Montpetit A, Cai YC, Lim S, McKenzie T, Farwell L, Rhodes B, Chad L, Hudson TJ, Sharpe A, Terhorst C, Greenwood CM, Wither J, Rioux JD (2008) Association of LY9 in UK and Canadian SLE families. Genes Immun 9:93–102

David M, Chen HE, Goelz S, Larner AC, Neel BG (1995) Differential regulation of the α/β interferon-stimulated Jak/Stat pathway by the SH2 domain-containing tyrosine phosphatase SHPTP1. Mol Cell Biol 15:7050–7058

Davies LC, Jenkins SJ, Allen JE, Taylor PR (2013) Tissue-resident macrophages. Nat Immunol 14:986–995

Dhodapkar MV, Steinman RM, Krasovsky J, Munz C, Bhardwaj N (2001) Antigen-specific inhibition of effector T cell function in humans after injection of immature dendritic cells. J Exp Med 193:233–238

Dong Z, Davidson D, Pérez-Quintero LA, Kurosaki T, Swat W, Veillette A (2012) The adaptor SAP controls NK Cell activation by regulating the enzymes Vav-1 and SHIP-1 and by enhancing conjugates with target cells. Immunity 36:974–985

Dubé N, Bourdeau A, Heinonen KM, Cheng A, Loy AL, Tremblay ML (2005) Genetic ablation of protein tyrosine phosphatase 1Baccelerates lymphomagenesis of p53-null mice through the regulation of B-cell development. Cancer Res 65(21):10088–10095.PMID:16267035

Elchebly M, Payette P, Michaliszyn E, Cromlish W, Collins S, Loy AL, Normandin D, Cheng A, Himms-Hagen J, Chan CC,Ramachandran C, Gresser MJ, Tremblay ML, Kennedy BP (1999) Increased insulin sensitivity and obesity resistance in mice lacking the protein tyrosine phosphatase-1B gene.Science 283(5407):1544–1548. PMID:10066179

Faure-Andre G, Vargas P, Yuseff MI, Heuze M, Diaz J, Lankar D et al (2008) Regulation of dendritic cell migration by CD74, the MHC class II-associated invariant chain. Science 322:1705–1710

Feldhammer M, Uetani N, Miranda-Saavedra D, Tremblay ML (2013) PTP1B: a simple enzyme for a complex world. Crit Rev Biochem Mol Biol. Sep-Oct 48(5):430–445

Fourcade J, Sun Z, Pagliano O, Guillaume P, Luescher IF, Sander C, Kirkwood JM, Olive D, Kuchroo V, Zarour HM (2012) CD8 T cells specific for tumor antigens can be rendered dysfunctional by the tumor microenvironment through upregulation of the inhibitory receptors BTLA and PD-1. Can Res 72:887–896

Frearson JA, Alexander DA (1998) The phosphotyrosine phosphatase SHP-2 participates in a multimeric signaling complex and regulates T cell receptor (TCR) coupling to the ras/mitogen-activated protein kinase (MAPK) pathway in Jurkat T cells. J Exp Med 187:1417–1426

Frey AB, Monu N (2008) Signaling defects in anti-tumor T cells. Immunol Rev 222:192–205

Gabrilovich D (2004) Mechanisms and functional significance of tumour-induced dendritic-cell defects. Nat Rev Immunol 4:941–952

Garg AD, Vara Perez M, Schaaf M, Agostinis P, Zitvogel L, Kroemer G, et al (2017) Trial watch: Dendritic cell-based anticancer immunotherapy. Onco Immunol e1328341

Golubovskaya V, Wu L (2016) Different subsets of T Cells, memory, effector functions, and CAR-T immunotherapy. Cancers 8:36

Grant L, Shearer K, Czopek A et al (2014) Myeloid-cell protein tyrosine phosphatase-1B deficiency in mice protects against high—fat diet and lipopolysaccharide induced inflammation, hyperinsulinemia and endotoxemia through an IL-10 STAT3-dependent mechanism. Diabetes 63:456–470

Gungabeesoon J, Tremblay ML, Uetani N (2016) Localizing PRL-2 expression and determining the effects of dietary Mg(2 +) on expression levels. Histochem Cell Biol July 146(1):99–111

Guntermann C, Alexander DR (2002) CTLA-4 suppresses proximal TCR signaling in resting human CD4 T cells by inhibiting ZAP-70 Tyr < sup > 319 </sup > phosphorylation: a potential role for tyrosine phosphatases. J Immunol 168:4420–4429

Hardy S, Uetani N, Wong N, Kostantin E, Labbé DP, Bégin LR, Mes-Masson A, Miranda-Saavedra D, Tremblay ML (2015) The protein tyrosine phosphatase PRL-2 interacts with the magnesium transporter CNNM3 to promote oncogenesis. Oncogene February 19 34 (8):986–995

Haymaker CL, Wu RC, Ritthipichai K, Bernatchez C, Forget MA, Chen JQ, Liu H, Wang E, Marincola F, Hwu P, Radvanyi LG (2015) BTLA marks a less-differentiated tumor-infiltrating lymphocyte subset in melanoma with enhanced survival properties. Oncoimmunology 4: e1014246

Heinonen KM, Bourdeau A, Doody KM, Tremblay ML (2009) Protein tyrosine phosphatases PTP-1B and TC-PTP play nonredundant roles in macrophage development and IFN-γ signaling. Proc Natl Acad Sci 106:9368–9372

Hoentjen F, Sartor RB, Ozaki M, Jobin C (2005) STAT3 regulates NF-kappaB recruitment to the IL-12p40 promoter in dendritic cells. Blood 105:689–696

Irie-Sasake J, Sasaki T, Matsumoto W, Opavsky A, Cheng M, Welstead G et al (2000) CD45 is a JAK phosphatase and negatively regulated cytokine receptor signaling. Nature 409:349–354

Julien SG, Dubé N, Hardy S, Tremblay ML (2011) Inside the human cancer tyrosine phosphatome. Nat Rev Cancer January 11(1):35–49

Kitamura H, Kamon H, Sawa S, Park SJ, Katunuma N, Ishihara K et al (2005) IL-6-STAT3 controls intracellular MHC class II alphabeta dimer level through cathepsin S activity in dendritic cells. Immunity 23:491–502

Kozicky LK, Sly LM (2015) Phosphatase regulation of macrophage activation. Semin Immunol 27:276–285

Lang R, Pauleau AL, Parganas E, Takahashi Y, Mages J, Ihle JN et al (2003) SOCS3 regulates the plasticity of gp130 signaling. Nat Immunol 4(6):546–550

Laouar Y, Welte T, Fu XY, Flavell RA (2003) STAT3 is required for Flt3L-dependent dendritic cell differentiation. Immunity 19:903–912

Latchman Y1, Wood CR, Chernova T, Chaudhary D, Borde M, Chernova I, Iwai Y, Long AJ, Brown JA, Nunes R, Greenfield EA, Bourque K, Boussiotis VA, Carter LL, Carreno BM, Malenkovich N, Nishimura H, Okazaki T, Honjo T, Sharpe AH, Freeman GJ (2001) PD-L2 is a second ligand for PD-1 and inhibits T cell activation. Nat Immunol 2:261–268

Latour S, Gish G, Helgason CD, Humphries RK, Pawson T, Veillette A (2001) Regulation of SLAM-mediated signal transduction by SAP, the X-linked lymphoproliferative gene product. Nat Immunol 2:681–690

Ledbetter JA, Tonks NK, Fischer EH, Clark EA (1988) CD45 regulates signal transduction and lymphocyte activation by specific association with receptor molecules on T or B cells. Proc Natl Acad Sci USA 85:8628–8632

Lee KM, Chuang E, Griffin M, Khattri R, Hong DK, Zhang W, Straus D, Samelson LE, Thompson CB, Bluestone JA (1998) Molecular basis of T cell inactivation by CTLA-4. Science 282:2263–2266

Li HS, Watowich SS (2013) Diversification of dendritic cell subsets: emerging roles for STAT proteins. Jak-Stat 2:e25112

Lim WA1, Pawson T (2010) Phosphotyrosine signaling: evolving a new cellular communication system. Cell September 3 142(5):661–667

Lissandrini D, Vermi W, Vezzalini M, Sozzani S, Facchetti F, Bellone G et al (2006) Receptor-type protein tyrosine phosphatase gamma (PTPgamma), a new identifier for myeloid dendritic cells and specialized macrophages. Blood 108:4223–4231

Li M, Xia P, Du Y et al (2014) T-cell immunoglobulin and ITIM Domain (TIGIT) receptor/poliovirus receptor (PVR) ligand engagement suppresses interferon-γ Production of natural killer cells via β-arrest in 2-mediated negative signaling. J Biol Chem 289:17647–17657

Liu X-G, Hou M, Liu Y (2017) TIGIT: a novel therapeutic target for tumor immunotherapy. Immunological Investigations 46:172–182

Lozano E, Dominguez-Villar M, Kuchroo V, Hafler DA (2012) The TIGIT/CD226 axis regulates human T cell function. J Immunol 188:3869–3875

Lu X, Malumbres R, Shields B, Jiang X, Sarosiek KA, Natkunam Y, Tiganis T, Lossos IS (2008) PTP1B is a negative regulator of interleukin 4–induced STAT6 signaling, Blood 112:4098–108

Lugo-Villarino G, Maldonado-Lopez R, Possemato R, Penaranda C, Glimcher LH (2003) T-bet is required for optimal production of IFN-gamma and antigen-specific T cell activation by dendritic cells. PNAS 100:7749–7754

Mantovani A, Marchesi D, Malesci A, Laghi L, Allavena P (2017) Tumour-associated macrophages as treatment targets in oncology. Nat Rev Clin Oncol 14(7):399–416

Marengère LE, Waterhouse P, Duncan GS, Mittrücker HW, Feng GS, Mak TW (1996) Regulation of T cell receptor signaling by tyrosine phosphatase SYP association with CTLA-4. Science 272:1170–1173

Medgyesi D, Hobeika E, Biesen R, Kollert F, Taddeo A, Voll RE, Hiepe F, Reth M (2014) The protein tyrosine phosphatase PTP1B is a negative regulator of CD40 and BAFF-R signaling and controls B cell autoimmunity. J Exp Med 211:427–440

Mills CD, Shearer J, Evans R, Caldwell MD (1992) Macrophage arginine metabolism and the inhibition or stimulation of cancer. J Immunol 149:2709–2714

Mills CD, Lenz LL, Harris RA (2016) A breakthrough: macrophage-directed cancer immunotherapy. Can Res 76(3):513–516

Murphy K, Weaver C (2017) Janeway's immunobiology, 9th edn. Garland Science, Taylor & Francis Group, pp 7, 78–79, 839, 841

Murphy K, Travers P, M. Walport M (2008) Janeway's Immunobiology. 7th edn. Garland Science, Taylor & Francis Group.888pp

Murray PJ, Wynn TA (2011) Protective and pathogenic functions of macrophage subsets. Nat Rev Immunol 11:723–737

Myers Michael P, Andersen Jannik N, Cheng Alan, Tremblay Michel L, Horvath Curt M, Parisien Jean-Patrick, Salmeen Annette, Barford David, Tonks Nicholas K (2001) TYK2 and JAK2 are substrates of protein-tyrosine phosphatase 1B. J Biol Chem 276:47771–47774

Nicolette CA, Healey D, Tcherepanova I, Whelton P, Monesmith T, Coombs L et al (2007) Dendritic cells for active immunotherapy: optimizing design and manufacture in order to develop commercially and clinically viable products. Vaccine 25(Suppl 2):B47–B60

Okazaki T, Maeda A, Nishimura H, Kurosaki T, Honjo T (2001) PD-1 immunoreceptor inhibits B cell receptor-mediated signaling by recruiting src homology 2-domain-containing tyrosine phosphatase 2 to phosphotyrosine. Proc Natl Acad Sci 98:13866–13871

Pandey R, Saxena M, Kapur R (2017) Role of SHP2 in hematopoiesis and leukemogenesis. Curr Opin Hematol July 24(4):307–313

Paulos CM, June CH (2010) Putting the brakes on BTLA in T cell–mediated cancer immunotherapy. J Clin Investig 120:76–80

Pawson T (2004) Specificity in signal transduction: from phosphotyrosine-SH2 domain interactions to complex cellular systems. Cell 116:191–203

Penafuerte C, Feldhammer M, Mills JR, Vinette V, Pike KA, Hall A et al (2017) Downregulation of PTP1B and TC-PTP phosphatases potentiate dendritic cell-based immunotherapy through IL-12/IFNγ signaling. OncoImmunology 6:e1321185

Pérez-Quintero LA, Roncagalli R, Guo H, Latour S, Davidson D, Veillette A (2014) EAT-2, a SAP-like adaptor, controls NK cell activation through phospholipase Cγ, Ca ++, and Erk, leading to granule polarization. J Exp Med 211:727–742

Pike KA, Hutchins AP, Vinette V, Théberge JF, Sabbagh L, Tremblay ML, et al (2014). Protein tyrosine phosphatase 1B is a regulator of the IL-10-induced transcriptional program in macrophages. Sci Signal 7(324): ra43

Pike KA, Tremblay ML (2016) TC-PTP and PTP1B: regulating JAK-STAT signaling, controlling lymphoid malignancies. Cytokine 82:52–57

Ramachandran IR, Song W, Lapteva N, Seethammagari M, Slawin KM, Spencer DM, Levitt JM (2011)The phosphatase SRC homology region 2 domain-containing phosphatase-1 is an intrinsic central regulator of dendritic cell function.J Immunol 186(7):3934–3945. PMID:21357539. doi:10.4049/jimmunol.1001675

Rothlin CV, Ghosh S, Zuniga EI, Oldstone MB, Lemke G (2007) TAM receptors are pleiotropic inhibitors of the innate immune response. Cell 131:1124–1136

Ruffell B, Coussens LM (2015) Macrophages and therapeutic resistance in cancer. Cancer Cell 13:462–472

Sathish JG1, Johnson KG, Fuller KJ, LeRoy FG, Meyaard L, Sims MJ, Matthews RJ (2001) Constitutive association of SHP-1 with Leukocyte-associated Ig-Like receptor-1 in human T cells. J Immunol 166:1763–1770

Schneider H, Smith X, Liu H, Bismuth G, Rudd CE (2008) CTLA-4 disrupts ZAP70 microcluster formation with reduced T cell/APC dwell times and calcium mobilization. Eur J Immunol 38:40–47

Shlapatska LM, Mikhalap SV, Berdova AG, Zelensky OM, Yun TJ, Nichols KE, Clark EA, Sidorenko SP (2001) CD150 Association with either the SH2-containing inositol phosphatase or the SH2-containing protein tyrosine phosphatase is regulated by the adaptor protein SH2D1A. J Immunol 166:5480–5487

Stitt TN, Conn G, Gore M, Lai C, Bruno J, Radziejewski C et al (1995) The anticoagulation factor protein S and its relative, Gas6, are ligands for the Tyro 3/Axl family of receptor tyrosine kinases. Cell 80:661–670

St-Pierre J, Tremblay ML (2012) Modulation of leptin resistance by protein tyrosine phosphatases. Cell Metab March 7 15(3):292–297

Stuible M, Dubé N, Tremblay ML (2008) PTP1B regulates cortactin tyrosine phosphorylation by targeting Tyr446. J Biol Chem June 6 283(23):15740–15746

Tao B, Jin W, Xu J, Liang Z, Yao J, Zhang Y et al (2014) Myeloid-specific disruption of tyrosine phosphatase Shp2 promotes alternative activation of macrophages and predisposes mice to pulmonary fibrosis. J Immunol 193(6):2801–2811

Tonks NK (2006) Protein tyrosine phosphatases: from genes, to function, to disease. Nat Rev Mol Cell Biol November 7(11):833–846

Tonks NK (2013) Protein tyrosine phosphatases–from housekeeping enzymes to master regulators of signal transduction. FEBS J January 280(2):346–378

Tonks NK, Charbonneau H, Diltz CD, Fischer EH, Walsh KA (1984) Demonstration that the leukocyte common antigen CD45 is a protein tyrosine phosphatase. Biochemistry 27:8695–8701

Uetani N, Hardy S, Gravel SP, Kiessling S, Pietrobon A, Wong NN, Chénard V, Cermakian N, St-Pierre J, Tremblay ML (2017) PRL2 links magnesium flux and sex-dependent circadian metabolic rhythms. JCI Insight July 6 2(13)

van Beek JJ, Wimmers F, Hato SV, de Vries IJ, Skold AE (2014) Dendritic cell cross talk with innate and innate-like effector cells in antitumor immunity: implications for DC vaccination. Crit Rev Immunol 34:517–536

Veillette A, Dong Z, Pérez-Quintero LA, Zhong MC, Cruz-Munoz ME (2009) Importance and mechanism of 'switch' function of SAP family adapters. Immunol Rev 232:229–239

Veillette A, Pérez-Quintero LA, Latour S (2013) X-linked lymphoproliferative syndromes and related autosomal recessive disorders. Curr Opin Allergy Clin Immunol 13:614–622

Wakamatsu E Mathis D, Benoist C (2013) Convergent and divergent effects of costimulatory molecules in conventional and regulatory CD4 + T cells. Proc Natl Acad Sci 110:1023–1028

Watanabe N, Gavrieli M et al (2003) BTLA is a lymphocyte inhibitory receptor with similarities to CTLA-4 and PD-1. Nat Immunol 4:670–679

Watson HA, Wehenkel S, Matthews J, Ager A (2016) SHP-1: the next checkpoint target for cancer immunotherapy? Biochem Soc Trans 44:356–362

Wherry EJ, Kurachi M (2015) Molecular and cellular insights into T cell exhaustion. Nat Rev Immunol 15:486–499

Wiede F, Shields BJ, Chew SH et al (2011) T cell protein tyrosine phosphatase attenuates T cell signaling to maintain tolerance in mice. J Clin Invest 121:4758–474

Wiede F, Sacirbegovic F, Leong YA, Yu D, Tiganis T (2017) PTPN2-deficiency exacerbates T follicular helper cell and B cell responses and promotes the development of autoimmunity. J Autoimmun 76:85–100

Wu, TH, Zhen Y, et al (2007) B and T lymphocyte attenuator interacts with CD3[zeta] and inhibits tyrosine phosphorylation of TCR[zeta] complex during T-cell activation. Immunol Cell Biol 85:590–595

Yokosuka T, Takamatsu M, et al (2012) Programmed cell death 1 forms negative costimulatory microclusters that directly inhibit T cell receptor signaling by recruiting phosphatase SHP2. J Exp Med 209:1201–1217

Xu D1, Qu CK (2008) Protein tyrosine phosphatases in the JAK/STAT pathway Front Biosci May 1 13:4925–4932

You-Ten KE1, Muise ES, Itié A, Michaliszyn E, Wagner J, Jothy S, Lapp WS, Tremblay ML (1997) Impaired bone marrow microenvironment and immune function in T cell protein tyrosine phosphatase–deficient mice. J Exper Med 186:683–693

Yu DH, Qu CK, Henegariu O, Lu X, Feng GS (1998) Protein-tyrosine phosphatase Shp-2 regulates cell spreading, migration, and focal adhesion. J Biol Chem 273:21125–21131

Zikherman J, Weiss A (2008) Alternative splicing of CD45: the tip of the iceberg. Immunity December 19 29(6):839–841

Immune Regulation by Ubiquitin Tagging as Checkpoint Code

Peng Zeng, Jieyu Ma, Runqing Yang and Yun-Cai Liu

Abstract The immune system is equipped with effective machinery to mobilize its activation to defend invading microorganisms, and at the same time, to refrain from attacking its own tissues to maintain immune tolerance. The balance of activation and tolerance is tightly controlled by diverse mechanisms, since breakdown of tolerance could result in disastrous consequences such as the development of autoimmune diseases. One of the mechanisms is by the means of protein ubiquitination, which involves the process of tagging a small peptide ubiquitin to protein substrates. E3 ubiquitin ligases are responsible for catalyzing the final step of ubiquitin–substrate conjugation by specifically recognizing substrates to determine their fates of degradation or functional modification. The ubiquitination process is reversible, which is carried out by deubiquitinating enzymes to release the ubiquitin molecule from the conjugated substrates. Protein ubiquitination and deubiquitination serve as checkpoint codes in many key steps of lymphocyte regulation including the development, activation, differentiation, and tolerance induction. In this chapter, we will discuss a few E3 ligases and deubiquitinating enzymes that are important in controlling immune responses, with emphasis on their roles in T cells.

Contents

1 Introduction 216
2 Cbl-b as a Key Regulator in T Cells 219

P. Zeng · J. Ma · R. Yang · Y.-C. Liu (✉)
Institute for Immunology, Tsinghua-Peking Center for Life Sciences, School of Medicine, Tsinghua University, Beijing 100084, China
e-mail: yuncail@hotmail.com

Y.-C. Liu
Division of Cell Biology, La Jolla Institute for Allergy and Immunology, La Jolla, CA 92037, USA

Current Topics in Microbiology and Immunology (2017) 410:215–248
DOI 10.1007/82_2017_64

Published Online: 20 September 2017

2.1 Cbl-b in T Cell Receptor Signaling 219
2.2 Cbl-b in T Cell Anergy 221
2.3 Cbl-b in T Cell Subsets 221
2.4 Cbl-b in Anti-tumor Immunity 222
3 The E3 Ligase Itch in T Cell Regulation 223
3.1 Itch in Th2 Cells 223
3.2 Itch in Tfh Differentiation 224
3.3 Itch in Th17 Differentiation 224
3.4 Itch in T Cell Anergy 224
4 Control of Treg Cells by Ubiquitination 225
4.1 Treg Cells in Immune Tolerance 225
4.2 VHL 226
4.3 SHARPIN 227
4.4 Itch 228
4.5 Cbl-b 228
4.6 GRAIL 229
4.7 TRAF6 229
5 Deubiquitinase A20 in Immune Regulation 230
5.1 A20 Functions in Lymphoid Cells 230
5.2 A20 Functions in Non-lymphoid Cells 231
5.3 The Role of A20 in TNF-Induced NFκB Activation 231
5.4 A20 Functions in Other Signaling Pathways 232
5.5 Regulation of A20 233
6 CYLD 233
6.1 CYLD in NFκB Signaling 233
6.2 CYLD in T Cell Development 234
6.3 CYLD in Lymphocyte Function 234
6.4 CYLD Function in Treg Cells 235
7 Concluding Remarks 235
References 237

1 Introduction

Immune tolerance is a state of hypo- or unresponsiveness of the adaptive immune system to self- or even nonself-antigens to avoid excessive immune responses, and is divided into central tolerance and peripheral tolerance (Sakaguchi et al. 2008). Central tolerance is induced during the processes of T and B cell development in the thymus and bone marrow, respectively. The major mechanisms to establish the central tolerance are through clonal deletion, receptor editing, and generation of naturally occurring or thymus-derived regulatory T (tTreg) cells (Abbas et al. 2013). In contrast, peripheral tolerance occurs after the mature T and B cells entering the peripheral lymphoid tissues, and keeps the self-reactive T and B cells which have escaped central tolerance under control. Several tolerance mechanisms are involved in this process, including ignorance, apoptosis, suppression by tTreg cells and peripherally induced regulatory T (pTreg) cells, anergy induction, or the balance of T helper (Th) cell subsets (Soyer et al. 2013; Walker and Abbas 2002).

Immune tolerance is the key mechanism to maintain tissue homeostasis. Deficits in central or peripheral tolerance can have disastrous consequences, such as the development of numerous autoimmune diseases such as type 1 diabetes, inflammatory bowel disease, multiple sclerosis, and systemic lupus erythematosus. However, immune tolerance can be counterintuitive in that tumor cells or pathogens can escape the immunological surveillance of the hosts (Zou 2006).

Protein ubiquitination is a highly ordered enzymatic cascade in which ubiquitin is tagged onto the lysine site of a substrate covalently (Park et al. 2014). This cascade involves E1, E2, and E3 enzymes. Initially, the ubiquitin activating enzyme E1 activates the ubiquitin by forming the high-energy isopeptide bond between E1 active cysteine and the ubiquitin C-terminal glycine residue (Fig. 1). Then, the activated ubiquitin is transferred onto E2 ubiquitin conjugating enzyme through similar isopeptide bond. Eventually, E3 ligases catalyze the isopeptide bond between the lysine of a substrate and active glycine of the ubiquitin. E3 ligases fall into three subfamilies: the homologous to the E6-associated protein C terminus (HECT) domain containing E3s; the really interesting new gene (RING) domain containing E3s; and the U-box containing E3s. The signature of HECT-type E3s is that they can transfer the ubiquitin from E2 onto themselves with isopeptide bond and then load the ubiquitin onto protein substrates. Unlike HECT-type E3s, the RING-type E3s directly promote the transferring of ubiquitin from E2 to substrates.

Ubiquitination has multiple functions that are dependent on different chain linkages (Komander and Rape 2012). There are at least 10 kinds of chain linkages: mono-ubiquitination, linear poly-ubiquitination, K6-, K11-, K27-, K29-, K33-, K48-, and K63-linked poly-ubiquitination, and mixed-linkage poly-ubiquitination depending on which lysine (or methionine) has been utilized for the isopeptide bond formation. K48-linkaged poly-ubiquitination marks the substrates for proteasomal degradation, whereas linear chains and K63-poly-ubiquitin chains can act as the platform to facilitate protein complex formation. K63-, mono-, and K33-linkaged poly-ubiquitination can also mediate the lysosomal-dependent proteolysis of substrates.

Ubiquitination is a reversible process that is carried out by the deubiquitinating enzymes (DUBs) (Sun 2008). Even though there are over 600 E3 ligases, ~100 DUBs are found in the human genome (Nijman et al. 2005). Based on their structure similarity, DUBs are divided into six families: ubiquitin C-terminal hydrolases (UCHs), ubiquitin-specific proteases (USPs), Machado-Joseph disease protein domain proteases (MJDs), ovarian-tumor proteases (OTUs), JAMM (JAB1/MPN/Mov34 metalloenzyme) (Nijman et al. 2005), and monocyte chemotactic protein-induced protein (MCPIP) family (Fraile et al. 2012). DUBs deconjugate the ubiquitin from substrates by specifically cleaving the isopeptide bond between them, or change the existing ubiquitin chains into another K-linked ubiquitin chains and alter the signaling on substrates (Wertz et al. 2004). Similar to ubiquitination, deubiquitination is a highly organized process that is regulated by

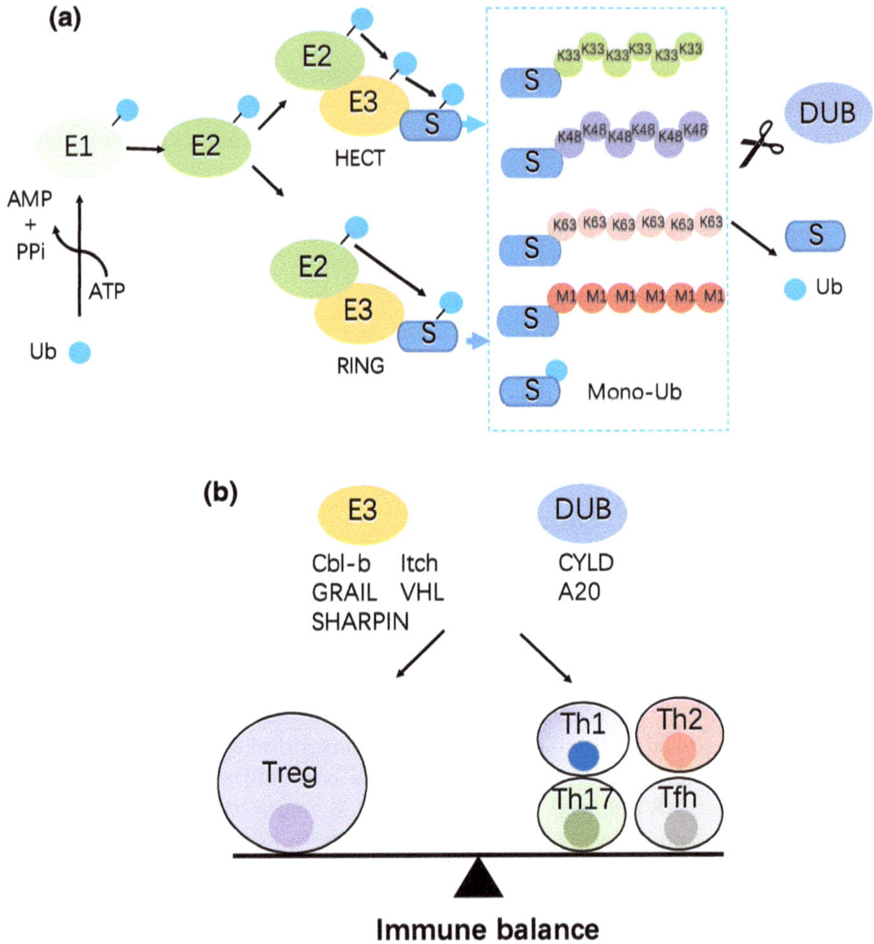

Fig. 1 Ubiquitin codes and the immune regulation, **a** The cascade of ubiquitination catalyzed by E1, E2, and E3 enzymes to form different forms of ubiquitin modification, which are reversed by deubiquitinating enzyme DUBs. **b** Selective E3 ligases and DUBs discussed in this chapter and their roles in controlling immune homeostasis by balancing Treg and T helper cells

many mechanisms at multiple levels such as transcriptional regulation or post-transcriptional modifications, subcellular localization, or protein–protein interactions.

Numerous studies have documented the critical roles of E3 ligases and DUBs in many aspects of immune regulation. In this chapter, we have selected some key enzymes to discuss their roles in ubiquitin–protein conjugation or deconjugation as checkpoint codes in immune regulation such as T cell activation and differentiation, anergy induction, and Treg development and function.

2 Cbl-b as a Key Regulator in T Cells

2.1 Cbl-b in T Cell Receptor Signaling

Cbl-b, derived from Casitas B-lineage lymphoma proto-oncogene-b, belongs to the family of RING-type E3 ligases that have three homologues including c-Cbl, Cbl-b, and Cbl-c. Cbl-b contains a tyrosine kinase binding (TKB) domain in the N terminus, a RING finger domain in the middle, and proline-rich domains and an ubiquitin-binding associated (UBA) domain in the C terminus. The TBK domain functions as a docking site for phosphorylated tyrosine-containing proteins such as Syk and Zap70 kinases. The RING finger domain binds to E2–ubiquitin complex for the ubiquitin conjugating onto substrates. The UBA domain on the C terminus can bind with ubiquitin and ubiquitin-like domains of proteins (Lutz-Nicoladoni et al. 2015; Loeser and Penninger 2007).

Cbl-b-deficient mice display hyper-proliferation of effector T cells and hyper-production of interleukin-2 (IL-2) even without CD28 costimulation (Bachmaier et al. 2000; Chiang et al. 2000). Loss of Cbl-b can rescue the dysfunction of $CD28^{-/-}$ T cells such as T cell proliferation, IL-2 production, and activation of GDP/GTP exchanger factor Vav1. Aged Cbl-b-deficient mice develop autoimmune responses such as autoantibody production, T and B lymphocyte infiltration into multiple tissues. Cbl-b suppresses Vav1-CDC42-Wiscott Aldrich syndrome protein (WASP) signaling pathway upon T cell receptor (TCR)/CD3 engagement (Krawczyk et al. 2000). Cbl-b-deficient T cells show TCR clustering upon anti-CD3 antibody treatment, as Cbl-b is the negative regulator for lipid raft aggregation and TCR clustering. However, Cbl-b deficiency does not restore the calcium mobilization and transactivation of nuclear factor of activated T cells, cytoplasmic 1 (NFATc1) after TCR stimulation. WASP is phosphorylated after the TCR engagement. Cbl-b, cooperating with c-Cbl, mediates the ubiquitination-dependent down-regulation of WASP after recognizing its tyrosine phosphorylation (Reicher et al. 2012; Watanabe et al. 2013). Cbl-b binds to and mediates non-proteolysis ubiquitination on Crk-L, which interrupts the interaction of Crk-L with C3G. As the GEF exchange activity of C3G is vital for Rap-1 activation, Cbl-b-deficient T cells show hyperactivation of the small GTPase Rap-1 and further enhance LFA-1 clustering which determines the cell adhesion strength (Zhang et al. 2003).

To figure out the specific mechanism that how Cbl-b negatively regulates Vav1 activation, we find that Cbl-b can bind with and catalyze the ubiquitination of the p85 regulatory subunit of phosphatidylinositol 3-kinase (PI3 K), which catalyzes PIP2 lipid into PIP3, and PIP3 promotes the GDP/GTP exchange activity of Vav1 by binding into its pleckstrin homology (PH) domain. The interaction between Cbl-b and PI3 K requires the C-terminal proline-rich region in Cbl-b and the SH3 domain in PI3 K subunit p85 (Fang et al. 2001). Moreover, the ubiquitination on p85 subunit of PI3 K mediated by Cbl-b does not lead to protein degradation, but blocks the recruitment of p85 toward CD28 as well as TCRζ. Cbl-b-deficient T cells uncouple CD28 costimulation from T cell proliferation and IL-2 production and these can be reversed by PI3 K inhibitor treatment (Fang et al. 2001). Moreover, Cbl-b cooperates with Itch to mediate the non-proteolytic K33-linkaged poly-ubiquitination on CD3ζ which dampens the association with and activation of Zap70 (Huang et al. 2010).

Meanwhile, CD28 costimulation stimulates the ubiquitination and protein degradation of Cbl-b to regulate the optimal T cell activation (Zhang et al. 2002). Besides, upon CD3 stimulation, SHP-1 binds with and dephosphorylates Cbl-b, and then blocks the ubiquitination of Cbl-b. CD28 costimulation dampens this interaction (Xiao et al. 2015). CTLA4-B7 interaction is essential for Cbl-b up-regulation at both transcriptional and translational levels after Cbl-b degradation induced by CD28 costimulation (Li et al. 2004).

By using fluorescence in situ hybridization assay to show protein tyrosine phosphorylation, Cbl-b protein and protein ubiquitination are colocalized with TCRζ in the immunological synapse (Wiedemann et al. 2005). In fact, Cbl-b and c-Cbl cooperatively promote TCR down-modulation upon TCR engagement as a comparison between wild-type and Cbl-b$^{-/-}$c-Cbl$^{F/F}$Lck^{cre} T cells (Naramura et al. 2002). NFκB signaling is hyperactivated in CD4^{+}CD8^{+} thymocytes in Cbl-b/c-Cbl double-knockout (DKO) mice. Pre-TCR protein level and pre-TCR signaling are also augmented in Cbl-b/c-Cbl DKO mice (Huang et al. 2006). The absence of Cbl-b and c-Cbl "licenses" both the CD4^{+} and CD8^{+} lineage T cell developments even in the MHC-I- and II-deficient mice. Recently, Cbl-b$^{F/F}$ mouse strain has been developed and Cbl-b$^{F/F}$c-Cbl$^{F/F}$CD4^{cre} mice display multi-organ autoimmune phenotype similar to the phenotype of Cbl-b$^{-/-}$c-Cbl$^{F/F}$Lck^{cre} mice (Goetz et al. 2016).

Loss of Cbl-b leads to the defect in the TGF-β-induced Foxp3 expression and effector T cells to Treg cell transition, which may be relevant to the unresponsiveness of TGF-β signaling and reduction of Smad2 phosphorylation (Wohlfert et al. 2006). Cbl-b-deficient effector T cells show in vitro resistance to the suppressive function of Treg cells and TGF-β, which contributes to the autoimmune response in Cbl-b-deficient mice (Wohlfert et al. 2004). Furthermore, Cbl-b inhibits TGF-β receptor signaling by interacting with SMAD7 and down-regulating SMAD7 through degradation (Gruber et al. 2013).

Cbl-b deficiency restores the defective antiviral antibody production and the formation of follicular dendritic cell (FDC) clusters in CD28-deficient mice. However, Cbl-b deficiency does not rescue the germinal center formation in CD28-deficient mice as loss of Cbl-b fails to rescue the cytokine production such as IL-4, IFN-γ, and the expression of activation surface markers like ICOS and OX40 (Krawczyk et al. 2005). In addition, loss of Cbl-b suppresses apoptosis in Th1 cells but not Th2 cells (Hanlon et al. 2005).

Although the E3 ligase activity of Cbl-b has been proved to be essential for its negative regulation of T cells in vivo (Paolino et al. 2011), Cbl-b negatively regulates T cell activation through blocking the association between Pten and Nedd4, preventing the ubiquitination on Pten mediated by Nedd4 and augmenting the production of PIP3 and Akt activation (Guo et al. 2012). Loss of Cbl-b also uncouples the activation of NFκB signaling but not the MAPKs, AP-1, and NFAT from CD28 costimulation. Cbl-b deficiency upregulates the activation of PI3K/Akt and PKCθ under the stimulation of CD3. Cbl-b inhibits the protein complex formation of PKCθ, CARMA-1, Bcl10, MALT-1, IKKβ, and IKKγ, which eventually regulates the activation of NFκB (Qiao et al. 2008). Cbl-b is shown to be ubiquitinated and negatively regulated by Nedd4 (Yang et al. 2008). The ubiquitination of Cbl-b upon the stimulation of T cell receptor is dependent on the kinase activity of PKCθ and Cbl-b deficiency can rescue the activation of PKCθ-deficient T cells (Gruber et al. 2009).

2.2 *Cbl-b in T Cell Anergy*

Anergy is a hyporesponsive state of T cells, which can be induced through TCR triggering without the appropriate costimulation. In vitro, anergy can be induced by stimulation of the T cells with anti-CD3 antibody, mitogen concanavalin A, ionomycin, peptides presented by chemically fixed antigen-presenting cells (APCs), inefficient APCs (such as small resting B cells), and even artificial planar lipid bilayers. In vivo, anergy is achieved by systemic administration of superantigens, adoptively transferring TCR transgenic T cells expressing cognate antigen as a self-antigen, or by delivering soluble peptide antigen into TCR transgenic mice (Zheng et al. 2008). Both mRNA and protein levels of Cbl-b are up-regulated in ionomycin-induced anergic T cells. Cbl-b knockout T cells are resistant to anergy induction (Heissmeyer et al. 2004). The enhanced accumulation of Cbl-b in central supramolecular activation clusters (cSMACs) is also observed (Doherty et al. 2010). Cbl-b induces the ubiquitination of PLCγ-1 and inhibits its phosphorylation (Jeon et al. 2004). The loss of Cbl-b prevents the anergy induction in vivo and in vitro. Moreover, repeated peptide challenge of Cbl-b-deficient mice leads to severe lethality, which also implies that Cbl-b-mediated anergy induction is vital for the tight control of the immune responses. In addition, Cbl-b catalyzes the mono-ubiquitination of CARMA-1 and prevents the CARMA-1/Bcl10/MALT-1 complex formation in anergic natural killer T cells upon stimulation with α-galactosylceramide (Kojo et al. 2009). As PKCθ is also modified with mono-ubiquitination in anergic T cells (Heissmeyer et al. 2004), deltex1, another E3 ligase, promotes mono-ubiquitination of PKCθ and thus causes the degradation of PKCθ by lysosomal pathway in anergic T cells and Cbl-b is also regulated by deltex1 through the degradation of PKCθ (Hsu et al. 2014).

2.3 *Cbl-b in T Cell Subsets*

Genetic deletion of Cbl-b causes the clonal expansion of $CD4^+$ T cells and promotes the IL-2, CD25, and CD71 expressions. The absence of Cbl-b restores the activation of PLCγ-1, AKT, and ERK (Zhang et al. 2008), and Cbl-b restrains the proliferation of self-reactive $CD4^+$ T cells (Hoyne et al. 2011; St Rose et al. 2009). Cbl-b deficiency promotes the Th2 and Th9 cell differentiations in the in vitro differentiation system (Qiao et al. 2014). Additionally, loss of Cbl-b leads to the elevated levels of STAT6 and IRF4. Cbl-b promotes the ubiquitination and degradation of STAT6 upon IL-4 ligation and TCR/CD28 costimulation. In the OVA/Alum-induced asthma mouse model, Cbl-b deficiency augments Th2- and Th9-related immune responses.

The proper function of cytotoxic T cells (CTLs) ensures the effective antiviral immune response and antitumor activity, whereas the dysfunctional CTLs can cause the virus infection-induced autoimmunity or exhaustion of CTLs in cancer surveillance. Inactivation of Cbl-b increases the cell surface expression levels of TCR and CD8, decreases antigen-stimulated down-modulation of cell surface TCR level, and enhances the IFN-γ production (Shamim et al. 2007). However, Cbl-b deficiency does not affect the CTLs-mediated cytotoxicity and clonal expansion of CTLs. After infection with moderate dose of LCMV, the down-regulation of virus-specific CTLs is delayed in Cbl-b-deficient mice compared with control mice (Ou et al. 2008). Moreover, in the LCMV infection-induced diabetes model, Cbl-b deficiency leads to higher incidences and faster kinetics of disease development (Gronski et al. 2004). Moreover, Cbl-b also plays an inhibitory role in the activation and cytotoxicity of $\gamma\delta$ T cells (Yin et al. 2013).

2.4 *Cbl-b in Anti-tumor Immunity*

Cbl-b-deficient mice show significant rejection of inoculated EG7 and EL4 lymphomas compared with wild-type mice. This efficient antitumor activity may due to CD28-independent hyper-proliferation, IL-2 and IFN-γ cytokine production, resistance against TGF-β treatment, and enhanced $CD8^+$ T cell infiltration in tumor tissues (Chiang et al. 2007). Moreover, adoptive transfer of Cbl-b knockout CTLs also effectively eradicates the inoculated EG7 tumors. Cbl-b inactivation confers rejection of TC-1 tumor cell line inoculation and ultraviolet B-induced skin malignancies, which is correlated with the resistance to Treg-mediated suppression and enhanced activation and tumor infiltration (Loeser et al. 2007). More strikingly, even after pre-injecting the TC-1 tumor cell line into Cbl-b-deficient mice and keeping the tumor cell-experienced mice for 1 year, these mice show more effective tumor rejection than the same aged un-experienced wild-type and Cbl-b-deficient mice after the inoculation of tenfold higher dose of TC-1 tumor cells. Even though adoptive transfer of Cbl-b-deficient $CD8^+$ T cells is effective for the tumor infiltration and eradication, they are not active for all tumor phenotypes, especially for the tumor with profound barriers (Yang et al. 2009). The antitumor activity of IL-7 is also partially involved the repressive effect of Cbl-b (Pellegrini et al. 2009). Blockade of the ligation of PD-L1/PD-1 on $CD8^+$ T cells decreases the expression of Cbl-b (Karwacz et al. 2011). Thus manipulating the expression of Cbl-b could be a promising strategy for the adoptive cell transfer therapy against multiple cancerous diseases (Hinterleitner et al. 2012; Shi et al. 2014; Stromnes et al. 2010; Zhou et al. 2014).

3 The E3 Ligase Itch in T Cell Regulation

Itch is a member of the HECT E3 ligase family. Itch is identified by genetic mapping of the mice-bearing mutations in agouti locus causing hair color changes, and one of the mice, called itchy mice, shows constant skin itching and multi-organ inflammation (Hustad et al. 1995; Perry et al. 1998). The Itch protein consists of a PKC-related C2 domain at N terminus, four WW domains in the middle, and a HECT ligase domain at the C terminus. The C2 domain of Itch can bind calcium ions and phospholipids, which is related to the cytosolic membrane location of Itch. The WW domains recognize and bind with proline-rich motifs such as PPxY and PPLP, which determine the substrate specificity. The HECT domain of Itch contains E2–ubiquitin-binding region, and a catalytic active cysteine residue, which mediates the transfer of the activated ubiquitin from the E2–ubiquitin complex and catalyzes the subsequent isopeptide bond formation between the lysine residue of the substrate and the glycine of ubiquitin (Aki et al. 2015).

3.1 Itch in Th2 Cells

Itch-deficient T cells are biased toward Th2 differentiation in vitro and Itch-deficient mice produce more IL-4 and Th2-biased IgG1 and IgE in vivo (Fang et al. 2002). Mechanistically, Itch binds with and mediates the ubiquitination-dependent degradation of JunB. Furthermore, upon anti-CD3/CD28 stimulation, JNK is activated by MEKK1 and mediates the phosphorylation-dependent activation of Itch and promotes Itch-mediated ubiquitination-dependent degradation of JunB and c-Jun (Gao et al. 2004). Itch is present in an inactive state via the binding of HECT domain with self-proline-rich region inside the WW domains. JNK induces the serine/threonine phosphorylation of Itch, which leads to the conformational change of Itch structure from an auto-inhibitory state to an activated state (Gallagher et al. 2006). The direct interaction of MEKK1 with Itch has also been reported to promote the E3 ligase activity of Itch (Enzler et al. 2009). Beyond the serine/threonine phosphorylation mediated by JNK, Itch can also be modified with tyrosine phosphorylation mediated by Fyn upon T cell stimulation (Yang et al. 2006). This modification blocks the recruitment of and subsequent degradation of JunB. Nedd4 family interacting protein 1 (Ndfip1), an E3 ligase adaptor protein, also promotes the ubiquitination and degradation of JunB by interacting with Itch (Oliver et al. 2006). In addition, Ndfip1 have Itch-dependent and Itch-independent roles in the immune responses (Ramon et al. 2011). In summary, Itch-mediated JunB ubiquitination and degradation is tightly regulated in T cells.

Itch-deficient mice are resistant to the high-fat diet-induced obesity (Marino et al. 2014). M2 macrophages induced by IL-4 resolve the obesity-related inflammation in adipose tissue. As Itch deficiency leads to Th2-biased T cell differentiation and high amount of IL-4 production, Th2/M2 macrophage polarization protects Itch-deficient mice from obesity-related insulin resistance and metabolic dysfunction.

3.2 Itch in Tfh Differentiation

Follicular T helper (Tfh) cells are one of the specialized T cell subsets which are localized in B cell follicles and provide help for B cells to promote their maturation, germinal center formation, Ig gene recombination, and high-affinity antibody production. Tfh cells are characterized by the expression of cell surface molecules including CXCR5, PD1, SAP, and ICOS, the cytokine IL-21, and the master transcription factor Bcl6 (Crotty 2011). We have found that Itch deficiency leads to the defective differentiation of Tfh cells and weakens germinal center responses and IgG responses upon LCMV infection (Xiao et al. 2014). Mechanistically, Itch mediates the ubiquitination and degradation of Foxo1. Consistently, Foxo1 plays a negative role in regulating Tfh differentiation (Stone et al. 2015).

3.3 Itch in Th17 Differentiation

Mice lacking the E3 ligase Itch develop spontaneous colitis and colitis-associated colorectal cancer (Kathania et al. 2016). The local source of cytokine IL-17 in the colonic lamina propria lymphocytes is mainly derived from Th17 cells, innate lymphoid cell 3 (ILC3), and $\gamma\delta$T cells. Itch can bind with RORγt, the master transcription factor of Th17, and ILC3 cells and promote the K48-linkaged ubiquitination and degradation of RORγt (Kathania et al. 2016). Thus, Th17 and ILC3 cells lacking Itch show elevated RORγt protein levels. Itch-deficient mice show augmented IL-17 production in all these cells, which eventually causes the colon inflammation. Ndfip1, which is the adaptor protein for Itch, is also shown to block the differentiation of Th17 cells (Layman et al. 2017). Mice lacking Ndfip1 show increased IL-17 production and augmented RORγt expression levels.

3.4 Itch in T Cell Anergy

Similar to Cbl-b, both mRNA and protein levels of Itch are elevated in ionomycin-induced anergic T cells (Heissmeyer et al. 2004). Ionomycin stimulation and homotypic cell adhesion induce cytoplasm-to-cell membrane translocation of

Itch, which promotes the mono-ubiquitination and lysosomal degradation of PLCγ-1. In addition, the degradation of PKCθ is increased in Itch-deficient T cells during anergy induction. Moreover, Itch-deficient mice are resistant to anergy induction in an airway inflammation model (Venuprasad et al. 2006). The elevated protein level of JunB in Itch-deficient T cells prevents the anergy induction and causes excessive Th2 cytokine production and severe airway inflammation in $Itch^{-/-}$ mice.

As described above, Itch cooperates with Cbl-b to regulate T cell activation via targeting the proximal TCR signaling (Huang et al. 2010). Briefly, Itch cooperates with Cbl-b and provokes the K33-linkaged poly-ubiquitination of CD3ζ, and then blocks the recruitment of Zap70 (Huang et al. 2010). Itch and Nedd4 also mediate the ubiquitination and lysosomal degradation of BCL10 upon T cell activation, and thus negatively regulate the NFκB signaling (Scharschmidt et al. 2004).

4 Control of Treg Cells by Ubiquitination

4.1 Treg Cells in Immune Tolerance

$CD4^+CD25^+$ Treg cells are defined as a subset of $CD4^+$ T cell populations, which play critical roles in immune tolerance; they are involved in preventing autoimmune responses and, at the same time, facilitating tumorigenesis and blocking beneficial antitumor immunity (Zou 2006). X-linked transcription factor Fork-head box P3 (Foxp3) is the master transcription factor for the development, maintenance, and function of Treg cells. Foxp3 deficiency or mutations results in severe autoimmunity and multiple organ inflammations in both mouse and human. Furthermore, retroviral transduction of Foxp3 in naïve $CD4^+$ T cells converts them toward a regulatory T cell phenotype, which highly express Treg surface markers CD25, cytotoxic T lymphocyte-associated antigen 4 (CTLA-4), and glucocorticoid-induced tumor necrosis factor receptor family-related gene (GITR) (Fontenot et al. 2003; Gavin et al. 2007; Hori et al. 2003).

As described above, Treg cells are divided into two subpopulations: tTreg cells, which are generated from thymus, and pTreg cells which are induced in the periphery from naïve $CD4^+$ T cells in the presence of IL-2 and TGF-β (Abbas et al. 2013). Both of them exhibit suppressive function in the immune response. Treg cells express several surface markers to exert their functions. For example, CTLA4 and lymphocyte activation gene 3 (LAG3) on Treg cell surface facilitate their interactions with DCs, leading to the inhibition of DC maturation, which in turn limits effector T cells responses (Wing et al. 2008; Liang et al. 2008); Neuropilin 1 (Nrp1) expressed by tTreg cells is reported to promote interaction duration between Treg cells and immature DCs, which restricts access of the effector cells to APCs (Sarris et al. 2008). Treg cell stability can also be maintained by Nrp1–Sema4a interaction (Delgoffe et al. 2013). In addition, by producing immunosuppressive cytokines IL-10, IL-35, and TGF-β (Collison et al. 2009; Murai et al. 2009;

Fahlen et al. 2005), tTreg cells can suppress conventional T cell activation. Thus, cytokine-dependent suppression is another way of Treg cells to control immune tolerance. Moreover, activated Treg cells can also kill the effector T cells in a granzyme- and perforin-dependent manner (Cao et al. 2007).

4.2 VHL

von Hippel–Lindau (VHL) is a component of VHL-elongin C/elongin B superfamily of E3 ubiquitin ligase complex that includes VHL, elongin B, and C, cullin2 and Ring box protein 1(Rbx1), and belongs to multi-subunit Ring-type E3 ligases (Kamura et al. 1999; Jackson et al. 2000). VHL, defined as a tumor suppressor, is mutated in most of the human sporadic renal cell carcinomas (Kaelin and Maher 1998). VHL plays a key role in cellular oxygen sensing by targeting hypoxia-inducible factor (HIF) for ubiquitination. Under normoxic conditions, HIF1α and HIF2α are hydroxylated by prolyl hydroxylase enzymes (PHDs), are then recognized by the VHL E3 ubiquitin ligase complex, and are targeted to ubiquitination and proteasomal degradation. Under hypoxic conditions, PHDs are inactive, thus resulting in HIFα stabilization, and the induction of the HIF-dependent transcriptional activity (Maxwell et al. 1999; Jaakkola et al. 2001; Ivan et al. 2001).

Th17 and pTreg arise from common $CD4^+$ T cells and share a common requirement of TGF-β for their induction. Two studies demonstrated that the oxygen sensor HIF1α plays an important role in regulating the balance of Th17 and Treg differentiation (Dang et al. 2011; Shi et al. 2011). Th17 differentiation induces the up-regulation of HIF1α in a STAT3-dependent manner, which, in turn, directly regulates the transcription of RORγt and drives the expression of the RORγt target genes such as the key effector cytokine IL-17A. Furthermore, HIF1α promotes Foxp3 degradation both under Th17- and pTreg skewing conditions, and lack of HIF1α enhances Treg differentiation (Dang et al. 2011). Similar results are shown that mTOR-dependent HIF1α expression is a metabolic checkpoint to regulate Th17 and Treg balances in the homeostasis (Shi et al. 2011). HIF1α deficiency protects mice from experimental autoimmune encephalomyelitis (EAE) and autoimmune neuroinflammation.

However, two other groups reported that HIF1α can also positively regulate Foxp3 transcription (Ben-Shoshan et al. 2008; Clambey et al. 2012). In vitro studies demonstrated that both HIF1α and Foxp3 are up-regulated under hypoxic conditions. Moreover, hypoxia maintains the suppressive function of $CD4^+CD25^+$ T cells. Retroviral transduction of HIF1α in vivo results in the elevated expression of Foxp3 in Treg cells (Ben-Shoshan et al. 2008). Another study showed that HIF1α directly binds to the Foxp3 promoter and hypoxia induces Fopx3 transcription in the presence of TGF-β (Clambey et al. 2012). In addition, the intrinsic expression of HIF1α in Treg cells is required for their suppressive function in controlling the inflammatory bowel disease.

VHL is the most well-studied E3 ligase component that regulates HIF1α degradation. Conditional deletion of VHL in $Foxp3^{+}$ Treg cells results in Th1-dominant autoimmunity (Lee et al. 2015); furthermore, VHL-deficient Treg cells fail to prevent the effector T cell-induced colitis, but convert them into IFN-γ producing Th1-like cells. HIF1α can directly bind to the promoter of IFN-γ to promote the pro-inflammatory cytokine production. Finally, ablation of IFN-γ or HIF1α in VHL-deficient Treg cells rescues the Th1 like status and enhances the suppressive function of Treg cells.

4.3 SHARPIN

The linear ubiquitin chain assembly complex (LUBAC) is composed of three subunits: the heme-oxidized iron-responsive element-binding protein 2 ubiquitin ligase-1 (HOIL-1), HOIL-1-interacting protein (HOIP), and a third SHANK-associated RH domain-interacting protein (SHARPIN) subunit. In the LUBAC complex, HOIP is the main catalytically active subunit, which promotes the generation of linear ubiquitin chains (Ikeda et al. 2011). The LUBAC complex participates in multiple signaling pathways and plays an important role in diverse physiological functions. Notably, the LUBAC complex generates linear ubiquitin chains to facilitate the interaction between receptor interacting protein 1 (RIP1) and IKK complex, leading to the activation of the downstream NFκB and MAPK pathways, and the expression of pro-inflammatory and prosurvival-related genes (Haas et al. 2009).

SHARPIN has a putative ubiquitin-binding NPL4 zinc-finger domain (NZF), which binds to ubiquitin, and an ubiquitin-like domain (UBL) which binds to HOIP (Gerlach et al. 2011). SHARPIN-deficient mice show diminished Treg numbers and attenuated Treg suppressive function (Park et al. 2016). SHARPIN is self-ubiquitinated in the NZF domain via the K63-linked ubiquitin chains, and then binds to TCRζ, leading to the inhibition between TCRζ and Zap70, and thus controls Treg development via thymic negative selection. Similar results are reported by another group (Redecke et al. 2016). Interestingly, SHARPIN-deficient Tregs exhibit robust IL-17 production (Park et al. 2016), a key pro-inflammatory cytokine of Th17. Therefore, SHARPIN may also regulate the balance between Treg and Th17 through regulating the TCR signaling.

In addition to SHARPIN, the two other RING-type E3 ligases HOIL1 and HOIP in the LUBAC complex also play a role in late thymocyte differentiation, Treg cell development, and homeostasis (Teh et al. 2016). By using the Rbck1 $^{fl/fl}$ $CD4^{cre}$, Rnf31 $^{fl/fl}$ $CD4^{cre}$, and $Sharpin^{-/-}$ mice, which delete HOIL1, HOIP, and SHARPIN, respectively, it was shown that the percentage and cell numbers of thymic $Foxp3^{+}$ Treg cells are markedly diminished in all three strains, demonstrating that all three LUBAC components are essential for Treg development. In addition, specific deletion of HOIP in $Foxp3^{+}$ Treg cells in Rnf31 $^{fl/fl}$ $Foxp3^{cre}$ mice suggested that HOIP acts in a cell-intrinsic manner to maintain peripheral Treg cell stability (Teh et al. 2016).

4.4 Itch

Itch-deficient naïve T cells are resistant to TGF-β-induced Treg differentiation and Foxp3 expression (Venuprasad et al. 2008). These deficient mice are unable to suppress airway inflammation. Mechanistic studies showed that Itch modulates the conjugation of mono-ubiquitination to the transcription factor TGF-β-inducible early gene 1 product (TIEG1), which enhances TIEG1 nuclear localization and transcriptional activation of Foxp3, and promotes Foxp3 expression. Thus, Itch mediates ubiquitin-dependent, proteasome-independent transcriptional activation pathway to positively regulate Treg differentiation and allergic responses.

Ndfip1 is needed for Itch to modulate the degradation of JunB, and this modification prevents JunB driving Th2 cytokine production (Oliver et al. 2006). Ndfip1 interacts with Itch after T cell stimulation and promotes Itch activation. Furthermore, Ndfip1$^{-/-}$ mice develop severe inflammatory disease. T cells derived from Ndfip1$^{-/-}$ mice are defective in converting into pTreg cells in vivo and in vitro and produce type 2 cytokine IL-4 (Beal et al. 2011). Eventually, TGF-β-dependent Ndfip1 expression promotes the degradation of JunB mediated by E3 ligase Itch and silences IL-4 production, thus promoting Foxp3 transcription and pTreg differentiation.

Itch not only regulates Treg cells differentiation but also mediates the balance of Treg cells and Th2 cells. Foxp3-restricted Itch-deficient mice display severe lung inflammation. In spite of the comparable frequency and intact suppressive activity of Itch-deficient Treg cells, this selective deficiency results in uncontrolled type 2 inflammation (Jin et al. 2013), due to the accumulation of Th2 key transcription factors GATA3 and STAT6 in Treg cells.

4.5 Cbl-b

As described earlier, Cbl-b has been implicated in multilayered T cell homeostasis. Our previous work identified that Cbl-b is required for Foxp3 transcription in pTreg differentiation through the PI3 K-AKT pathway to regulate Foxo1 and Foxo3a phosphorylation (Harada et al. 2010). Cbl-b was later reported to partially rescue the defective development of tTreg cells in CD28$^{-/-}$ mice. The authors demonstrated that Cbl-b binds to Foxp3 together with Stub1, a U-box domain type E3 ubiquitin ligase, to target Foxp3 to ubiquitination and degradation (Zhao et al. 2015), suggesting a direct role for Cbl-b to modulate tTreg differentiation.

4.6 GRAIL

Gene related to anergy in lymphocytes (GRAIL) encoded by Rnf128 is a RING-type E3 ligase which is important for the T cell anergy induction and immune tolerance (Anandasabapathy et al. 2003; Heissmeyer and Rao 2004; Heissmeyer et al. 2004). GRAIL is unregulated in both tTreg and pTreg cells. Furthermore, forced expression of GRAIL protein in $CD4^{+}$T cells promotes them to convert into a regulatory phenotype in the absence of detectable Foxp3 expression (MacKenzie et al. 2007). GRAIL-deficient mice are resistant to immune tolerance induction and are sensitive to autoimmune diseases. Naïve T cells lacking GRAIL are hyperactivated and over-proliferative, whereas Treg cells from the GRAIL-deficient mice are defective in maintaining the suppressive capacity, associating with the increased IL-21 production. At the molecular level, GRAIL modulates the ubiquitin-dependent degradation of TCR–CD3 complex and regulates the NFATc1 transcription factor expression, which is important for IL-21 production (Nurieva et al. 2010). These data indicated that the GRAIL acts as an essential regulator of T cell tolerance by regulating naïve T cell anergy and maintaining Treg cell stability and suppressive function.

4.7 TRAF6

Tumor necrosis factor (TNF) receptor-associated factor 6 (TRAF6) has an N-terminal RING finger domain and acts as an adaptor protein to mediate protein–protein interaction through processing the K63-linked non-proteolytic poly-ubiquitination. TRAF6 regulates multiple signaling pathways such as NFκB, MAPK, or PI3K-AKT pathway, or pathways downstream of the TNF receptor superfamily, the Toll-like receptor (TLR) family, TGF-β receptor, IL-17 receptor, and TCR signaling (Walsh et al. 2015).

TRAF6-deficient mice exhibit thymic atrophy, severe osteopetrosis, defective lymph node organogenesis, dramatically reduced Treg cell numbers in thymus, and compromised development of medullary thymic epithelial cells (mTECs) (Naito et al. 1999; Akiyama et al. 2005), indicating the critical role for TRAF6 in the regulation of central self-tolerance and autoimmune diseases. In addition, T cell-specific deletion of TRAF6 in the $TRAF6^{fl/fl}$ $CD4^{cre}$ mice develops systemic inflammatory disease characterized by hyperactivation of $CD4^{+}$ T cells, but unimpaired Treg cell development (King et al. 2006). However, TRAF6-deficient T cells are resistant to suppression by normal Treg cells and exhibit hyperactivation of PI3K-AKT pathway compared with wild-type T cells (King et al. 2006). In addition, TRAF6-deficient T cells are resistant to anergy induction in vivo and in vitro (King et al. 2008), further indicating that TRAF6 is essential for peripheral tolerance induction.

However, another study demonstrated that TRAF6-deficient mice show a huge decrease in Treg numbers in thymus, but with unchanged Foxp3 expression (Shimo et al. 2011). Moreover, in vivo mixed fetal liver transfer assay indicates that TRAF6 deficiency specifically impairs the development of $Foxp3^{+}$Treg cells from hematopoietic cells in the recipient thymus. However, TRAF6 does not affect the pTreg polarization from naïve T cells. Similarly, TRAF6-deficient T cells also generate normal numbers of pTreg cells under pTreg skewing conditions (Cejas et al. 2010). Therefore, TRAF6 may have the biological function only in tTreg development but not in pTreg cells.

To further analyze the cell-intrinsic function of this E3 ligase in Treg development and function, TRAF6 $^{fl/fl}$ $Foxp3^{cre}$ mice are generated (Muto et al. 2013). These mice develop allergic skin diseases, arthritis, and lymphadenopathy characterized by hyperactivation of Th2 cells and impaired suppressive function of Treg cells in vivo. Moreover, TRAF6-deficient Treg cells are more easily converted into non-Treg cells when adoptively transferred into $Rag1^{-/-}$ recipient mice. These data suggested that TRAF6 is essential for maintaining Treg stability through inhibiting Th2-related gene expression.

5 Deubiquitinase A20 in Immune Regulation

5.1 A20 Functions in Lymphoid Cells

A20 is first linked to TNF and IL-1 signaling termination (Song et al. 1996). Later, a genetic study confirmed its roles in downstream of TNF, IL-1, and LPS signals (Lee et al. 2000). The importance of A20 is further demonstrated by $A20^{-/-}$ mice which show severe inflammation in multi-organs and hypersensitivity to NFκB activation in immune cells (Lee et al. 2000). Multiple immune cell functions are regulated by A20. Lack of A20 impairs T cell survival and proliferation, partly due to the pro-necroptotic RIPK1–RIPK3 complex accumulation in cytoplasm (Onizawa et al. 2015). Mice with exclusive ablation of A20 in B cells display increased GC formation, excessive production of autoantibodies, and pro-inflammatory cytokine IL-6 (Chu et al. 2011, 2012; Tavares et al. 2010; Hovelmeyer et al. 2011). Additionally, the levels of IL-4, IgE in the serum, and the cell numbers of mast cells and eosinophils are significantly increased upon sensitization in mice lacking A20 in B cells (Li et al. 2016).

A20 is located in a susceptibility locus related to inflammatory bowel disease (Lee et al. 2000). Interestingly, mice lacking A20 both in intestine epithelial cells (IECs) and in myeloid cells, but not in either alone, spontaneously develop colitis (Vereecke et al. 2014). In addition, IECs derived from mice deficient in A20 in myeloid cells show much more apoptosis in a cytokine-dependent manner, which may suggest that A20 assists IECs to resist cell death induced by excessive cytokines from A20-deficient myeloid cells.

Lymphocyte priming by APCs is one of the important steps to control immune responses. A20 has been found to properly regulate the priming actions on macrophages or DCs. A20 $^{fl/fl}$Cd11c-cre mice show aberrant activation of lymphocytes and colitis development (Hammer et al. 2011). The raised level of activation markers like CD80 86 and CD40 and the abnormal production of pro-inflammatory cytokines are associated with A20 deficiency in macrophages or DCs (Kool et al. 2011; Song et al. 2008), which in turn promote effector $CD4^{+}$ T cell differentiation (Wang et al. 2012; Song et al. 2008). What is surprising is that disrupting A20 in DCs alone resembles the feature of systemic lupus erythematous in mice (Kool et al. 2011), which show elevated double-strand DNA antibody production and splenomegaly. Constitutive NFκB and JNK signaling activations in DCs may be responsible for their survival and inappropriately activation of other lymphocytes (Kool et al. 2011).

A20 also plays a crucial role in maintaining the hematopoietic stem cell pool, which is illuminated by different genetic models (Nakagawa et al. 2015; Nagamachi et al. 2014). Lack of A20 causes anemia and lymphopenia, and increases IFN-γ production, which is identified as the major cause of inflammation in $A20^{-/-}$ mice (Nakagawa et al. 2015).

5.2 A20 Functions in Non-lymphoid Cells

A20 also plays an important anti-inflammatory role in various non-lymphoid cells. It has been shown that A20 expression on lung epithelial cells is linked to DC recruitment and type-2 immune responses during chronic endotoxin intake (Schuijs et al. 2015). In addition, A20-silencing in THP1 cells and enterocytes disrupts LPS-induced tolerance (Hu et al. 2014; Wang et al. 2009). Moreover, A20 protects mice from the damage of EAE, since A20 deficiency in neuroectodermal cells increased immune cell infiltration and pro-inflammatory gene expression (Wang et al. 2013). The anti-inflammatory function of A20 is also elucidated in pristine-induced lupus nephritis mice model, in which A20 overexpression alleviates the lupus inflammation and injury in the kidney (Li et al. 2015).

5.3 The Role of A20 in TNF-Induced NFκB Activation

Molecular studies have revealed that A20 is a bipartite ubiquitin-editing enzyme with DUB and E3 ligase functions. Both OUT and ZF domains are important for A20 to terminate the TNF-induced NFκB activation (Wertz et al. 2004; Lu et al. 2013). Upon TNF stimulation, A20 is recruited into TNF receptor complex and relieves the K63 ubiquitin chains from RIP1 by its OUT domain. Next, A20 utilizes its ZF4 domain to catalyze K48-linked poly-ubiquitin chain formation on RIP1 to promote the degradation of RIP1 (Wertz et al. 2004).

RING finger protein 11(RNF11), Itch, and Tax1-binding protein 1 (TAX1BP1) have been reported as the adaptor proteins for A20 complex (Shembade et al. 2008, 2009). The two ZF domains and PPXY motifs on TAX1BP1 provide a platform for RIP1, Itch, and A20 interaction (Shembade et al. 2008). The human T cell leukemia virus type I oncoprotein Tax serves as a negative regulator of A20 by disrupting A20 and its adaptor protein interaction. The effect of Tax on A20 is dependent on its association with cell adhesion molecule CADM1, which initiates the phosphorylation on TAX1BP1 by IKKα and then interferes the association of TAX1BP1 with A20, thus promoting NFκB signaling (Pujari et al. 2015).

Another substrate of A20 is NFκB essential modulator (NEMO). A20 is recruited to NEMO and LUBAC in a TNF-inducible manner. It then prevents NEMO binding to LUBAC utilizing its ZF7 motif, which has high affinity to linear ubiquitin chains (Verhelst et al. 2012). Rhomboid domain containing 3 (Rhbdd3) facilitates A20 recruitment to NEMO via K27-linked poly-ubiquitin chains at its K268 residue (Liu et al. 2014). Breaking the link between NEMO and A20 causes severe inflammation. A NEMO truncation, which is defective in binding to A20, triggers skin and intestine inflammation. $CD14^+$ and $CD4^+$ cells derived from NEMO-truncated patients produce higher pro-inflammatory factors IL-1β and TNF (Zilberman-Rudenko et al. 2016).

5.4 A20 Functions in Other Signaling Pathways

TNF or TNF receptor 1 deficiency fails to mitigate the spontaneous inflammation in A20 knockout mice, suggesting that A20 plays a key role in TNF-independent pro-inflammatory signals (Boone et al. 2004). Several evidences support that A20 mediates TLR signaling transduction: A20-deficient macrophages are hyperreactive to TLR ligand stimulation (Boone et al. 2004); TLR4-MyD88 signaling activation is partly responsible for arthritis pathogenesis in mice with A20 deficiency in myeloid cells ($A20^{myel-KO}$ mice) (Matmati et al. 2011); MyD88 deficiency rescues the cachexia, T cell hyperactivation, and lethality phenotype in $A20^{-/-}$ mice (Turer et al. 2008). At the molecular level, A20 deubiquitylates TRAF6 and then terminates the TLR signaling (Boone et al. 2004). More detailed mechanistic study demonstrates that by promoting the ubiquitination and degradation of E2 enzymes Ubc13 and UbcH5c, A20 and its adaptor protein TAX1BP1 inhibit E2/E3 complex formation, and thus inactivate the E3 ligase activities of TRAF2, TRAF6, and cellular inhibitor of apoptosis protein-1 (cIAP1), and mediate TLR signaling termination (Shembade et al. 2010). Similar to $A20^{-/-}$ mice, patients with germ line mutation in A20 develop systemic inflammation, characterized by higher pro-inflammatory cytokine production and spontaneous NLRP3 inflammasome activation (Zhou et al. 2016). A20 restricts pro-IL-1β ubiquitination to suppress NLRP3 inflammasome activity (Duong et al. 2015).

Besides that, A20 regulates nucleotide-binding oligomerization domain containing 2 (NOD2) signaling. A20-deficient cells are hypersensitive to muramyl dipeptide stimuli (Hitotsumatsu et al. 2008). Moreover, JAK-STAT signaling is important for enteritis development in A20$^{myel-KO}$ mice (De Wilde et al. 2017).

In addition to its ubiquitin-editing ability to attenuate inflammatory responses, A20 can also exert catalytic-independent activity to regulate signaling transduction. Through binding to cIAP1 by ZF7 motif, A20 breaks the association between TRAF2 and TRAF3, which abolishes the stability of NFκB-inducing kinase NIK, an important regulator of non-canonical pathway(Yamaguchi et al. 2013).

5.5 Regulation of A20

The activity of A20 is monitored by different mechanisms including transcriptional gene expression, post-transcriptional modification, and proteasome-mediated degradation. At the transcription level, A20 gene expression is repressed by transcription repressor DREAM through binding to downstream regulatory elements, but is enhanced by the transcription factor USF1 (Tiruppathi et al. 2014); Orphan nuclear receptor ERRα strengthens A20 promoter activity (Yuk et al. 2015); Reactive oxygen species boosts A20 expression by increasing the H3K4me3 modification on its promoter region (Li et al. 2013). At the post-transcription level, RNA-binding protein 1 (RC3H1) down-regulates the level of A20 by targeting A20 UTR region (Murakawa et al. 2015). Phosphorylation at A20 serine 381 residue by IKKβ elevates its cleavage ability for the K63 poly-ubiquitin chains (Wertz et al. 2015; Hutti et al. 2007). At the regulation of protein level, A20 is cleaved by protease MALT1 upon TCR stimulation (Duwel et al. 2009; Coornaert et al. 2008).

6 CYLD

6.1 CYLD in NFκB Signaling

Cylindromatosis (CYLD) is originally discovered as a tumor suppressor product and is broadly mutated in familial cylindromatosis, multiple familial trichoepithelioma, and Brooke–Spiegler syndrome (Bowen et al. 2005). Structural studies have demonstrated that CYLD encompasses three cytoskeletal-associated protein-glycine-conserved (CAP–GLY) domains and one deubiquitinase catalytic motif (Bignell et al. 2000). In vitro biochemical studies show that CYLD prefers to cleave K63-linked and linear, rather than K48-linked, poly-ubiquitin chains (Komander et al. 2009). The substrates of CYLD are very diverse, such as NEMO, RIPK2, and Bcl3 (Massoumi et al. 2006; Kovalenko et al. 2003; Wex et al. 2015), with most of them being involved in NFκB signaling. Hence, CYLD has been pointed out to be an important factor in coordinating NFκB signaling.

6.2 CYLD in T Cell Development

$CYLD^{-/-}$ mice show reduced numbers of mature $CD4^+$ and $CD8^+$ thymocytes and peripheral T cells (Reiley et al. 2006). The effect of CYLD on thymocyte development starts from the CD4 and CD8 double-positive stage, because thymocyte development at double-negative stage shows no obvious change. Lack of CYLD impairs proximal TCR signaling in that effective Lck and Zap70 association requires CYLD to trim the K48 and Lys 63-linked poly-ubiquitin chains from Lck. In the absence of CYLD, the accumulation of ubiquitin chains on the active form of Lck results in reduced Zap70 phosphorylation by Lck (Reiley et al. 2006).

Mice harboring a catalytically inactive mutation of CYLD display similar defective T cell phenotype as $CYLD^{-/-}$ mice. However, CYLD-truncated mice do not exhibit defective proximal TCR signaling (Tsagaratou et al. 2010). Instead, constitutive NFκB signaling activation might be the major cause for the effect on T cell development, since ablation of NEMO rescues the defect (Tsagaratou et al. 2010). Paradoxically, mice with exon 2 and 3 deletions of CYLD display normal T cell development (Zhang et al. 2006). The reasons for these inconsistent and even contradictory findings remain unknown.

CYLD also affects T cell development through an indirect pathway by regulating the development of mTECs, which serve as the gatekeeper to get rid of auto-reactive T cells. In mice with a short-splice variant of CYLD (sCYLD), which lacks exons 7 and 8 and loses the binding sites for TRAF2 and NEMO, their mTECs fail to develop from immature into mature stage and cause the impairment of T cell negative selection (Reissig et al. 2015).

6.3 CYLD in Lymphocyte Function

CYLD is an important molecule in maintaining T cell activation and homeostasis. Lack of CYLD has been connected with hyperreactive activation and excessive pro-inflammatory cytokine production in T cells (Reissig et al. 2012; Reiley et al. 2007). Transferring $CYLD^{-/-}$ T cells into $Rag^{-/-}$ recipient mice aggravates colitis development. The mechanism of CYLD in controlling T cell activation is by inhibiting TAK-1 ubiquitination, which abolishes the downstream JNK and IKKβ signaling pathway (Reiley et al. 2007). Another indirect evidence suggesting the role of CYLD in regulating T cell activation is that T cell activation is impaired in MALT-1 deficient T cells (Mc Guire et al. 2013). In the absence of MALT1 proteolytic activity, CYLD may persist in cells and control T cell activation.

CYLD-deficient mice display decreased immature NKT cells, due to the survival defect of immature NKT cells (Lee et al. 2010). CYLD-deficient NKT cells exhibit defective IL-7 response and ICOS expression, both of which are important for NKT survival (Lee et al. 2010). In addition, constitutive IKKβ activation and IκBα degradation in CYLD-deficient B cells result in B cell hyperplasia, lymphoid organ enlargement, and accumulation of marginal zone B cells (Jin et al. 2007). Mice carrying sCYLD mutation also display B cell hyperplasia phenotype like CYLD$^{-/-}$ mice. CYLD$^{-/-}$ B cells highly express activation markers such as CD21, CD23, CD80, and CD86 (Jin et al. 2007), whereas sCYLD-containing B cells show high content of nuclear Bcl-3 and increased levels of Bcl-2 expression (Hovelmeyer et al. 2007). Furthermore, DCs derived from sCYLD mice contain a larger amount of nuclear RelB and express elevated costimulatory molecules and pro-inflammatory cytokines, but the mechanisms of sCYLD regulation in DCs still need further investigation (Bros et al. 2010).

6.4 CYLD Function in Treg Cells

CYLD negatively regulates Treg generation in the thymus and peripheral lymphoid tissues. Treg cells carrying the sCYLD mutant exhibit reduced CD25 and CTLA-4 expression and partly lose their suppressive function (Reissig et al. 2012). Consistent with this observation, Treg cells from sCYLD mice fail to inhibit the colitis development initiated by transferring naïve T cells into RAG$^{-/-}$ recipients (Reissig et al. 2012). However, it is also shown that CYLD mutation with exon4 deletion only affects pTreg cell generation, but not influences their suppressive function in vitro (Zhao et al. 2011). The inducible pTreg cells from CYLD-deficient CD4^{+} T cells display strong p38, TAK-1, and AP-1 activation upon TGF-β stimulation. By deubiquitinating the K63 chain at K360 and 374 residues of Smad7, CYLD negatively modulates TAK-1 activation and downstream AP-1 signaling and thus controls the development of Treg cells (Zhao et al. 2011).

7 Concluding Remarks

In this chapter, we have discussed a few E3 ligases and DUBs in the regulation of immune responses. These modifying enzymes catalyze differential ubiquitin chain formation on their protein substrates via using various lysine residues of the ubiquitin moiety, which serves as checkpoint codes during T cell development and differentiation (Fig. 2). There have been significant progresses in this field of research, including the discoveries of many profound immunological phenotypes due to the genetic deficiency or mutation of these enzymes, the manifestation of affected T cells including Th1, 2, 9, and 17, Tfh, and Treg cell subtypes, or at different stages or states including the early development in the thymus, naïve T cell activation, anergy induction, or T cell subset differentiation. Particularly, the ubiquitin system plays an essential role in tolerance induction to maintain immune homeostasis, and dys-regulation of immune tolerance can cause harmful responses to self-tissues, but benefits antitumor or antiviral immunity. Molecular and biochemical studies have helped to identify target proteins for ubiquitin conjugation or

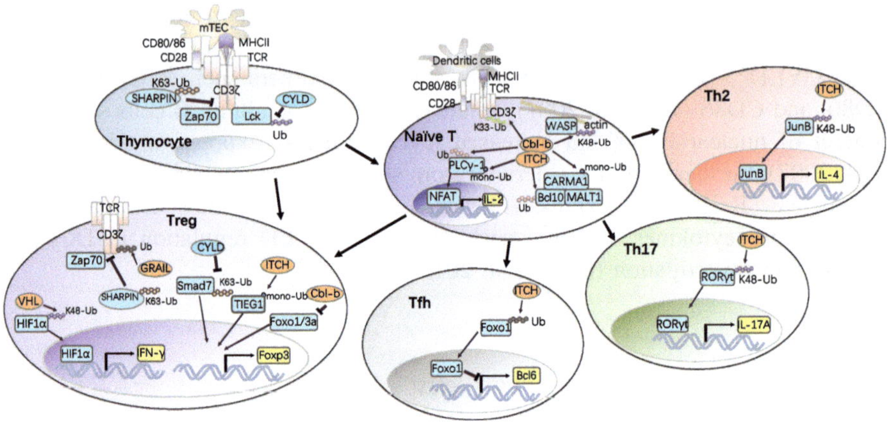

Fig. 2 Checkpoint regulation of T cells by E3 ligases and DUBs. Some of the key E3 ligases including Cbl-b, Itch, GRAIL, SHARPIN, VHL, and the deubiquitinase CYLD are presented. They catalyze the ubiquitin conjugation or deconjugation by targeting protein substrates for differential ubiquitin chain formation. Such modifications serve as codes at different checkpoints of T cell development, activation, and differentiation

deconjugation, which further elucidate the molecular mechanisms by which E3 ligases or DUBs control the development and function of lymphocytes.

Looking forward, it becomes obvious that we face many obstacles and challenges. For one thing, we are still at a very early stage of deciphering the complexity of the ubiquitin system, given the enormous numbers of both E3 ligases and DUBs, with only a fraction of them having been characterized so far. In addition, it is still unclear what determines the specificity of substrate recruitment, the fate of a ubiquitinated substrate, and the crosstalk with other post-translational modifications. In the context of immune regulation, we would be eager to know how a particular E3 ligase or DUB functions under conditions of different stimuli or in various environmental cues, or what will be the impact of such regulation on the magnitude of a particular immune response. In addition, systemic analysis using cutting-edge proteomic techniques, large-scale gene-silencing, gene expression and epigenetic profiling, and deep sequencing technology will allow us to gain a global view of the networks controlled by ubiquitination or deubiquitination, in combination with different models of inflammation and tumorigenesis. Importantly, how the observations in murine models will be implicated in humans in health and diseased conditions will require much more efforts for investigation. Further, elucidating the detailed cellular and molecular mechanisms will eventually help to identify potential targets for therapeutic intervention in human immunologic diseases and cancer.

Acknowledgements This work is supported by funding from Tsinghua-Peking center of life sciences, NSFC 81630041,MOST YFC0903900, NIH RO1AI123398, and R21AI122258.

References

Abbas AK, Benoist C, Bluestone JA, Campbell DJ, Ghosh S, Hori S, Jiang S, Kuchroo VK, Mathis D, Roncarolo MG, Rudensky A, Sakaguchi S, Shevach EM, Vignali DA, Ziegler SF (2013) Regulatory T cells: recommendations to simplify the nomenclature. Nat Immunol 14 (4):307–308. doi:10.1038/ni.2554

Aki D, Zhang W, Liu YC (2015) The E3 ligase Itch in immune regulation and beyond. Immunol Rev 266(1):6–26. doi:10.1111/imr.12301

Akiyama T, Maeda S, Yamane S, Ogino K, Kasai M, Kajiura F, Matsumoto M, Inoue J (2005) Dependence of self-tolerance on TRAF6-directed development of thymic stroma. Science 308 (5719):248–251. doi:10.1126/science.1105677

Anandasabapathy N, Ford GS, Bloom D, Holness C, Paragas V, Seroogy C, Skrenta H, Hollenhorst M, Fathman CG, Soares L (2003) GRAIL: An E3 ubiquitin ligase that inhibits cytokine gene transcription is expressed in anergic CD4(+) T cells. Immunity 18(4):535–547. doi:10.1016/S1074-7613(03)00084-0

Bachmaier K, Krawczyk C, Kozieradzki I, Kong YY, Sasaki T, Oliveira-dos-Santos A, Mariathasan S, Bouchard D, Wakeham A, Itie A, Le J, Ohashi PS, Sarosi I, Nishina H, Lipkowitz S, Penninger JM (2000) Negative regulation of lymphocyte activation and autoimmunity by the molecular adaptor Cbl-b. Nature 403(6766):211–216. doi:10.1038/35003228

Beal AM, Ramos-Hernandez N, Riling CR, Nowelsky EA, Oliver PM (2011) TGF-beta induces the expression of the adaptor Ndfip1 to silence IL-4 production during iTreg cell differentiation. Nat Immunol 13(1):77–85. doi:10.1038/ni.2154

Ben-Shoshan J, Maysel-Auslender S, Mor A, Keren G, George J (2008) Hypoxia controls CD4 + CD25 + regulatory T-cell homeostasis via hypoxia-inducible factor-1alpha. Eur J Immunol 38(9):2412–2418. doi:10.1002/eji.200838318

Bignell GR, Warren W, Seal S, Takahashi M, Rapley E, Barfoot R, Green H, Brown C, Biggs PJ, Lakhani SR, Jones C, Hansen J, Blair E, Hofmann B, Siebert R, Turner G, Evans DG, Schrander-Stumpel C, Beemer FA, van Den Ouweland A, Halley D, Delpech B, Cleveland MG, Leigh I, Leisti J, Rasmussen S (2000) Identification of the familial cylindromatosis tumour-suppressor gene. Nat Genet 25(2):160–165. doi:10.1038/76006

Boone DL, Turer EE, Lee EG, Ahmad R-C, Wheeler MT, Tsui C, Hurley P, Chien M, Chai S, Hitotsumatsu O (2004) The ubiquitin-modifying enzyme A20 is required for termination of toll-like receptor responses. Nat Immunol 5(10):1052–1060

Bowen S, Gill M, Lee DA, Fisher G, Geronemus RG, Vazquez ME, Celebi JT (2005) Mutations in the CYLD gene in Brooke-Spiegler syndrome, familial cylindromatosis, and multiple familial trichoepithelioma: lack of genotype-phenotype correlation. J Invest Dermatol 124(5):919–920. doi:10.1111/j.0022-202X.2005.23688.x

Bros M, Dexheimer N, Besche V, Masri J, Trojandt S, Hovelmeyer N, Reissig S, Massoumi R, Grabbe S, Waisman A, Reske-Kunz AB (2010) Mutated cylindromatosis gene affects the functional state of dendritic cells. Eur J Immunol 40(10):2848–2857. doi:10.1002/eji.200939285

Cao X, Cai SF, Fehniger TA, Song J, Collins LI, Piwnica-Worms DR, Ley TJ (2007) Granzyme B and perforin are important for regulatory T cell-mediated suppression of tumor clearance. Immunity 27(4):635–646. doi:10.1016/j.immuni.2007.08.014

Cejas PJ, Walsh MC, Pearce EL, Han D, Harms GM, Artis D, Turka LA, Choi Y (2010) TRAF6 inhibits Th17 differentiation and TGF-beta-mediated suppression of IL-2. Blood 115 (23):4750–4757. doi:10.1182/blood-2009-09-242768

Chiang YJ, Kole HK, Brown K, Naramura M, Fukuhara S, Hu RJ, Jang IK, Gutkind JS, Shevach E, Gu H (2000) Cbl-b regulates the CD28 dependence of T-cell activation. Nature 403 (6766):216–220. doi:10.1038/35003235

Chiang JY, Jang IK, Hodes R, Gu H (2007) Ablation of Cbl-b provides protection against transplanted and spontaneous tumors. J Clin Invest 117(4):1029–1036. doi:10.1172/JCI29472

Chu Y, Vahl JC, Kumar D, Heger K, Bertossi A, Wojtowicz E, Soberon V, Schenten D, Mack B, Reutelshofer M, Beyaert R, Amann K, van Loo G, Schmidt-Supprian M (2011) B cells lacking the tumor suppressor TNFAIP3/A20 display impaired differentiation and hyperactivation and cause inflammation and autoimmunity in aged mice. Blood 117(7):2227–2236. doi:10.1182/blood-2010-09-306019

Chu Y, Soberon V, Glockner L, Beyaert R, Massoumi R, van Loo G, Krappmann D, Schmidt-Supprian M (2012) A20 and CYLD do not share significant overlapping functions during B cell development and activation. J Immunol 189(9):4437–4443

Clambey ET, McNamee EN, Westrich JA, Glover LE, Campbell EL, Jedlicka P, de Zoeten EF, Cambier JC, Stenmark KR, Colgan SP, Eltzschig HK (2012) Hypoxia-inducible factor-1 alpha-dependent induction of FoxP3 drives regulatory T-cell abundance and function during inflammatory hypoxia of the mucosa. Proc Natl Acad Sci USA 109(41):E2784–E2793. doi:10.1073/pnas.1202366109

Collison LW, Pillai MR, Chaturvedi V, Vignali DA (2009) Regulatory T cell suppression is potentiated by target T cells in a cell contact, IL-35- and IL-10-dependent manner. J Immunol 182(10):6121–6128. doi:10.4049/jimmunol.0803646

Coornaert B, Baens M, Heyninck K, Bekaert T, Haegman M, Staal J, Sun L, Chen ZJ, Marynen P, Beyaert R (2008) T cell antigen receptor stimulation induces MALT1 paracaspase-mediated cleavage of the NF-kappaB inhibitor A20. Nat Immunol 9(3):263–271. doi:10.1038/ni1561

Crotty S (2011) Follicular Helper CD4 T Cells (T-FH). Annu Rev Immunol 29(29):621–663

Dang EV, Barbi J, Yang HY, Jinasena D, Yu H, Zheng Y, Bordman Z, Fu J, Kim Y, Yen HR, Luo W, Zeller K, Shimoda L, Topalian SL, Semenza GL, Dang CV, Pardoll DM, Pan F (2011) Control of T(H)17/T(reg) balance by hypoxia-inducible factor 1. Cell 146(5):772–784. doi:10.1016/j.cell.2011.07.033

De Wilde K, Martens A, Lambrecht S, Jacques P, Drennan MB, Debusschere K, Govindarajan S, Coudenys J, Verheugen E, Windels F, Catrysse L, Lories R, McGonagle D, Beyaert R, van Loo G, Elewaut D (2017) A20 inhibition of STAT1 expression in myeloid cells: a novel endogenous regulatory mechanism preventing development of enthesitis. Ann Rheum Dis 76 (3):585–592. doi:10.1136/annrheumdis-2016-209454

Delgoffe GM, Woo SR, Turnis ME, Gravano DM, Guy C, Overacre AE, Bettini ML, Vogel P, Finkelstein D, Bonnevier J, Workman CJ, Vignali DA (2013) Stability and function of regulatory T cells is maintained by a neuropilin-1-semaphorin-4a axis. Nature 501(7466):252–256. doi:10.1038/nature12428

Doherty M, Osborne DG, Browning DL, Parker DC, Wetzel SA (2010) Anergic CD4 + T cells form mature immunological synapses with enhanced accumulation of c-Cbl and Cbl-b. J Immunol 184(7):3598–3608. doi:10.4049/jimmunol.0902285

Duong BH, Onizawa M, Oses-Prieto JA, Advincula R, Burlingame A, Malynn BA, Ma A (2015) A20 restricts ubiquitination of pro-interleukin-1beta protein complexes and suppresses NLRP3 inflammasome activity. Immunity 42(1):55–67. doi:10.1016/j.immuni.2014.12.031

Duwel M, Welteke V, Oeckinghaus A, Baens M, Kloo B, Ferch U, Darnay BG, Ruland J, Marynen P, Krappmann D (2009) A20 negatively regulates T cell receptor signaling to NF-kappaB by cleaving Malt1 ubiquitin chains. J Immunol 182(12):7718–7728. doi:10.4049/jimmunol.0803313

Enzler T, Chang X, Facchinetti V, Melino G, Karin M, Su B, Gallagher E (2009) MEKK1 binds HECT E3 ligase Itch by its amino-terminal RING motif to regulate Th2 cytokine gene expression. J Immunol 183(6):3831–3838. doi:10.4049/jimmunol.0803412

Fahlen L, Read S, Gorelik L, Hurst SD, Coffman RL, Flavell RA, Powrie F (2005) T cells that cannot respond to TGF-beta escape control by CD4(+)CD25(+) regulatory T cells. J Exp Med 201(5):737–746. doi:10.1084/jem.20040685

Fang D, Wang HY, Fang N, Altman Y, Elly C, Liu YC (2001) Cbl-b, a RING-type E3 ubiquitin ligase, targets phosphatidylinositol 3-kinase for ubiquitination in T cells. J Biol Chem 276 (7):4872–4878. doi:10.1074/jbc.M008901200

Fang D, Elly C, Gao B, Fang N, Altman Y, Joazeiro C, Hunter T, Copeland N, Jenkins N, Liu YC (2002) Dysregulation of T lymphocyte function in itchy mice: a role for Itch in TH2 differentiation. Nat Immunol 3(3):281–287. doi:10.1038/ni763

Fontenot JD, Gavin MA, Rudensky AY (2003) Foxp3 programs the development and function of CD4 + CD25 + regulatory T cells. Nat Immunol 4(4):330–336. doi:10.1038/ni904

Fraile JM, Quesada V, Rodriguez D, Freije JM, Lopez-Otin C (2012) Deubiquitinases in cancer: new functions and therapeutic options. Oncogene 31(19):2373–2388. doi:10.1038/onc.2011.443

Gallagher E, Gao M, Liu YC, Karin M (2006) Activation of the E3 ubiquitin ligase Itch through a phosphorylation-induced conformational change. Proc Natl Acad Sci USA 103(6):1717–1722

Gao M, Labuda T, Xia Y, Gallagher E, Fang D, Liu YC, Karin M (2004) Jun turnover is controlled through JNK-dependent phosphorylation of the E3 ligase itch. Science 306(5694):271–275

Gavin MA, Rasmussen JP, Fontenot JD, Vasta V, Manganiello VC, Beavo JA, Rudensky AY (2007) Foxp3-dependent programme of regulatory T-cell differentiation. Nature 445 (7129):771–775. doi:10.1038/nature05543

Gerlach B, Cordier SM, Schmukle AC, Emmerich CH, Rieser E, Haas TL, Webb AI, Rickard JA, Anderton H, Wong WW, Nachbur U, Gangoda L, Warnken U, Purcell AW, Silke J, Walczak H (2011) Linear ubiquitination prevents inflammation and regulates immune signalling. Nature 471(7340):591–596. doi:10.1038/nature09816

Goetz B, An W, Mohapatra B, Zutshi N, Iseka F, Storck MD, Meza J, Sheinin Y, Band V, Band H (2016) A novel CBL-Bflox/flox mouse model allows tissue-selective fully conditional CBL/CBL-B double-knockout: CD4-Cre mediated CBL/CBL-B deletion occurs in both T-cells and hematopoietic stem cells. Oncotarget 7(32):51107–51123

Gronski MA, Boulter JM, Moskophidis D, Nguyen LT, Holmberg K, Elford AR, Deenick EK, Kim HO, Penninger JM, Odermatt B, Gallimore A, Gascoigne NR, Ohashi PS (2004) TCR affinity and negative regulation limit autoimmunity. Nat Med 10(11):1234–1239. doi:10.1038/nm1114

Gruber T, Hermann-Kleiter N, Hinterleitner R, Fresser F, Schneider R, Gastl G, Penninger JM, Baier G (2009) PKC-theta modulates the strength of T cell responses by targeting Cbl-b for ubiquitination and degradation. Sci Signal 2(76):ra30. doi:10.1126/scisignal.2000046

Gruber T, Hinterleitner R, Hermann-Kleiter N, Meisel M, Kleiter I, Wang CM, Viola A, Pfeifhofer-Obermair C, Baier G (2013) Cbl-b mediates TGF beta sensitivity by downregulating inhibitory SMAD7 in primary T cells. J Mol Cell Biol 5(6):358–368

Guo H, Qiao G, Ying H, Li Z, Zhao Y, Liang Y, Yang L, Lipkowitz S, Penninger JM, Langdon WY, Zhang J (2012) E3 ubiquitin ligase Cbl-b regulates Pten via Nedd4 in T cells independently of its ubiquitin ligase activity. Cell Rep 1(5):472–482. doi:10.1016/j.celrep.2012.04.008

Haas TL, Emmerich CH, Gerlach B, Schmukle AC, Cordier SM, Rieser E, Feltham R, Vince J, Warnken U, Wenger T, Koschny R, Komander D, Silke J, Walczak H (2009) Recruitment of the linear ubiquitin chain assembly complex stabilizes the TNF-R1 signaling complex and is required for TNF-mediated gene induction. Mol Cell 36(5):831–844. doi:10.1016/j.molcel.2009.10.013

Hammer GE, Turer EE, Taylor KE, Fang CJ, Advincula R, Oshima S, Barrera J, Huang EJ, Hou B, Malynn BA, Reizis B, DeFranco A, Criswell LA, Nakamura MC, Ma A (2011) Expression of A20 by dendritic cells preserves immune homeostasis and prevents colitis and spondyloarthritis. Nat Immunol 12(12):1184–1193. doi:10.1038/ni.2135

Hanlon A, Jang S, Salgame P (2005) Cbl-b differentially regulates activation-induced apoptosis in T helper 1 and T helper 2 cells. Immunology 116(4):507–512. doi:10.1111/j.1365-2567.2005.02252.x

Harada Y, Harada Y, Elly C, Ying G, Paik JH, DePinho RA, Liu YC (2010) Transcription factors Foxo3a and Foxo1 couple the E3 ligase Cbl-b to the induction of Foxp3 expression in induced regulatory T cells. J Exp Med 207(7):1381–1391. doi:10.1084/jem.20100004

Heissmeyer V, Rao A (2004) E3 ligases in T cell anergy–turning immune responses into tolerance. Sci STKE 2004(241):pe29. doi:10.1126/stke.2412004pe29

Heissmeyer V, Macian F, Im SH, Varma R, Feske S, Venuprasad K, Gu H, Liu YC, Dustin ML, Rao A (2004) Calcineurin imposes T cell unresponsiveness through targeted proteolysis of signaling proteins. Nat Immunol 5(3):255–265

Hinterleitner R, Gruber T, Pfeifhofer-Obermair C, Lutz-Nicoladoni C, Tzankov A, Schuster M, Penninger JM, Loibner H, Lametschwandtner G, Wolf D, Baier G (2012) Adoptive transfer of siRNA Cblb-silenced CD8 + T lymphocytes augments tumor vaccine efficacy in a B16 melanoma model. PLoS ONE 7(9):e44295. doi:10.1371/journal.pone.0044295

Hitotsumatsu O, Ahmad RC, Tavares R, Wang M, Philpott D, Turer EE, Lee BL, Shiffin N, Advincula R, Malynn BA, Werts C, Ma A (2008) The ubiquitin-editing enzyme A20 restricts nucleotide-binding oligomerization domain containing 2-triggered signals. Immunity 28 (3):381–390. doi:10.1016/j.immuni.2008.02.002

Hori S, Nomura T, Sakaguchi S (2003) Control of regulatory T cell development by the transcription factor Foxp3. Science 299(5609):1057–1061. doi:10.1126/science.1079490

Hovelmeyer N, Wunderlich FT, Massoumi R, Jakobsen CG, Song J, Worns MA, Merkwirth C, Kovalenko A, Aumailley M, Strand D, Bruning JC, Galle PR, Wallach D, Fassler R, Waisman A (2007) Regulation of B cell homeostasis and activation by the tumor suppressor gene CYLD. J Exp Med 204(11):2615–2627. doi:10.1084/jem.20070318

Hovelmeyer N, Reissig S, Xuan NT, Adams-Quack P, Lukas D, Nikolaev A, Schluter D, Waisman A (2011) A20 deficiency in B cells enhances B-cell proliferation and results in the development of autoantibodies. Eur J Immunol 41(3):595–601. doi:10.1002/eji.201041313

Hoyne GF, Flening E, Yabas M, Teh C, Altin JA, Randall K, Thien CB, Langdon WY, Goodnow CC (2011) Visualizing the role of Cbl-b in control of islet-reactive CD4 T cells and susceptibility to type 1 diabetes. J Immunol 186(4):2024–2032. doi:10.4049/jimmunol.1002296

Hsu TS, Hsiao HW, Wu PJ, Liu WH, Lai MZ (2014) Deltex1 promotes protein kinase Ctheta degradation and sustains Casitas B-lineage lymphoma expression. J Immunol 193(4):1672–1680. doi:10.4049/jimmunol.1301416

Hu J, Wang G, Liu X, Zhou L, Jiang M, Yang L (2014) A20 is critical for the induction of Pam3CSK4-tolerance in monocytic THP-1 cells. PLoS ONE 9(1):e87528. doi:10.1371/journal.pone.0087528

Huang F, Kitaura Y, Jang I, Naramura M, Kole HH, Liu L, Qin H, Schlissel MS, Gu H (2006) Establishment of the major compatibility complex-dependent development of CD4 + and CD8 + T cells by the Cbl family proteins. Immunity 25(4):571–581. doi:10.1016/j.immuni.2006.08.021

Huang H, Jeon MS, Liao L, Yang C, Elly C, Yates JR 3rd, Liu YC (2010) K33-linked polyubiquitination of T cell receptor-zeta regulates proteolysis-independent T cell signaling. Immunity 33(1):60–70. doi:10.1016/j.immuni.2010.07.002

Hustad CM, Perry WL, Siracusa LD, Rasberry C, Cobb L, Cattanach BM, Kovatch R, Copeland NG, Jenkins NA (1995) Molecular genetic characterization of six recessive viable alleles of the mouse agouti locus. Genetics 140(1):255–265

Hutti JE, Turk BE, Asara JM, Ma A, Cantley LC, Abbott DW (2007) IkappaB kinase beta phosphorylates the K63 deubiquitinase A20 to cause feedback inhibition of the NF-kappaB pathway. Mol Cell Biol 27(21):7451–7461. doi:10.1128/MCB.01101-07

Ikeda F, Deribe YL, Skanland SS, Stieglitz B, Grabbe C, Franz-Wachtel M, van Wijk SJ, Goswami P, Nagy V, Terzic J, Tokunaga F, Androulidaki A, Nakagawa T, Pasparakis M, Iwai K, Sundberg JP, Schaefer L, Rittinger K, Macek B, Dikic I (2011) SHARPIN forms a linear ubiquitin ligase complex regulating NF-kappaB activity and apoptosis. Nature 471 (7340):637–641. doi:10.1038/nature09814

Ivan M, Kondo K, Yang H, Kim W, Valiando J, Ohh M, Salic A, Asara JM, Lane WS, Kaelin WG Jr (2001) HIFalpha targeted for VHL-mediated destruction by proline hydroxylation: implications for O2 sensing. Science 292(5516):464–468. doi:10.1126/science.1059817

Jaakkola P, Mole DR, Tian YM, Wilson MI, Gielbert J, Gaskell SJ, von Kriegsheim A, Hebestreit HF, Mukherji M, Schofield CJ, Maxwell PH, Pugh CW, Ratcliffe PJ (2001)

Targeting of HIF-alpha to the von Hippel-Lindau ubiquitylation complex by O2-regulated prolyl hydroxylation. Science 292(5516):468–472. doi:10.1126/science.1059796

Jackson PK, Eldridge AG, Freed E, Furstenthal L, Hsu JY, Kaiser BK, Reimann JD (2000) The lore of the RINGs: substrate recognition and catalysis by ubiquitin ligases. Trends Cell Biol 10 (10):429–439. doi:10.1016/S0962-8924(00)01834-1

Jeon MS, Atfield A, Venuprasad K, Krawczyk C, Sarao R, Elly C, Yang C, Arya S, Bachmaier K, Su L, Bouchard D, Jones R, Gronski M, Ohashi P, Wada T, Bloom D, Fathman CG, Liu YC, Penninger JM (2004) Essential role of the E3 ubiquitin ligase Cbl-b in T cell anergy induction. Immunity 21(2):167–177. doi:10.1016/j.immuni.2004.07.013

Jin W, Reiley WR, Lee AJ, Wright A, Wu X, Zhang M, Sun SC (2007) Deubiquitinating enzyme CYLD regulates the peripheral development and naive phenotype maintenance of B cells. J Biol Chem 282(21):15884–15893. doi:10.1074/jbc.M609952200

Jin HS, Park Y, Elly C, Liu YC (2013) Itch expression by Treg cells controls Th2 inflammatory responses. J Clin Invest 123(11):4923–4934. doi:10.1172/JCI69355

Kaelin WG, Jr., Maher ER (1998) The VHL tumour-suppressor gene paradigm. Trends Genet 14 (10):423–426. doi:10.1016/S0168-9525(98)01558-3

Kamura T, Koepp DM, Conrad MN, Skowyra D, Moreland RJ, Iliopoulos O, Lane WS, Kaelin WG, Jr., Elledge SJ, Conaway RC, Harper JW, Conaway JW (1999) Rbx1, a component of the VHL tumor suppressor complex and SCF ubiquitin ligase. Science 284 (5414):657–661. doi:10.1126/science.284.5414.657

Karwacz K, Bricogne C, MacDonald D, Arce F, Bennett CL, Collins M, Escors D (2011) PD-L1 co-stimulation contributes to ligand-induced T cell receptor down-modulation on CD8 + T cells. EMBO Mol Med 3(10):581–592. doi:10.1002/emmm.201100165

Kathania M, Khare P, Zeng M, Cantarel B, Zhang H, Ueno H, Venuprasad K (2016) Itch inhibits IL-17-mediated colon inflammation and tumorigenesis by ROR-gammat ubiquitination. Nat Immunol 17(8):997–1004. doi:10.1038/ni.3488

King CG, Kobayashi T, Cejas PJ, Kim T, Yoon K, Kim GK, Chiffoleau E, Hickman SP, Walsh PT, Turka LA, Choi Y (2006) TRAF6 is a T cell-intrinsic negative regulator required for the maintenance of immune homeostasis. Nat Med 12(9):1088–1092. doi:10.1038/nm1449

King CG, Buckler JL, Kobayashi T, Hannah JR, Bassett G, Kim T, Pearce EL, Kim GG, Turka LA, Choi Y (2008) Cutting edge: requirement for TRAF6 in the induction of T cell anergy. J Immunol 180(1):34–38

Kojo S, Elly C, Harada Y, Langdon WY, Kronenberg M, Liu YC (2009) Mechanisms of NKT cell anergy induction involve Cbl-b-promoted monoubiquitination of CARMA1. Proc Natl Acad Sci USA 106(42):17847–17851. doi:10.1073/pnas.0904078106

Komander D, Rape M (2012) The ubiquitin code. Annu Rev Biochem 81:203–229. doi:10.1146/annurev-biochem-060310-170328

Komander D, Reyes-Turcu F, Licchesi JD, Odenwaelder P, Wilkinson KD, Barford D (2009) Molecular discrimination of structurally equivalent Lys 63-linked and linear polyubiquitin chains. EMBO Rep 10(5):466–473. doi:10.1038/embor.2009.55

Kool M, van Loo G, Waelput W, De Prijck S, Muskens F, Sze M, van Praet J, Branco-Madeira F, Janssens S, Reizis B, Elewaut D, Beyaert R, Hammad H, Lambrecht BN (2011) The ubiquitin-editing protein A20 prevents dendritic cell activation, recognition of apoptotic cells, and systemic autoimmunity. Immunity 35(1):82–96. doi:10.1016/j.immuni.2011.05.013

Kovalenko A, Chable-Bessia C, Cantarella G, Israel A, Wallach D, Courtois G (2003) The tumour suppressor CYLD negatively regulates NF-kappaB signalling by deubiquitination. Nature 424 (6950):801–805. doi:10.1038/nature01802

Krawczyk C, Bachmaier K, Sasaki T, Jones RG, Snapper SB, Bouchard D, Kozieradzki I, Ohashi PS, Alt FW, Penninger JM (2000) Cbl-b is a negative regulator of receptor clustering and raft aggregation in T cells. Immunity 13(4):463–473

Krawczyk CM, Jones RG, Atfield A, Bachmaier K, Arya S, Odermatt B, Ohashi PS, Penninger JM (2005) Differential control of CD28-regulated in vivo immunity by the E3 ligase Cbl-b. J Immunol 174(3):1472–1478

Layman AAK, Sprout SL, Phillips D, Oliver PM (2017) Ndfip1 restricts Th17 cell potency by limiting lineage stability and proinflammatory cytokine production. Sci Rep 7:39649. doi:10.1038/srep39649

Lee EG, Boone DL, Chai S, Libby SL, Chien M, Lodolce JP, Ma A (2000) Failure to regulate TNF-induced NF-kappaB and cell death responses in A20-deficient mice. Science 289 (5488):2350–2354

Lee AJ, Zhou X, Chang M, Hunzeker J, Bonneau RH, Zhou D, Sun SC (2010) Regulation of natural killer T-cell development by deubiquitinase CYLD. EMBO J 29(9):1600–1612. doi:10.1038/emboj.2010.31

Lee JH, Elly C, Park Y, Liu YC (2015) E3 Ubiquitin Ligase VHL Regulates Hypoxia-Inducible Factor-1alpha to Maintain Regulatory T Cell Stability and Suppressive Capacity. Immunity 42 (6):1062–1074. doi:10.1016/j.immuni.2015.05.016

Li D, Gal I, Vermes C, Alegre ML, Chong AS, Chen L, Shao Q, Adarichev V, Xu X, Koreny T, Mikecz K, Finnegan A, Glant TT, Zhang J (2004) Cutting edge: Cbl-b: one of the key molecules tuning CD28- and CTLA-4-mediated T cell costimulation. J Immunol 173 (12):7135–7139

Li Y, Zhang P, Wang C, Han C, Meng J, Liu X, Xu S, Li N, Wang Q, Shi X (2013) Immune responsive gene 1 (IRG1) promotes endotoxin tolerance by increasing A20 expression in macrophages through reactive oxygen species. J Biol Chem 288(23):16225–16234

Li M, Shi X, Qian T, Li J, Tian Z, Ni B, Hao F (2015) A20 overexpression alleviates pristine-induced lupus nephritis by inhibiting the NF-kappaB and NLRP3 inflammasome activation in macrophages of mice. Int J Clin Exp Med 8(10):17430–17440

Li MY, Zhu M, Linghu EQ, Feng F, Zhu B, Wu C, Guo MZ (2016) Interleukin-13 suppresses interleukin-10 via inhibiting A20 in peripheral B cells of patients with food allergy. Oncotarget 7(48):79914–79924. doi:10.18632/oncotarget.13107

Liang B, Workman C, Lee J, Chew C, Dale BM, Colonna L, Flores M, Li N, Schweighoffer E, Greenberg S, Tybulewicz V, Vignali D, Clynes R (2008) Regulatory T cells inhibit dendritic cells by lymphocyte activation gene-3 engagement of MHC class II. J Immunol 180(9):5916–5926

Liu J, Han C, Xie B, Wu Y, Liu S, Chen K, Xia M, Zhang Y, Song L, Li Z, Zhang T, Ma F, Wang Q, Wang J, Deng K, Zhuang Y, Wu X, Yu Y, Xu T, Cao X (2014) Rhbdd3 controls autoimmunity by suppressing the production of IL-6 by dendritic cells via K27-linked ubiquitination of the regulator NEMO. Nat Immunol 15(7):612–622. doi:10.1038/ni.2898

Loeser S, Penninger JM (2007) Regulation of peripheral T cell tolerance by the E3 ubiquitin ligase Cbl-b. Semin Immunol 19(3):206–214. doi:10.1016/j.smim.2007.02.004

Loeser S, Loser K, Bijker MS, Rangachari M, van der Burg SH, Wada T, Beissert S, Melief CJ, Penninger JM (2007) Spontaneous tumor rejection by cbl-b-deficient CD8 + T cells. J Exp Med 204(4):879–891. doi:10.1084/jem.20061699

Lu TT, Onizawa M, Hammer GE, Turer EE, Yin Q, Damko E, Agelidis A, Shifrin N, Advincula R, Barrera J, Malynn BA, Wu H, Ma A (2013) Dimerization and ubiquitin mediated recruitment of A20, a complex deubiquitinating enzyme. Immunity 38(5):896–905. doi:10.1016/j.immuni.2013.03.008

Lutz-Nicoladoni C, Wolf D, Sopper S (2015) Modulation of Immune Cell Functions by the E3 Ligase Cbl-b. Front Oncol 5:58. doi:10.3389/fonc.2015.00058

MacKenzie DA, Schartner J, Lin J, Timmel A, Jennens-Clough M, Fathman CG, Seroogy CM (2007) GRAIL is up-regulated in CD4 + CD25 + T regulatory cells and is sufficient for conversion of T cells to a regulatory phenotype. J Biol Chem 282(13):9696–9702. doi:10.1074/jbc.M604192200

Marino A, Menghini R, Fabrizi M, Casagrande V, Mavilio M, Stoehr R, Candi E, Mauriello A, Moreno-Navarrete JM, Gomez-Serrano M, Peral B, Melino G, Lauro R, Fernandez Real JM, Federici M (2014) ITCH deficiency protects from diet-induced obesity. Diabetes 63 (2):550–561. doi:10.2337/db13-0802

Massoumi R, Chmielarska K, Hennecke K, Pfeifer A, Fassler R (2006) Cyld inhibits tumor cell proliferation by blocking Bcl-3-dependent NF-kappaB signaling. Cell 125(4):665–677. doi:10.1016/j.cell.2006.03.041

Matmati M, Jacques P, Maelfait J, Verheugen E, Kool M, Sze M, Geboes L, Louagie E, Mc Guire C, Vereecke L, Chu Y, Boon L, Staelens S, Matthys P, Lambrecht BN, Schmidt-Supprian M, Pasparakis M, Elewaut D, Beyaert R, van Loo G (2011) A20 (TNFAIP3) deficiency in myeloid cells triggers erosive polyarthritis resembling rheumatoid arthritis. Nat Genet 43(9):908–912. doi:10.1038/ng.874

Maxwell PH, Wiesener MS, Chang GW, Clifford SC, Vaux EC, Cockman ME, Wykoff CC, Pugh CW, Maher ER, Ratcliffe PJ (1999) The tumour suppressor protein VHL targets hypoxia-inducible factors for oxygen-dependent proteolysis. Nature 399(6733):271–275. doi:10.1038/20459

Mc Guire C, Wieghofer P, Elton L, Muylaert D, Prinz M, Beyaert R, van Loo G (2013) Paracaspase MALT1 deficiency protects mice from autoimmune-mediated demyelination. J Immunol 190(6):2896–2903. doi:10.4049/jimmunol.1201351

Murai M, Turovskaya O, Kim G, Madan R, Karp CL, Cheroutre H, Kronenberg M (2009) Interleukin 10 acts on regulatory T cells to maintain expression of the transcription factor Foxp3 and suppressive function in mice with colitis. Nat Immunol 10(11):1178–1184. doi:10.1038/ni.1791

Murakawa Y, Hinz M, Mothes J, Schuetz A, Uhl M, Wyler E, Yasuda T, Mastrobuoni G, Friedel CC, Dölken L (2015) RC3H1 post-transcriptionally regulates A20 mRNA and modulates the activity of the IKK/NF-[kappa] B pathway. Nat Commun 6

Muto G, Kotani H, Kondo T, Morita R, Tsuruta S, Kobayashi T, Luche H, Fehling HJ, Walsh M, Choi Y, Yoshimura A (2013) TRAF6 is essential for maintenance of regulatory T cells that suppress Th2 type autoimmunity. PLoS ONE 8(9):e74639. doi:10.1371/journal.pone.0074639

Nagamachi A, Nakata Y, Ueda T, Yamasaki N, Ebihara Y, Tsuji K, Honda Z, Takubo K, Suda T, Oda H, Inaba T, Honda H (2014) Acquired deficiency of A20 results in rapid apoptosis, systemic inflammation, and abnormal hematopoietic stem cell function. PLoS ONE 9(1): e87425. doi:10.1371/journal.pone.0087425

Naito A, Azuma S, Tanaka S, Miyazaki T, Takaki S, Takatsu K, Nakao K, Nakamura K, Katsuki M, Yamamoto T, Inoue J (1999) Severe osteopetrosis, defective interleukin-1 signalling and lymph node organogenesis in TRAF6-deficient mice. Genes Cells 4(6):353–362. doi:10.1046/j.1365-2443.1999.00265.x

Nakagawa MM, Thummar K, Mandelbaum J, Pasqualucci L, Rathinam CV (2015) Lack of the ubiquitin-editing enzyme A20 results in loss of hematopoietic stem cell quiescence. J Exp Med 212(2):203–216. doi:10.1084/jem.20132544

Naramura M, Jang IK, Kole H, Huang F, Haines D, Gu H (2002) c-Cbl and Cbl-b regulate T cell responsiveness by promoting ligand-induced TCR down-modulation. Nat Immunol 3 (12):1192–1199. doi:10.1038/ni855

Nijman SM, Luna-Vargas MP, Velds A, Brummelkamp TR, Dirac AM, Sixma TK, Bernards R (2005) A genomic and functional inventory of deubiquitinating enzymes. Cell 123(5):773–786. doi:10.1016/j.cell.2005.11.007

Nurieva RI, Zheng S, Jin W, Chung Y, Zhang Y, Martinez GJ, Reynolds JM, Wang SL, Lin X, Sun SC, Lozano G, Dong C (2010) The E3 ubiquitin ligase GRAIL regulates T cell tolerance and regulatory T cell function by mediating T cell receptor-CD3 degradation. Immunity 32 (5):670–680. doi:10.1016/j.immuni.2010.05.002

Oliver PM, Cao X, Worthen GS, Shi P, Briones N, MacLeod M, White J, Kirby P, Kappler J, Marrack P, Yang B (2006) Ndfip1 protein promotes the function of itch ubiquitin ligase to prevent T cell activation and T helper 2 cell-mediated inflammation. Immunity 25(6):929–940. doi:10.1016/j.immuni.2006.10.012

Onizawa M, Oshima S, Schulze-Topphoff U, Oses-Prieto JA, Lu T, Tavares R, Prodhomme T, Duong B, Whang MI, Advincula R (2015) The ubiquitin-modifying enzyme A20 restricts ubiquitination of the kinase RIPK3 and protects cells from necroptosis. Nat Immunol 16 (6):618–627

Ou R, Zhang M, Huang L, Moskophidis D (2008) Control of virus-specific CD8 + T-cell exhaustion and immune-mediated pathology by E3 ubiquitin ligase Cbl-b during chronic viral infection. J Virol 82(7):3353–3368. doi:10.1128/JVI.01350-07

Paolino M, Thien CB, Gruber T, Hinterleitner R, Baier G, Langdon WY, Penninger JM (2011) Essential role of E3 ubiquitin ligase activity in Cbl-b-regulated T cell functions. J Immunol 186 (4):2138–2147. doi:10.4049/jimmunol.1003390

Park Y, Jin HS, Aki D, Lee J, Liu YC (2014) The ubiquitin system in immune regulation. Adv Immunol 124:17–66. doi:10.1016/B978-0-12-800147-9.00002-9

Park Y, Jin HS, Lopez J, Lee J, Liao L, Elly C, Liu YC (2016) SHARPIN controls regulatory T cells by negatively modulating the T cell antigen receptor complex. Nat Immunol 17(3):286–296. doi:10.1038/ni.3352

Pellegrini M, Calzascia T, Elford AR, Shahinian A, Lin AE, Dissanayake D, Dhanji S, Nguyen LT, Gronski MA, Morre M, Assouline B, Lahl K, Sparwasser T, Ohashi PS, Mak TW (2009) Adjuvant IL-7 antagonizes multiple cellular and molecular inhibitory networks to enhance immunotherapies. Nat Med 15(5):528–536. doi:10.1038/nm.1953

Perry WL, Hustad CM, Swing DA, O'Sullivan TN, Jenkins NA, Copeland NG (1998) The itchy locus encodes a novel ubiquitin protein ligase that is disrupted in a(18H) mice. Nat Genet 18 (2):143–146

Pujari R, Hunte R, Thomas R, van der Weyden L, Rauch D, Ratner L, Nyborg JK, Ramos JC, Takai Y, Shembade N (2015) Human T-cell leukemia virus type 1 (HTLV-1) tax requires CADM1/TSLC1 for inactivation of the NF-kappaB inhibitor A20 and constitutive NF-kappaB signaling. PLoS Pathog 11(3):e1004721. doi:10.1371/journal.ppat.1004721

Qiao GL, Li ZP, Molinero L, Alegre ML, Ying HY, Sun ZM, Penninger JM, Zhang J (2008) T-Cell receptor-induced NF-kappa B activation is negatively regulated by E3 ubiquitin ligase Cbl-b. Mol Cell Biol 28(7):2470–2480

Qiao G, Ying H, Zhao Y, Liang Y, Guo H, Shen H, Li Z, Solway J, Tao E, Chiang YJ, Lipkowitz S, Penninger JM, Langdon WY, Zhang J (2014) E3 ubiquitin ligase Cbl-b suppresses proallergic T cell development and allergic airway inflammation. Cell Rep 6 (4):709–723. doi:10.1016/j.celrep.2014.01.012

Ramon HE, Riling CR, Bradfield J, Yang B, Hakonarson H, Oliver PM (2011) The ubiquitin ligase adaptor Ndfip1 regulates T cell-mediated gastrointestinal inflammation and inflammatory bowel disease susceptibility. Mucosal Immunol 4(3):314–324. doi:10.1038/mi.2010.69

Redecke V, Chaturvedi V, Kuriakose J, Hacker H (2016) SHARPIN controls the development of regulatory T cells. Immunology 148(2):216–226. doi:10.1111/imm.12604

Reicher B, Joseph N, David A, Pauker MH, Perl O, Barda-Saad M (2012) Ubiquitylation-dependent negative regulation of WASp Is essential for actin cytoskeleton dynamics. Mol Cell Biol 32(15):3153–3163

Reiley WW, Zhang M, Jin W, Losiewicz M, Donohue KB, Norbury CC, Sun SC (2006) Regulation of T cell development by the deubiquitinating enzyme CYLD. Nat Immunol 7 (4):411–417. doi:10.1038/ni1315

Reiley WW, Jin W, Lee AJ, Wright A, Wu X, Tewalt EF, Leonard TO, Norbury CC, Fitzpatrick L, Zhang M, Sun SC (2007) Deubiquitinating enzyme CYLD negatively regulates the ubiquitin-dependent kinase Tak1 and prevents abnormal T cell responses. J Exp Med 204 (6):1475–1485. doi:10.1084/jem.20062694

Reissig S, Hovelmeyer N, Weigmann B, Nikolaev A, Kalt B, Wunderlich TF, Hahn M, Neurath MF, Waisman A (2012) The tumor suppressor CYLD controls the function of murine regulatory T cells. J Immunol 189(10):4770–4776. doi:10.4049/jimmunol.1201993

Reissig S, Hovelmeyer N, Tang Y, Weih D, Nikolaev A, Riemann M, Weih F, Waisman A (2015) The deubiquitinating enzyme CYLD regulates the differentiation and maturation of thymic medullary epithelial cells. Immunol Cell Biol 93(6):558–566. doi:10.1038/icb.2014.122

Sakaguchi S, Yamaguchi T, Nomura T, Ono M (2008) Regulatory T cells and immune tolerance. Cell 133(5):775–787. doi:10.1016/j.cell.2008.05.009

Sarris M, Andersen KG, Randow F, Mayr L, Betz AG (2008) Neuropilin-1 expression on regulatory T cells enhances their interactions with dendritic cells during antigen recognition. Immunity 28(3):402–413. doi:10.1016/j.immuni.2008.01.012

Scharschmidt E, Wegener E, Heissmeyer V, Rao A, Krappmann D (2004) Degradation of Bcl10 induced by T-cell activation negatively regulates NF-kappa B signaling. Mol Cell Biol 24 (9):3860–3873

Schuijs MJ, Willart MA, Vergote K, Gras D, Deswarte K, Ege MJ, Madeira FB, Beyaert R, van Loo G, Bracher F (2015) Farm dust and endotoxin protect against allergy through A20 induction in lung epithelial cells. Science 349(6252):1106–1110

Shamim M, Nanjappa SG, Singh A, Plisch EH, LeBlanc SE, Walent J, Svaren J, Seroogy C, Suresh M (2007) Cbl-b regulates antigen-induced TCR down-regulation and IFN-gamma production by effector CD8 T cells without affecting functional avidity. J Immunol 179 (11):7233–7243

Shembade N, Harhaj NS, Parvatiyar K, Copeland NG, Jenkins NA, Matesic LE, Harhaj EW (2008) The E3 ligase Itch negatively regulates inflammatory signaling pathways by controlling the function of the ubiquitin-editing enzyme A20. Nat Immunol 9(3):254–262. doi:10.1038/ni1563

Shembade N, Parvatiyar K, Harhaj NS, Harhaj EW (2009) The ubiquitin-editing enzyme A20 requires RNF11 to downregulate NF-kappaB signalling. EMBO J 28(5):513–522. doi:10.1038/emboj.2008.285

Shembade N, Ma A, Harhaj EW (2010) Inhibition of NF-κB signaling by A20 through disruption of ubiquitin enzyme complexes. Science 327(5969):1135–1139

Shi LZ, Wang R, Huang G, Vogel P, Neale G, Green DR, Chi H (2011) HIF1alpha-dependent glycolytic pathway orchestrates a metabolic checkpoint for the differentiation of TH17 and Treg cells. J Exp Med 208(7):1367–1376. doi:10.1084/jem.20110278

Shi ZD, Li XF, Hao L, Zhao Y, Wang YX, Dong BZ, Chen WH, Zhang ZG, Wang YM, Fu Q, Han CH, Li S (2014) Cbl-b gene silencing in splenic T lymphocytes as a therapeutic strategy to target the prostate cancer RM-1 cell tumors in immune competent mice. Eur Rev Med Pharmacol Sci 18(24):3819–3830

Shimo Y, Yanai H, Ohshima D, Qin J, Motegi H, Maruyama Y, Hori S, Inoue J, Akiyama T (2011) TRAF6 directs commitment to regulatory T cells in thymocytes. Genes Cells 16 (4):437–447. doi:10.1111/j.1365-2443.2011.01500.x

Song HY, Rothe M, Goeddel DV (1996) The tumor necrosis factor-inducible zinc finger protein A20 interacts with TRAF1/TRAF2 and inhibits NF-kappaB activation. Proc Natl Acad Sci USA 93(13):6721–6725

Song X-T, Kabler KE, Shen L, Rollins L, Huang XF, Chen S-Y (2008) A20 is an antigen presentation attenuator, and its inhibition overcomes regulatory T cell–mediated suppression. Nat Med 14(3):258–265

Soyer OU, Akdis M, Ring J, Behrendt H, Crameri R, Lauener R, Akdis CA (2013) Mechanisms of peripheral tolerance to allergens. Allergy 68(2):161–170. doi:10.1111/all.12085

St Rose MC, Qui HZ, Bandyopadhyay S, Mihalyo MA, Hagymasi AT, Clark RB, Adler AJ (2009) The E3 ubiquitin ligase Cbl-b regulates expansion but not functional activity of self-reactive CD4 T cells. J Immunol 183(8):4975–4983

Stone EL, Pepper M, Katayama CD, Kerdiles YM, Lai CY, Emslie E, Lin YC, Yang E, Goldrath AW, Li MO, Cantrell DA, Hedrick SM (2015) ICOS coreceptor signaling inactivates the transcription factor FOXO1 to promote Tfh cell differentiation. Immunity 42(2):239–251. doi:10.1016/j.immuni.2015.01.017

Stromnes IM, Blattman JN, Tan X, Jeevanjee S, Gu H, Greenberg PD (2010) Abrogating Cbl-b in effector CD8(+) T cells improves the efficacy of adoptive therapy of leukemia in mice. J Clin Invest 120(10):3722–3734. doi:10.1172/JCI41991

Sun SC (2008) Deubiquitylation and regulation of the immune response. Nat Rev Immunol 8 (7):501–511. doi:10.1038/nri2337

Tavares RM, Turer EE, Liu CL, Advincula R, Scapini P, Rhee L, Barrera J, Lowell CA, Utz PJ, Malynn BA, Ma A (2010) The ubiquitin modifying enzyme A20 restricts B cell survival and prevents autoimmunity. Immunity 33(2):181–191. doi:10.1016/j.immuni.2010.07.017

Teh CE, Lalaoui N, Jain R, Policheni AN, Heinlein M, Alvarez-Diaz S, Sheridan JM, Rieser E, Deuser S, Darding M, Koay HF, Hu Y, Kupresanin F, O'Reilly LA, Godfrey DI, Smyth GK, Bouillet P, Strasser A, Walczak H, Silke J, Gray DH (2016) Linear ubiquitin chain assembly complex coordinates late thymic T-cell differentiation and regulatory T-cell homeostasis. Nat Commun 7:13353. doi:10.1038/ncomms13353

Tiruppathi C, Soni D, Wang DM, Xue J, Singh V, Thippegowda PB, Cheppudira BP, Mishra RK, Debroy A, Qian Z, Bachmaier K, Zhao YY, Christman JW, Vogel SM, Ma A, Malik AB (2014) The transcription factor DREAM represses the deubiquitinase A20 and mediates inflammation. Nat Immunol 15(3):239–247. doi:10.1038/ni.2823

Tsagaratou A, Trompouki E, Grammenoudi S, Kontoyiannis DL, Mosialos G (2010) Thymocyte-specific truncation of the deubiquitinating domain of CYLD impairs positive selection in a NF-kappaB essential modulator-dependent manner. J Immunol 185(4):2032–2043. doi:10.4049/jimmunol.0903919

Turer EE, Tavares RM, Mortier E, Hitotsumatsu O, Advincula R, Lee B, Shifrin N, Malynn BA, Ma A (2008) Homeostatic MyD88-dependent signals cause lethal inflamMation in the absence of A20. J Exp Med 205(2):451–464. doi:10.1084/jem.20071108

Venuprasad K, Elly C, Gao M, Salek-Ardakani S, Harada Y, Luo JL, Yang C, Croft M, Inoue K, Karin M, Liu YC (2006) Convergence of Itch-induced ubiquitination with MEKK1-JNK signaling in Th2 tolerance and airway inflammation. J Clin Invest 116(4):1117–1126. doi:10.1172/JCI26858

Venuprasad K, Huang H, Harada Y, Elly C, Subramaniam M, Spelsberg T, Su J, Liu YC (2008) The E3 ubiquitin ligase Itch regulates expression of transcription factor Foxp3 and airway inflammation by enhancing the function of transcription factor TIEG1. Nat Immunol 9(3):245–253. doi:10.1038/ni1564

Vereecke L, Vieira-Silva S, Billiet T, van Es JH, Mc Guire C, Slowicka K, Sze M, van den Born M, De Hertogh G, Clevers H (2014) A20 controls intestinal homeostasis through cell-specific activities. Nat Commun 5

Verhelst K, Carpentier I, Kreike M, Meloni L, Verstrepen L, Kensche T, Dikic I, Beyaert R (2012) A20 inhibits LUBAC-mediated NF-kappaB activation by binding linear polyubiquitin chains via its zinc finger 7. EMBO J 31(19):3845–3855. doi:10.1038/emboj.2012.240

Walker LS, Abbas AK (2002) The enemy within: keeping self-reactive T cells at bay in the periphery. Nat Rev Immunol 2(1):11–19. doi:10.1038/nri701

Walsh MC, Lee J, Choi Y (2015) Tumor necrosis factor receptor- associated factor 6 (TRAF6) regulation of development, function, and homeostasis of the immune system. Immunol Rev 266(1):72–92. doi:10.1111/imr.12302

Wang J, Ouyang Y, Guner Y, Ford HR, Grishin AV (2009) Ubiquitin-editing enzyme A20 promotes tolerance to lipopolysaccharide in enterocytes. J Immunol 183(2):1384–1392. doi:10.4049/jimmunol.0803987

Wang L, Hong B, Jiang X, Jones L, Chen SY, Huang XF (2012) A20 controls macrophage to elicit potent cytotoxic CD4(+) T cell response. PLoS ONE 7(11):e48930. doi:10.1371/journal.pone.0048930

Wang X, Deckert M, Xuan NT, Nishanth G, Just S, Waisman A, Naumann M, Schluter D (2013) Astrocytic A20 ameliorates experimental autoimmune encephalomyelitis by inhibiting NF-kappaB- and STAT1-dependent chemokine production in astrocytes. Acta Neuropathol 126(5):711–724. doi:10.1007/s00401-013-1183-9

Watanabe Y, Sasahara Y, Ramesh N, Massaad MJ, Yeng Looi C, Kumaki S, Kure S, Geha RS, Tsuchiya S (2013) T-cell receptor ligation causes Wiskott-Aldrich syndrome protein degradation and F-actin assembly downregulation. J Allergy Clin Immunol 132(3):648–655 e641. doi:10.1016/j.jaci.2013.03.046

Wertz IE, O'Rourke KM, Zhou H, Eby M, Aravind L, Seshagiri S, Wu P, Wiesmann C, Baker R, Boone DL, Ma A, Koonin EV, Dixit VM (2004) De-ubiquitination and ubiquitin ligase domains of A20 downregulate NF-kappaB signalling. Nature 430(7000):694–699. doi:10.1038/nature02794

Wertz IE, Newton K, Seshasayee D, Kusam S, Lam C, Zhang J, Popovych N, Helgason E, Schoeffler A, Jeet S, Ramamoorthi N, Kategaya L, Newman RJ, Horikawa K, Dugger D, Sandoval W, Mukund S, Zindal A, Martin F, Quan C, Tom J, Fairbrother WJ, Townsend M, Warming S, DeVoss J, Liu J, Dueber E, Caplazi P, Lee WP, Goodnow CC, Balazs M, Yu K, Kolumam G, Dixit VM (2015) Phosphorylation and linear ubiquitin direct A20 inhibition of inflammation. Nature 528(7582):370–375. doi:10.1038/nature16165

Wex K, Schmid U, Just S, Wang X, Wurm R, Naumann M, Schluter D, Nishanth G (2015) Receptor-interacting protein Kinase-2 inhibition by CYLD impairs antibacterial immune responses in macrophages. Front Immunol 6:650. doi:10.3389/fimmu.2015.00650

Wiedemann A, Muller S, Favier B, Penna D, Guiraud M, Delmas C, Champagne E, Valitutti S (2005) T-cell activation is accompanied by an ubiquitination process occurring at the immunological synapse. Immunol Lett 98(1):57–61

Wing K, Onishi Y, Prieto-Martin P, Yamaguchi T, Miyara M, Fehervari Z, Nomura T, Sakaguchi S (2008) CTLA-4 control over Foxp3 + regulatory T cell function. Science 322 (5899):271–275. doi:10.1126/science.1160062

Wohlfert EA, Callahan MK, Clark RB (2004) Resistance to CD4(+)CD25(+) regulatory T cells and TGF-beta in Cbl-b(-/-) mice. J Immunol 173(2):1059–1065

Wohlfert EA, Gorelik L, Mittler R, Flavell RA, Clark RB (2006) Cutting edge: deficiency in the E3 ubiquitin ligase Cbl-b results in a multifunctional defect in T cell TGF-beta sensitivity in vitro and in vivo. J Immunol 176(3):1316–1320

Xiao NM, Eto D, Elly C, Peng GY, Crotty S, Liu YC (2014) The E3 ubiquitin ligase Itch is required for the differentiation of follicular helper T cells. Nat Immunol 15(7):657–666

Xiao Y, Qiao G, Tang J, Tang R, Guo H, Warwar S, Langdon WY, Tao L, Zhang J (2015) Protein tyrosine phosphatase SHP-1 modulates T cell responses by controlling Cbl-b degradation. J Immunol 195(9):4218–4227. doi:10.4049/jimmunol.1501200

Yamaguchi N, Oyama M, Kozuka-Hata H, Inoue J-i (2013) Involvement of A20 in the molecular switch that activates the non-canonical NF-κB pathway. Scientific Reports 3

Yang C, Zhou W, Jeon MS, Demydenko D, Harada Y, Zhou H, Liu YC (2006) Negative regulation of the E3 ubiquitin ligase itch via Fyn-mediated tyrosine phosphorylation. Mol Cell 21(1):135–141. doi:10.1016/j.molcel.2005.11.014

Yang B, Gay DL, MacLeod MK, Cao X, Hala T, Sweezer EM, Kappler J, Marrack P, Oliver PM (2008) Nedd4 augments the adaptive immune response by promoting ubiquitin-mediated degradation of Cbl-b in activated T cells. Nat Immunol 9(12):1356–1363. doi:10.1038/ni.1670

Yang TM, Martin ML, Nielsen JS, Milne K, Wall EM, Lin W, Watson PH, Nelson BH (2009) Mammary tumors with diverse immunological phenotypes show differing sensitivity to adoptively transferred CD8 + T cells lacking the Cbl-b gene. Cancer Immunol Immunother 58 (11):1867–1877

Yin S, Zhang J, Mao Y, Hu Y, Cui L, Kang N, He W (2013) Vav1-phospholipase C-gamma1 (Vav1-PLC-gamma1) pathway initiated by T cell antigen receptor (TCRgammadelta) activation is required to overcome inhibition by ubiquitin ligase Cbl-b during gammadeltaT cell cytotoxicity. J Biol Chem 288(37):26448–26462. doi:10.1074/jbc.M113.484600

Yuk JM, Kim TS, Kim SY, Lee HM, Han J, Dufour CR, Kim JK, Jin HS, Yang CS, Park KS, Lee CH, Kim JM, Kweon GR, Choi HS, Vanacker JM, Moore DD, Giguere V, Jo EK (2015) Orphan nuclear receptor ERRalpha controls macrophage metabolic signaling and A20 expression to negatively regulate TLR-induced inflammation. Immunity 43(1):80–91. doi:10.1016/j.immuni.2015.07.003

Zhang J, Bardos T, Li DD, Gal I, Vermes C, Xu JY, Mikecz K, Finnegan A, Lipkowitz S, Glant TT (2002) Cutting edge: Regulation of T cell activation threshold by CD28 costimulation through targeting Cbl-b for ubiquitination. J Immunol 169(5):2236–2240

Zhang W, Shao Y, Fang D, Huang J, Jeon MS, Liu YC (2003) Negative regulation of T cell antigen receptor-mediated Crk-L-C3G signaling and cell adhesion by Cbl-b. J Biol Chem 278 (26):23978–23983. doi:10.1074/jbc.M212671200

Zhang J, Stirling B, Temmerman ST, Ma CA, Fuss IJ, Derry JM, Jain A (2006) Impaired regulation of NF-kappaB and increased susceptibility to colitis-associated tumorigenesis in CYLD-deficient mice. J Clin Invest 116(11):3042–3049. doi:10.1172/JCI28746

Zhang R, Zhang N, Mueller DL (2008) Casitas B-lineage lymphoma b inhibits antigen recognition and slows cell cycle progression at late times during CD4 + T cell clonal expansion. J Immunol 181(8):5331–5339

Zhao Y, Guo H, Qiao G, Zucker M, Langdon WY, Zhang J (2015) E3 ubiquitin ligase Cbl-b regulates thymic-derived CD4 + CD25 + regulatory T cell development by targeting Foxp3 for ubiquitination. J Immunol 194(4):1639–1645. doi:10.4049/jimmunol.1402434

Zhao Y, Thornton AM, Kinney MC, Ma CA, Spinner JJ, Fuss IJ, Shevach EM, Jain A (2011) The deubiquitinase CYLD targets Smad7 protein to regulate transforming growth factor beta (TGF-beta) signaling and the development of regulatory T cells. J Biol Chem 286(47):40520–40530. doi:10.1074/jbc.M111.292961

Zheng Y, Zha Y, Gajewski TF (2008) Molecular regulation of T-cell anergy. EMBO Rep 9(1):50–55. doi:10.1038/sj.embor.7401138

Zhou SK, Chen WH, Shi ZD, Wang SP, Li L, Wen XF, Wang YM (2014) Silencing the expression of Cbl-b enhances the immune activation of T lymphocytes against RM-1 prostate cancer cells in vitro. J Chin Med Assoc 77(12):630–636. doi:10.1016/j.jcma.2014.03.008

Zhou Q, Wang H, Schwartz DM, Stoffels M, Park YH, Zhang Y, Yang D, Demirkaya E, Takeuchi M, Tsai WL, Lyons JJ, Yu X, Ouyang C, Chen C, Chin DT, Zaal K, Chandrasekharappa SC, E PH, Yu Z, Mullikin JC, Hasni SA, Wertz IE, Ombrello AK, Stone DL, Hoffmann P, Jones A, Barham BK, Leavis HL, van Royen-Kerkof A, Sibley C, Batu ED, Gul A, Siegel RM, Boehm M, Milner JD, Ozen S, Gadina M, Chae J, Laxer RM, Kastner DL, Aksentijevich I (2016) Loss-of-function mutations in TNFAIP3 leading to A20 haploinsufficiency cause an early-onset autoinflammatory disease. Nat Genet 48(1):67–73. doi:10.1038/ng.3459

Zilberman-Rudenko J, Shawver LM, Wessel AW, Luo Y, Pelletier M, Tsai WL, Lee Y, Vonortas S, Cheng L, Ashwell JD, Orange JS, Siegel RM, Hanson EP (2016) Recruitment of A20 by the C-terminal domain of NEMO suppresses NF-kappaB activation and autoinflammatory disease. Proc Natl Acad Sci USA 113(6):1612–1617. doi:10.1073/pnas.1518163113

Zou W (2006) Regulatory T cells, tumour immunity and immunotherapy. Nat Rev Immunol 6 (4):295–307. doi:10.1038/nri1806

MicroRNA in Immune Regulation

Cheng-Jang Wu and Li-Fan Lu

Abstract The immune system protects us from enormously diverse microbial pathogens but needs to be tightly regulated to avoid deleterious immune-mediated inflammation and tissue damage. A wide range of molecular determinants and cellular components work in concert to control the magnitude and duration of a given immune response. In the past decade, microRNAs (miRNAs), a major class of small non-coding RNA species, have been extensively studied as key molecular players in immune regulation. In this chapter, we will discuss how miRNAs function as negative regulators to restrict innate and adaptive immune responses. Moreover, we will review the current reports regarding miRNAs in human immunological diseases. Finally, we will also address the emerging roles of other non-coding RNAs, long non-coding RNAs (lncRNAs) in particular, in the regulation of the immune system.

Contents

1 Introduction 250
2 miRNAs in the Immune System 250
3 miRNAs as Negative Regulators of Immune Responses 252
   3.1 miRNA in Adaptive Immunity 252
   3.2 miRNA in Innate Immunity 254

C.-J. Wu · L.-F. Lu (✉)
Division of Biological Sciences, University of California, San Diego, La Jolla, CA 92093, USA
e-mail: lifanlu@ucsd.edu

L.-F. Lu
Moores Cancer Center, University of California, San Diego, La Jolla, CA 92093, USA

L.-F. Lu
Center for Microbiome Innovation, University of California, San Diego, La Jolla, CA 92093, USA

Current Topics in Microbiology and Immunology (2017) 410:249–268
DOI 10.1007/82_2017_65

Published Online: 13 September 2017

4 miRNAs in Human Immunological Diseases ... 256
4.1 miRNA in Autoimmunity and Hypersensitivity Diseases ... 256
4.2 miRNA in Blood Cancer ... 258
5 Other Non-coding RNAs ... 259
6 Concluding Remarks ... 261
References ... 261

1 Introduction

Since the first microRNA (miRNA), *lin-4*, was discovered in *Caenorhabditis elegans* by the laboratories of Victor Ambros and Gary Ruvkun in 1993 (Lee et al. 1993; Wightman et al. 1993), these small non-coding RNA species have been extensively studied for their roles in post-transcriptional regulation of gene expression. Like other protein encoding genes, primary miRNA transcripts are first transcribed in the nucleus but are sequentially processed by a microprocessor complex, which contains the RNase III Drosha and the double-stranded RNA binding protein DGCR8 in mammals and Pasha in other species, into a characteristic ~60 nucleotide stem-loop structure, named precursor-miRNA (pre-miRNA) (Lee et al. 2003; Han et al. 2006; Denli et al. 2004; Gregory et al. 2004). The pre-miRNAs are then exported via exportin 5 (Lund et al. 2004; Yi et al. 2003) into the cytoplasm, where the hairpin of pre-miRNA is further processed by the evolutionarily conserved RNase III enzyme Dicer to generate mature miRNAs (Bernstein et al. 2001; Hutvagner et al. 2001). These mature miRNAs are capable of being incorporated into the RNA-induced silencing complex (RISC) where they interact with the core component Argonaute protein (Hutvagner and Zamore 2002; Lingel et al. 2003; Liu et al. 2004). Once assembling with mature miRNA and engaging the targets, active RISCs recognize complementary messenger RNA (mRNA) transcripts for degradation or translational silencing (Jing et al. 2005; Song et al. 2004; Liu et al. 2004). By modulating the expression level of their target genes, miRNAs have been found to regulate almost all biological processes including cell growth, differentiation, and apoptosis in both plants and animals (Ameres and Zamore 2013). As is the case for non-immune cells, the role of miRNA-mediated gene regulation in the immune system in recent years has become a focus of intense investigation.

2 miRNAs in the Immune System

Even though our understanding of miRNA biogenesis and its overall cellular function was initially established based on discoveries in nematodes and plants (Lee et al. 1993; Baulcombe 2004), accumulating evidence has demonstrated that miRNAs also have a crucial role in controlling all aspects of immune responses (Mehta and Baltimore 2016). The importance of miRNA in immune cells was first demonstrated in studies where key components in miRNA biogenesis were disrupted. To this end, conditional deletion of Dicer during early B cell development leads to a dramatic block at the pro- to pre-B transition (Koralov et al. 2008). Subsequent studies using mice with Dicer ablated specifically in peripheral activated B cells further demonstrated that miRNAs are also required for germinal center B cell formation and the generation of the antibody diversity (Xu et al. 2012). Interestingly, unlike what has been reported in B cell differentiation, Dicer ablation in early T cell progenitors did not exhibit any substantial alterations in thymic T cell development with the exception of reduced thymic cellularity (Cobb et al. 2005). Nevertheless, the specificity of the role of miRNA in peripheral T cells is much more apparent, as T cells lacking Dicer failed to differentiate into multiple helper T cell lineages and exhibited aberrant effector T cell function (Muljo et al. 2005). Moreover, when Dicer or Drosha ablation was restricted to regulatory T (Treg) cell lineage, mice developed highly aggressive autoimmunity comparable to those devoid of a functional Foxp3 gene (Zhou et al. 2008; Liston et al. 2008; Chong et al. 2008), suggesting an indispensable role of miRNA in controlling Treg cell biology.

To date, more than hundreds of miRNAs have been reported to be differentially expressed in immune cells. Distinct miRNA signatures not only were found in individual immune cell lineages but could also be detected in the same cell subsets that are in different developmental stages. For example, whereas miR-139 is highly expressed in Pro- and Pre-B cells, elevated levels of miR-28, miR-320 and miR-148a are detected in germinal center (GC) B cells and plasma cells, respectively (Kuchen et al. 2010). Moreover, expression of miRNAs in immune cells can also be dynamically regulated in response to a variety of stimuli, such as antigens recognized by T or B cell receptors, proinflammatory cytokines, and microbial components that trigger Toll-like receptors (O'Connell et al. 2007; Taganov et al. 2006; Cobb et al. 2006). To this end, a recent study reported that T cell activation induces proteasome-mediated degradation of Argonaute, and subsequently causes a global down-regulation of mature miRNAs (Bronevetsky et al. 2013). It was suggested that activation-induced miRNA down-regulation confers effector functions to helper T cells via relaxing the repression of genes that direct T cell differentiation. Finally, hierarchical clustering analysis of miRNA profiling clearly separated cells of the immune system from other tissues. Taken together, these results implied certain miRNAs might play a specific role in controlling development and effector functions of the immune system (Kuchen et al. 2010), and that miRNAs need to be tightly regulated as aberrant expression of miRNAs often leads to dysregulated innate and adaptive immunity.

3 miRNAs as Negative Regulators of Immune Responses

3.1 *miRNA in Adaptive Immunity*

While studies of mice with B or T cell-specific deletion of the entire miRNA pathway seemed to suggest that miRNAs generally play a positive role in promoting adaptive immunity as discussed in the previous section, many miRNAs have been identified as important negative regulators in restricting B and T cell responses (Fig. 1). For example, miR-150, a miRNA that is predominantly expressed in mature B cells was shown to control multiple B cell populations through regulating the expression level of transcription factor c-Myb (Xiao et al. 2007).

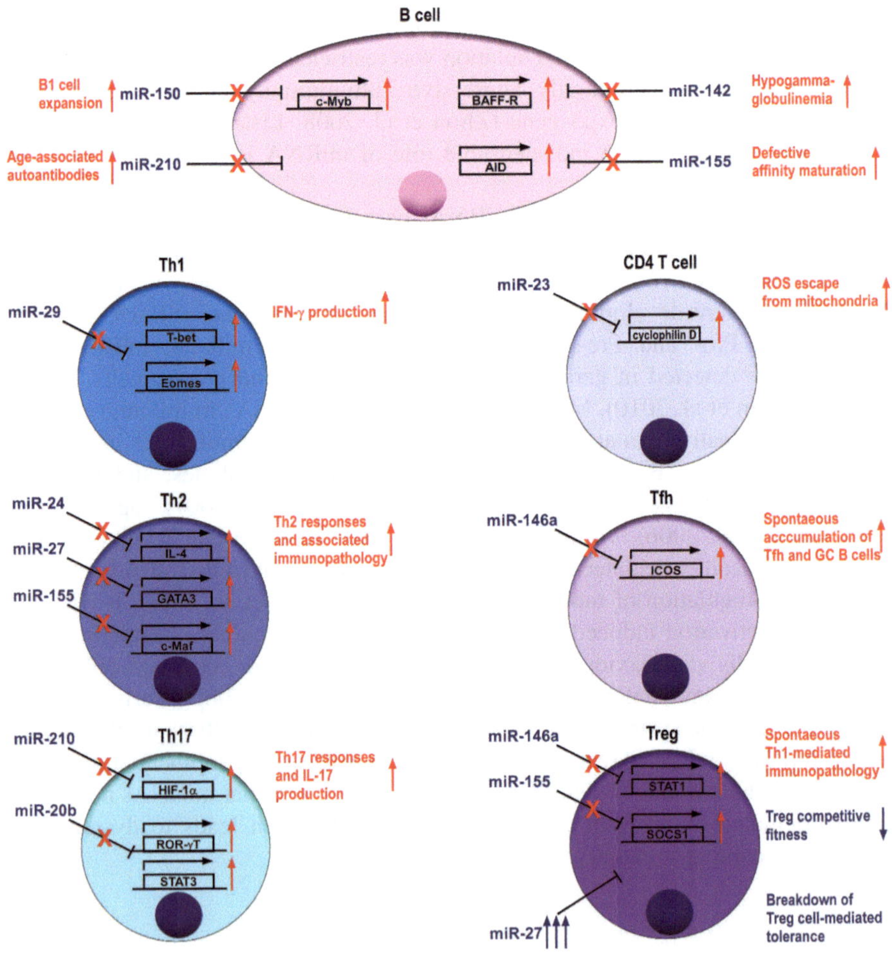

Fig. 1 miRNAs negatively regulate adaptive immune cells. Specific miRNAs expressed by B or T cells repress key target genes that are involved in adaptive immune responses. Tfh cell, T follicular helper cell; Th cell, T helper cell; Treg cell, regulatory T cell

Genetic ablation of miR-150 resulted in the expansion of B1 cells, one of the subsets of mature B cells, in spleen and peritoneal cavity. While the numbers of follicular B cells were not significantly altered, upon immunization elevated antibody responses could be easily detected (Xiao et al. 2007). Compared to miR-150, miR-210, a miRNA that is highly induced upon B cell activation, appears to play an even larger role in functioning as a negative feedback regulator to restrain B cell responses; deletion of miR-210 leads to the development of age-associated autoantibodies (Mok et al. 2013). In contrast, while miR-142 was shown to target B cell-activating factor receptor (BAFF-R), a molecule that is critical for B cell proliferation and survival, hypogammaglobulinemia phenotype was detected in mice devoid of miR-142 despite having increased follicular B cell numbers (Kramer et al. 2015). Similarly, despite the fact that activation-induced cytidine deaminase (AID), a potent enzyme critical for somatic hypermutation and class-switch recombination (CSR), has been shown to be a bona fide miR-155 target where lack of AID regulation by miR-155 led to defective affinity maturation (Teng et al. 2008; Dorsett et al. 2008), miR-155-deficient mice actually exhibited reduced germinal center function and failed to generate high-affinity IgG1 antibodies (Rodriguez et al. 2007). Together, these results suggested that other miR-142 or miR-155 targets are likely responsible for their respective effects on humoral immunity and further demonstrate the complex nature of miRNA-mediated immune regulation.

Like B cell, activated T cells also express increased level of miR-155; mice devoid of miR-155 displayed increased lung airway remodeling, and miR-155-deficient $CD4^+$ T cell cells are intrinsically biased toward Th2 differentiation in vitro. Mechanistically, it was shown that miR-155 can inhibit Th2 responses through modulating the level of c-Maf, a transcription factor known to promote Th2 immunity (Rodriguez et al. 2007). In addition to miR-155, we and others have recently demonstrated that miR-24 and miR-27, two members of the miR-23 cluster family, collaboratively limit Th2 responses and associated immune pathology through targeting IL-4, GATA3 as well as other Th2-related genes in both direct and indirect manners (Cho et al. 2016; Pua et al. 2016). While miR-23 does not seem to play any role in Th2 regulation, it is indispensable for restraining activation-induced necrosis of $CD4^+$ T cells by enforcing intracellular reactive oxygen species (ROS) equilibrium through targeting cyclophilin D, a regulator of ROS escape from mitochondria (Zhang et al. 2016). In addition to Th2 regulation, miRNAs have also been implicated in regulating other Th lineages. To this end, miR-29 was shown to control Th1 responses by repressing multiple genes associated with Th1 differentiation and function including both T-bet and Eomes, two transcription factors known to induce IFN-γ production and IFN-γ itself (Ma et al. 2011; Steiner et al. 2011). On the other hand, hypoxia-induced miR-210 was reported to negatively regulate Th17 responses through restricting the expression of HIF-1α, a key transcription factor that promotes Th17 polarization under limited oxygen (Wang et al. 2014). Similarly, Th17 differentiation and the resultant pathogenesis of experimental autoimmune encephalomyelitis (EAE), a mouse model of multiple sclerosis, could be suppressed by miR-20b via targeting RORγt and STAT3, two key Th17 transcription factors (Zhu et al. 2014). Finally, miR-146a has highly induced in T follicular helper (Tfh) cells, a specialized Th cell subset required for humoral immunity, and can act as a post-transcriptional brake to control Tfh cell and corresponding GC B cell responses by regulating ICOS-ICOSL axis (Pratama et al. 2015).

In addition to its role in Tfh cells, miR-146a has also been shown to function as a key molecular regulator to confer suppressor function to Treg cells. In the absence of miR-146a-mediated regulation of STAT1, mice succumbed to spontaneous IFN-γ-dependent Th1-mediated immunopathology (Lu et al. 2010). On the other hand, albeit dispensable for Treg cell suppressor function, Foxp3-dependent miR-155 ensures Treg cell competitive fitness through targeting SOCS1 (Lu et al. 2009, 2015). Moreover, our recent work also demonstrated that miR-27 controls multiple aspects of Treg cell biology and suggests that excessive expression of miR-27 in Treg cells resulted in a breakdown of Treg cell-mediated immunological tolerance (Cruz et al. 2017). Together with the aforementioned studies in which the entire miRNA pathway was ablated in Treg cells, these results suggested that miRNAs are able to mediate their regulatory effects on the immune system indirectly through maintaining optimal Treg cell function and homeostasis (Zhou et al. 2008; Liston et al. 2008; Chong et al. 2008).

3.2 miRNA in Innate Immunity

Similar to what was described in the adaptive immune system, significant progress has been made over the past decade in characterizing individual miRNAs that control the function of innate immune cells (Fig. 2). Among them, miRNA-mediated regulation of myeloid cell function is best characterized. Both miR-146 and miR-155 were identified in macrophages in response to LPS activation (Taganov et al. 2006; O'Connell et al. 2007). Between these two miRNA, miR-146a serves as a negative regulator to limit inflammatory responses by targeting TRAF6 and IRAK1 (Taganov et al. 2006; Boldin et al. 2011). During virus infection, miR-146a expression is also upregulated in macrophages in RIG-I-dependent manner, and its function negatively regulates type I interferon (IFN) production through repressing IRAK2 (Hou et al. 2009). While miR-155 is generally considered as a positive player in mediating inflammatory responses as it can directly repress SOCS1 and SHIP1 (Androulidaki et al. 2009; O'Connell et al. 2009), two known inhibitory molecules affecting multiple signaling pathways, miR-155 can also down-modulate TLR/IL-1 inflammatory pathway via targeting TAB2 (Ceppi et al. 2009). Moreover, protein kinase Akt1, which is activated by LPS in macrophages, positively regulates let-7e, a miRNA that can inhibit TLR4 expression to control endotoxin sensitivity and tolerance (Androulidaki et al. 2009). In addition to direct TLR4 targeting, miR-21, another miRNA induced by LPS can also act as a negative feedback regulator of TLR4 signaling by targeting PDCD4, a proinflammatory protein required for LPS-induced death (Sheedy et al. 2010). Finally, miRNAs can also restrict DC function by directly suppressing the

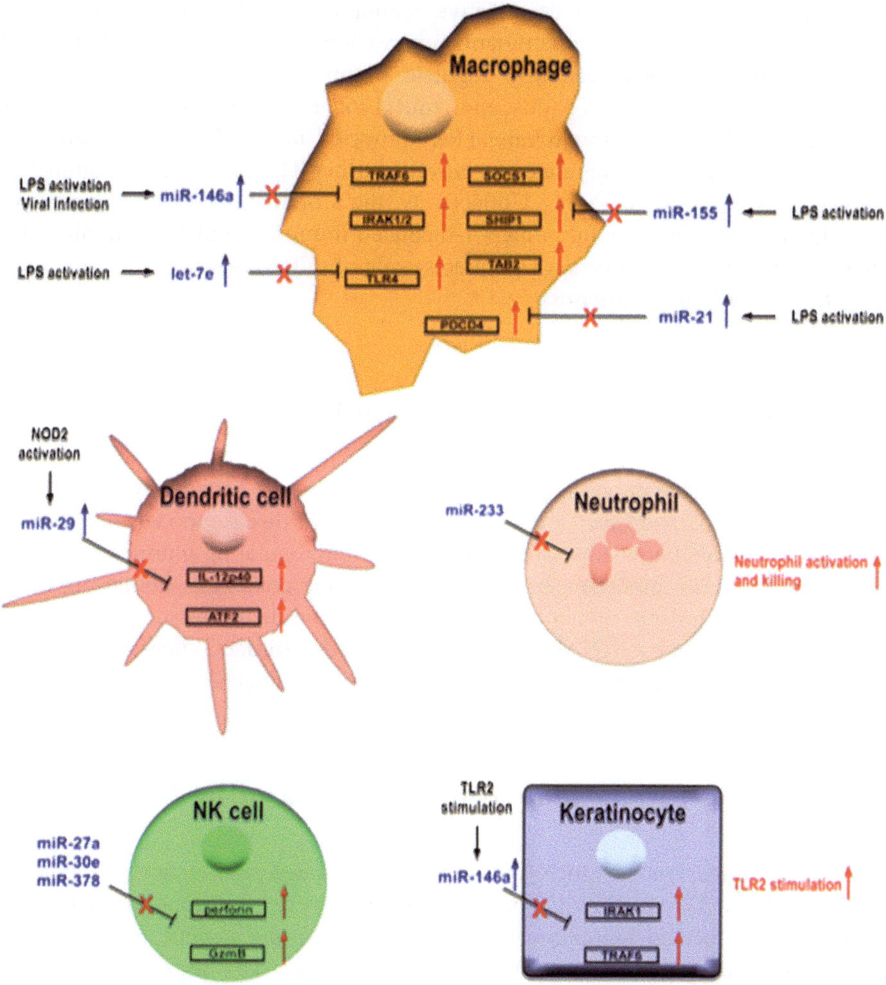

Fig. 2 miRNAs negatively regulate innate immune cells. For miRNAs expressed in different innate immune cell populations, their key target genes are shown. NK, natural killer

production of proinflammatory cytokines. To this end, NOD2-induced miR-29 expression in human dendritic cells (DCs) was shown to inhibit IL-23 expression by repressing IL-12p40 directly and IL-23p19 indirectly via targeting ATF2 (Brain et al. 2013).

Besids myeloid cell subsets, much experimental evidence has also pointed to miRNAs as important negative regulators for other innate immune cells. For example, miR-223 has been shown to function as a cell intrinsic negative modulator of neutrophil activation and killing. Mice lack of miR-223 developed spontaneous inflammatory lung pathology and exhibited exaggerated tissue damage upon LPS treatment (Johnnidis et al. 2008). Moreover, miR-27a*, miR-378, and miR-30e have all been shown to act as negative regulators of NK cell cytotoxicity by silencing perforin and granzyme (GzmB) expression (Kim et al. 2011; Wang et al. 2012). Finally, miRNAs can also regulate immune responses by targeting in non-innate immune cells. To this end, miR-146a expression was shown to be induced by TLR2 stimulation in human keratinocytes. miR-146a can then serve as a potent negative feedback regulator to prevent further TLR2-induced inflammatory responses (Meisgen et al. 2014). Collectively, the aforementioned studies not only provide important insights into miRNA-mediated immune regulation but also offer the molecular basis to understand the precise role of miRNAs in the pathogenesis of human immunological diseases.

4 miRNAs in Human Immunological Diseases

As summarized in the previous sections, studies employing genetically manipulated mice with in vitro experimental approaches and in vivo disease models have helped us to gain valuable knowledge of miRNA function in regulating immune responses. However, it is also important to study miRNA in the context of human immunological diseases. Indeed, numerous studies have shown the correlation between the expression of several miRNAs and immune related disorders, such as autoimmunity, hypersensitive diseases, and hematopoietic malignancies.

4.1 miRNA in Autoimmunity and Hypersensitivity Diseases

Abnormal expression of miRNAs has been associated with many autoimmune diseases (Table 1). One of the best examples is systemic lupus erythematosus (SLE), a multifaceted autoimmune disease with a strong genetic predisposition, characterized by enhanced type I interferon signaling. To this end, it has been reported that peripheral blood mononuclear cells (PBMCs) isolated from SLE patients express reduced miR-146a. The amount of miR-146a was shown to negatively correlate with the clinical disease activity and type I interferon levels in patients. It was suggested that lack of miR-146a-mediated regulation of STAT1 and IRF5 led to the excessive production of type I interferon (Tang et al. 2009). Moreover, sequencing analysis of single-nucleotide polymorphisms (SNPs) in SLE patients identified a genetic variant in the miR-146a promoter region that is functionally significant in downregulating the expression of miR-146a by altering its binding affinity for Ets-1 (Luo et al. 2011). In addition to miR-146a, diminished expressions of both miR-125a and miR-155 were also reported in patients with SLE (Zhao et al. 2010; Lashine et al. 2015). While elevated level of RANTES in the absence of optimal miR-125a-mediated regulation was considered to promote the disease, increased expression of protein phosphatase 2A (PP2A) in juvenile SLE patients with reduced miR-155 was thought to be responsible for enhanced

Table 1 miRNAs involved in human immunological diseases

Immune disease	miRNA	Target
Systemic lupus	miR-21	RASGRP1
Erythematosus	miR-125a	KLF13
	miR-146a	IRF5, STAT1, Ets-1
	miR-148a	DNMT1
	miR-155	PP2A
Multiple sclerosis	miR-27b	BMI1, TGFBR1, and SMAD4
	miR-326	Ets-1
Type 1 diabetes	miR-326	Ets-1, VDR
Rheumatoid arthritis	miR-146a	FAF1
	miR-155	MMP-3
	miR-223	IGF-1
Psoriasis	miR-203	SOCS-3
Asthma	miR-19a	PTEN, SOCS1, and TNFAIP3

pathogenesis. On the other hand, elevated levels of miR-21 and miR-148a were detected in circulating $CD4^+$ T cells from SLE patients. Both of these two miRNAs are able to upregulate autoimmune-associated methylation-sensitive genes such as CD70 and LFA-1 through promoting DNA hypomethylation and by repressing the expression of RASGRP1 and DNA methyltransferase 1 (DNMT1), respectively (Pan et al. 2010).

Like SLE, multiple sclerosis (MS) has also been linked to the aberrant expression of many miRNAs. For example, the expression level of miR-326 has been shown to be highly correlated with disease severity in MS patients (Du et al. 2009). Mechanistically, miR-326 promotes Th17 cell induction through targeting Ets-1, a known negative regulator of Th17 differentiation. Moreover, increased miR-27, miR-128, as well as miR-340, were detected in $CD4^+$ T cells of patients with MS. It was shown that through repressing BMI1, a molecule that stabilizes GATA3, miR-27b inhibits Th2 differentiation and promotes proinflammatory Th1 autoimmune responses (Guerau-de-Arellano et al. 2011). Following up this study, miR-27 was further shown to dampen TGFβ signaling, leading to impaired Treg cells and enhanced susceptibility to developing multiple sclerosis (Severin et al. 2016). Besides these two autoimmune diseases, abnormal miRNA expression has also been associated with other autoimmune inflammation including rheumatoid arthritis (RA), psoriasis and type I diabetes (Li et al. 2010; Stanczyk et al. 2008; Lu et al. 2014; Sonkoly et al. 2007; Sebastiani et al. 2011).

In addition to autoimmunity, dysregulated miRNAs also contribute to the development or pathogenesis of hypersensitivity diseases like asthma. miRNA profiling analysis of human airway-infiltrating T cells revealed that miR-19a, a member of the miR-17-92 cluster, was greatly upregulated in $CD4^+$ T cells isolated from asthmatic airways compared with cells from healthy subjects. Mechanistically, miR-19a was shown to repress multiple genes including PTEN, SOCS1, and TNFAIP3, leading to specific augmentation of Th2 responses and associated immune pathology (Simpson et al. 2014). Together, more and more studies have revealed a causative role of miRNA in the development of many human immunological diseases.

4.2 *miRNA in Blood Cancer*

As one of the major functions of miRNAs is to regulate cell differentiation and proliferation, it is not surprising that when dysregulated miRNAs can drive the development of malignancies of the immune system (Table 2). miRNAs can either act as tumor suppressors or function as oncomirs to promote or prevent tumorigenesis. As tumor suppressors, miR-15a and miR-16 were shown to inhibit the development of B cell chronic lymphocytic leukemias (B-CLL) through targeting Bcl-2, and that in more than 50% of B-CLL patients a region encoding miR-15a and miR-16 was found to be deleted (Cimmino et al. 2005; Calin et al. 2002). miR-28 was also identified as a tumor suppressor and is significantly downregulated in Burkitt lymphoma. Oncogene Myc was shown to negatively regulate miR-28 expression leading to the uncontrolled proliferation of certain B cell subsets (Schneider et al. 2014). Another example is miR-29b whose expression is deregulated in primary acute myelogeneous leukemia (AML). Restoration of miR-29b in AML cell lines was able to induce apoptosis and dramatically reduce tumorigenicity, pointing to a clear tumor suppressor role (Garzon et al. 2009).

In contrast to the aforementioned roles of miRNA in preventing tumorigenesis, many miRNAs also exhibit oncogenic activity in hematologic malignancies. For example, the miR-17-92 cluster, which is located in a region of DNA that is frequently amplified in human B cell lymphomas, has been shown to promote malignancy of immune cells (He et al. 2005; Tagawa and Seto 2005; Inomata et al. 2009; Lu et al. 2010). In addition to miR-17-92 cluster, miR-155, another

Table 2 miRNAs involved in human blood cancer

miRNA	Blood cancer	Target
Tumor suppressors		
miR-15a/miR16-1	B cell chronic lymphocytic leukemia	Bcl-2
miR-28	Burkitt lymphoma	MAD2L1, BAG1
miR-29b	Acute myelogeneous leukemia	MCL-1
OncomiRs		
miR-17-92	B cell lymphomas	c-myc
miR-125b	Myeloid and B cell leukemia	IRF4
miR-155	Burkitt's lymphoma, diffuse large B cell lymphomas, Hodgkin's lymphomas, and NK cell lymphoma	PTEN, PDCD4, and SHIP1
miR-223	T cell acute lymphoblastic leukemia	FBXW7

well-characterized oncomir, has also been shown to be expressed at higher levels in many different types of B cell lymphomas including Hodgkin's lymphoma, DLBCL, Burkitt's lymphoma as well as NK cell lymphoma (Eis et al. 2005; Kluiver et al. 2005; van den Berg et al. 2003; Metzler et al. 2004; Yamanaka et al. 2009). On the other hand, miR-223 was shown to be upregulated in human T cell acute lymphoblastic leukemia (T-ALL) in a TAL1-dependent manner (Mansour et al. 2013). It is thought that miR-223 promotes T-ALL through repressing a tumor suppressor, FBXW7. Finally, elevated levels of the oncomir, miR-125b, has been reported in a variety of human neoplastic blood disorders and could potentially induce myeloid- and B cell leukemia by inhibiting IRF4 (So et al. 2014). The miRNA signatures identified from these clinical studies not only provide great value as prognostic parameters for cancer progression but might also serve as potential novel therapeutic targets to treat human hematopoietic malignancies.

5 Other Non-coding RNAs

Like miRNAs, many other non-coding RNA (ncRNA) species including long non-coding RNAs (lncRNAs) have also been identified as important gene regulators in the immune system (Fig. 3). Since the first lncRNA H19 was reported in 1990 (Brannan et al. 1990), extensive investigation in lncRNA-mediated gene regulation has demonstrated that lncRNAs can regulate gene expression in various biological processes, including immune responses (Ponting et al. 2009). LncRNAs are transcribed by RNA polymerase II, 5'-capped, polyadenylated, and undergo splicing similar to that for mRNAs (Guttman et al. 2009). They can function both in *cis* to regulate the gene expression in close genomic proximity at the site of transcription, or in *trans* to target distant transcriptional activators or repressors (Ponting et al. 2009). Moreover, since lncRNAs usually contain multiple modular domains that can either interact with proteins or form complementary pairs with nucleotides, these molecules could connect DNA, RNA, and proteins and be involved in nearly all stages of gene regulation (Guttman and Rinn 2012).

A Recent report suggests that up to two third of transcribed genes across all cell types in humans are classified as lncRNAs (Iyer et al. 2015). In T cells, a lncRNA, NRON (non-coding RNA, repressor of NFAT) was shown to form a large cytoplasmic RNA-protein scaffold complex that can repress the transcriptional activation of NFAT-responsive genes via regulating NFAT nuclear trafficking (Willingham et al. 2005; Sharma et al. 2011). Moreover, another lncRNA, lncRNA-CD244, whose expression has been shown to be driven by CD244 signaling upon Tuberculosis (TB) infection, is able to inhibit cytokine production by $CD8^+$ T cells by mediating histone H3K27 trimethylation at promoter regions of IFN-γ and TNF-α (Wang et al. 2015). On the other hand, in Th17 cells, rather than inhibiting their effector function, a lncRNA, lncRNA Rmrp was shown to interact with a complex of RORγt and an RNA helicase, DDX5, to activate RORγt-dependent Th17-relative gene transcription (Huang et al. 2015).

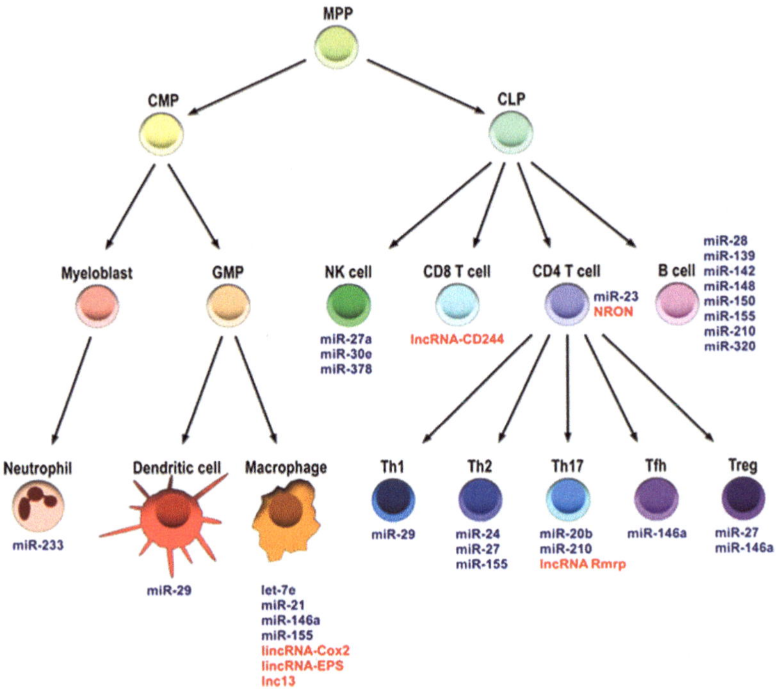

Fig. 3 Non-coding RNAs in the immune system. The schematic describes miRNAs (*blue font*) and lncRNAs (*red font*) that have key roles in controlling immune responses. CLP, common lymphoid progenitor; CMP, common myeloid progenitor; GMP, granulocyte–monocyte progenitor; MPP, multipotent progenitor

As for the role of lncRNAs in the innate immune cells, a TLR signaling induced intergenic lncRNA, lncRNA-Cox2, was reported to repress the expression of many critical inflammatory genes in macrophages through forming a complex with heterogeneous nuclear ribonucleoprotein (hnRNP) A/B and A2/B1 (Carpenter et al. 2013). In addition to lncRNA-Cox2, lncRNA-EPS was also recently identified as a key regulator in controlling macrophage inflammatory responses. Mice with lncRNA-EPS deficiency exhibited enhanced inflammation and lethality upon LPS challenges. Mechanistically, lncRNA-EPS was shown to limit the expression of immune response genes (IRGs) by controlling nucleosome positioning through interacting with hnRNP L (Atianand et al. 2016).

Finally, the link between lncRNA and human autoimmune diseases has also been demonstrated. Genome-wide association studies (GWAS) of celiac disease patients identified five SNPs in a region encoding a lncRNA, lnc13. Biopsies from celiac disease patients appeared to have substantially lower amounts of lnc13 compared with healthy donors. Further studies have shown that lnc13 is primarily expressed in the nucleus of human macrophages from the lamina propria. Like the aforementioned lncRNAs, lnc13 was also shown to repress many inflammatory genes through interaction with a hnRNP, hnRNP D in particular, as well as Hdac1 and chromatin. It was thus suggested that decreased levels of lnc13 in intestinal tissue from patients with celiac disease likely contributes to the observed inflammation in this autoimmune disorder (Castellanos-Rubio et al. 2016). Despite the great efforts made to study lncRNA biology, unlike miRNAs, it remains a major challenge to functionally evaluate a lncRNA as the sequence of the transcript lends no insight into how it may actually work within a given cell type. Nevertheless, it is evident that lncRNAs exhibit important regulatory functions in controlling immune responses as well as other biological processes.

6 Concluding Remarks

Over a decade of intense scrutiny into the role of miRNAs in the immune system, there is little doubt that miRNAs function as crucial gene modulators that would impact almost all facets of immune responses in both physiological and pathological settings. While miRNAs do not completely turn off (or in some cases, turn on) the expression of their targets, they can act by repressing genes involved in positive-feedback regulatory circuits or by regulating a set of genes that are in a shared pathway or protein complex. As such, even relatively small changes in gene expression introduced by miRNAs could cause major biological consequences. From impairment of immune functions to the pathogenesis of a variety of immunological diseases, the fact that dysregulation of individual miRNAs in the immune system has repeatedly been demonstrated to have profound physiological effects further supports this notion. Moreover, beyond gaining further molecular insights into miRNA-mediated gene regulation in immunological research, recent advances in modulating miRNA function by miRNA mimics or antisense oligonucleotides have shown promise in miRNA-targeted therapeutics. With increasing knowledge of miRNA biology and the development of novel approaches for efficient delivery of miRNA modifying agents to specific immune cell subsets, we are confident that manipulating miRNA pathways will soon become a viable option to treat a wide array of human immunological diseases.

References

Ameres SL, Zamore PD (2013) Diversifying microRNA sequence and function. Nat Rev Mol Cell Biol 14:475–488

Androulidaki A, Iliopoulos D, Arranz A, Doxaki C, Schworer S, Zacharioudaki V, Margioris AN, Tsichlis PN, Tsatsanis C (2009) The kinase Akt1 controls macrophage response to lipopolysaccharide by regulating microRNAs. Immunity 31:220–231

Atianand MK, Hu W, Satpathy AT, Shen Y, Ricci EP, Alvarez-Dominguez JR, Bhatta A, Schattgen SA, McGowan JD, Blin J, Braun JE, Gandhi P, Moore MJ, Chang HY, Lodish HF, Caffrey DR, Fitzgerald KA (2016) A long noncoding RNA lincRNA-EPS acts as a transcriptional brake to restrain inflammation. Cell 165:1672–1685

Baulcombe D (2004) RNA silencing in plants. Nature 431:356–363

Bernstein E, Caudy AA, Hammond SM, Hannon GJ (2001) Role for a bidentate ribonuclease in the initiation step of RNA interference. Nature 409:363–366

Boldin MP, Taganov KD, Rao DS, Yang L, Zhao JL, Kalwani M, Garcia-Flores Y, Luong M, Devrekanli A, Xu J, Sun G, Tay J, Linsley PS, Baltimore D (2011) miR-146a is a significant brake on autoimmunity, myeloproliferation, and cancer in mice. J Exp Med 208:1189–1201

Brain O, Owens BM, Pichulik T, Allan P, Khatamzas E, Leslie A, Steevels T, Sharma S, Mayer A, Catuneanu AM, Morton V, Sun MY, Jewell D, Coccia M, Harrison O, Maloy K, Schonefeldt S, Bornschein S, Liston A, Simmons A (2013) The intracellular sensor NOD2 induces microRNA-29 expression in human dendritic cells to limit IL-23 release. Immunity 39:521–536

Brannan CI, Dees EC, Ingram RS, Tilghman SM (1990) The product of the H19 gene may function as an RNA. Mol Cell Biol 10:28–36

Bronevetsky Y, Villarino AV, Eisley CJ, Barbeau R, Barczak AJ, Heinz GA, Kremmer E, Heissmeyer V, McManus MT, Erle DJ, Rao A, Ansel KM (2013) T cell activation induces proteasomal degradation of argonaute and rapid remodeling of the microRNA repertoire. J Exp Med 210:417–432

Calin GA, Dumitru CD, Shimizu M, Bichi R, Zupo S, Noch E, Aldler H, Rattan S, Keating M, Rai K, Rassenti L, Kipps T, Negrini M, Bullrich F, Croce CM (2002) Frequent deletions and down-regulation of micro- RNA genes miR15 and miR16 at 13q14 in chronic lymphocytic leukemia. Proc Natl Acad Sci USA 99:15524–15529

Carpenter S, Aiello D, Atianand MK, Ricci EP, Gandhi P, Hall LL, Byron M, Monks B, Henry-Bezy M, Lawrence JB, O'Neill LA, Moore MJ, Caffrey DR, Fitzgerald KA (2013) A long noncoding RNA mediates both activation and repression of immune response genes. Science 341:789–792

Castellanos-Rubio A, Fernandez-Jimenez N, Kratchmarov R, Luo X, Bhagat G, Green PH, Schneider R, Kiledjian M, Bilbao JR, Ghosh S (2016) A long noncoding RNA associated with susceptibility to celiac disease. Science 352:91–95

Ceppi M, Pereira PM, Dunand-Sauthier I, Barras E, Reith W, Santos MA, Pierre P (2009) MicroRNA-155 modulates the interleukin-1 signaling pathway in activated human monocyte-derived dendritic cells. Proc Natl Acad Sci USA 106:2735–2740

Cho S, Wu CJ, Yasuda T, Cruz LO, Khan AA, Lin LL, Nguyen DT, Miller M, Lee HM, Kuo ML, Broide DH, Rajewsky K, Rudensky AY, Lu LF (2016) miR-23 approximately 27 approximately 24 clusters control effector T cell differentiation and function. J Exp Med 213:235–249

Chong MM, Rasmussen JP, Rudensky AY, Littman DR (2008) The RNAseIII enzyme Drosha is critical in T cells for preventing lethal inflammatory disease. J Exp Med 205:2005–2017

Cimmino A, Calin GA, Fabbri M, Iorio MV, Ferracin M, Shimizu M, Wojcik SE, Aqeilan RI, Zupo S, Dono M, Rassenti L, Alder H, Volinia S, Liu CG, Kipps TJ, Negrini M, Croce CM (2005) miR-15 and miR-16 induce apoptosis by targeting BCL2. Proc Natl Acad Sci U S A 102:13944–13949

Cobb BS, Nesterova TB, Thompson E, Hertweck A, O'Connor E, Godwin J, Wilson CB, Brockdorff N, Fisher AG, Smale ST, Merkenschlager M (2005) T cell lineage choice and differentiation in the absence of the RNase III enzyme Dicer. J Exp Med 201:1367–1373

Cobb BS, Hertweck A, Smith J, O'Connor E, Graf D, Cook T, Smale ST, Sakaguchi S, Livesey FJ, Fisher AG, Merkenschlager M (2006) A role for Dicer in immune regulation. J Exp Med 203:2519–2527

Cruz LO, Hashemifar SS, Wu CJ, Cho S, Nguyen DT, Lin LL, Khan AA, Lu LF (2017) Excessive expression of miR-27 impairs Treg-mediated immunological tolerance. J Clin Invest 127:530–542

Denli AM, Tops BB, Plasterk RH, Ketting RF, Hannon GJ (2004) Processing of primary microRNAs by the microprocessor complex. Nature 432:231–235

Dorsett Y, McBride KM, Jankovic M, Gazumyan A, Thai TH, Robbiani DF, di Virgilio M, B RS-M, Heidkamp G, Schwickert TA, Eisenreich T, Rajewsky K, Nussenzweig MC (2008)

MicroRNA-155 suppresses activation-induced cytidine deaminase-mediated Myc-Igh translocation. Immunity 28:630–638

Du C, Liu C, Kang J, Zhao G, Ye Z, Huang S, Li Z, Wu Z, Pei G (2009) MicroRNA miR-326 regulates TH-17 differentiation and is associated with the pathogenesis of multiple sclerosis. Nat Immunol 10:1252–1259

Eis PS, Tam W, Sun L, Chadburn A, Li Z, Gomez MF, Lund E, Dahlberg JE (2005) Accumulation of miR-155 and BIC RNA in human B cell lymphomas. Proc Natl Acad Sci USA 102:3627–3632

Garzon R, Heaphy CE, Havelange V, Fabbri M, Volinia S, Tsao T, Zanesi N, Kornblau SM, Marcucci G, Calin GA, Andreeff M, Croce CM (2009) MicroRNA 29b functions in acute myeloid leukemia. Blood 114:5331–5341

Gregory RI, Yan KP, Amuthan G, Chendrimada T, Doratotaj B, Cooch N, Shiekhattar R (2004) The microprocessor complex mediates the genesis of microRNAs. Nature 432:235–240

Guerau-De-arellano M, Smith KM, Godlewski J, Liu Y, Winger R, Lawler SE, Whitacre CC, Racke MK, Lovett-Racke AE (2011) Micro-RNA dysregulation in multiple sclerosis favours pro-inflammatory T-cell-mediated autoimmunity. Brain 134:3578–3589

Guttman M, Rinn JL (2012) Modular regulatory principles of large non-coding RNAs. Nature 482:339–346

Guttman M, Amit I, Garber M, French C, Lin MF, Feldser D, Huarte M, Zuk O, Carey BW, Cassady JP, Cabili MN, Jaenisch R, Mikkelsen TS, Jacks T, Hacohen N, Bernstein BE, Kellis M, Regev A, Rinn JL, Lander ES (2009) Chromatin signature reveals over a thousand highly conserved large non-coding RNAs in mammals. Nature 458:223–227

Han J, Lee Y, Yeom KH, Nam JW, Heo I, Rhee JK, Sohn SY, Cho Y, Zhang BT, Kim VN (2006) Molecular basis for the recognition of primary microRNAs by the Drosha-DGCR8 complex. Cell 125:887–901

He L, Thomson JM, Hemann MT, Hernando-Monge E, Mu D, Goodson S, Powers S, Cordon-Cardo C, Lowe SW, Hannon GJ, Hammond SM (2005) A microRNA polycistron as a potential human oncogene. Nature 435:828–833

Hou J, Wang P, Lin L, Liu X, Ma F, An H, Wang Z, Cao X (2009) MicroRNA-146a feedback inhibits RIG-I-dependent Type I IFN production in macrophages by targeting TRAF6, IRAK1, and IRAK2. J Immunol 183:2150–2158

Huang W, Thomas B, Flynn RA, Gavzy SJ, Wu L, Kim SV, Hall JA, Miraldi ER, Ng CP, Rigo F, Meadows S, Montoya NR, Herrera NG, Domingos AI, Rastinejad F, Myers RM, Fuller-Pace FV, Bonneau R, Chang HY, Acuto O, Littman DR (2015) DDX5 and its associated lncRNA Rmrp modulate TH17 cell effector functions. Nature 528:517–522

Hutvagner G, Zamore PD (2002) A microRNA in a multiple-turnover RNAi enzyme complex. Science 297:2056–2060

Hutvagner G, McLachlan J, Pasquinelli AE, Balint E, Tuschl T, Zamore PD (2001) A cellular function for the RNA-interference enzyme Dicer in the maturation of the let-7 small temporal RNA. Science 293:834–838

Inomata M, Tagawa H, Guo YM, Kameoka Y, Takahashi N, Sawada K (2009) MicroRNA-17-92 down-regulates expression of distinct targets in different B-cell lymphoma subtypes. Blood 113:396–402

Iyer MK, Niknafs YS, Malik R, Singhal U, Sahu A, Hosono Y, Barrette TR, Prensner JR, Evans JR, Zhao S, Poliakov A, Cao X, Dhanasekaran SM, Wu YM, Robinson DR, Beer DG, Feng FY, Iyer HK, Chinnaiyan AM (2015) The landscape of long noncoding RNAs in the human transcriptome. Nat Genet 47:199–208

Jing Q, Huang S, Guth S, Zarubin T, Motoyama A, Chen J, di Padova F, Lin SC, Gram H, Han J (2005) Involvement of microRNA in AU-rich element-mediated mRNA instability. Cell 120:623–634

Johnnidis JB, Harris MH, Wheeler RT, Stehling-Sun S, Lam MH, Kirak O, Brummelkamp TR, Fleming MD, Camargo FD (2008) Regulation of progenitor cell proliferation and granulocyte function by microRNA-223. Nature 451:1125–1129

Kim TD, Lee SU, Yun S, Sun HN, Lee SH, Kim JW, Kim HM, Park SK, Lee CW, Yoon SR, Greenberg PD, Choi I (2011) Human microRNA-27a* targets Prf1 and GzmB expression to regulate NK-cell cytotoxicity. Blood 118:5476–5486

Kluiver J, Poppema S, de Jong D, Blokzijl T, Harms G, Jacobs S, Kroesen BJ, van den Berg A (2005) BIC and miR-155 are highly expressed in Hodgkin, primary mediastinal and diffuse large B cell lymphomas. J Pathol 207:243–249

Koralov SB, Muljo SA, Galler GR, Krek A, Chakraborty T, Kanellopoulou C, Jensen K, Cobb BS, Merkenschlager M, Rajewsky N, Rajewsky K (2008) Dicer ablation affects antibody diversity and cell survival in the B lymphocyte lineage. Cell 132:860–874

Kramer NJ, Wang WL, Reyes EY, Kumar B, Chen CC, Ramakrishna C, Cantin EM, Vonderfecht SL, Taganov KD, Chau N, Boldin MP (2015) Altered lymphopoiesis and immunodeficiency in miR-142 null mice. Blood 125:3720–3730

Kuchen S, Resch W, Yamane A, Kuo N, Li Z, Chakraborty T, Wei L, Laurence A, Yasuda T, Peng S, Hu-Li J, Lu K, Dubois W, Kitamura Y, Charles N, Sun HW, Muljo S, Schwartzberg PL, Paul WE, O'Shea J, Rajewsky K, Casellas R (2010) Regulation of microRNA expression and abundance during lymphopoiesis. Immunity 32:828–839

Lashine YA, Salah S, Aboelenein HR, Abdelaziz AI (2015) Correcting the expression of miRNA-155 represses PP2Ac and enhances the release of IL-2 in PBMCs of juvenile SLE patients. Lupus 24:240–247

Lee RC, Feinbaum RL, Ambros V (1993) The C. elegans heterochronic gene lin-4 encodes small RNAs with antisense complementarity to lin-14. Cell 75:843–854

Lee Y, Ahn C, Han J, Choi H, Kim J, Yim J, Lee J, Provost P, Radmark O, Kim S, Kim VN (2003) The nuclear RNase III Drosha initiates microRNA processing. Nature 425:415–419

Li J, Wan Y, Guo Q, Zou L, Zhang J, Fang Y, Zhang J, Zhang J, Fu X, Liu H, Lu L, Wu Y (2010) Altered microRNA expression profile with miR-146a upregulation in CD4 + T cells from patients with rheumatoid arthritis. Arthritis Res Ther 12:R81

Lingel A, Simon B, Izaurralde E, Sattler M (2003) Structure and nucleic-acid binding of the Drosophila Argonaute 2 PAZ domain. Nature 426:465–469

Liston A, Lu LF, O'Carroll D, Tarakhovsky A, Rudensky AY (2008) Dicer-dependent microRNA pathway safeguards regulatory T cell function. J Exp Med 205:1993–2004

Liu J, Carmell MA, Rivas FV, Marsden CG, Thomson JM, Song JJ, Hammond SM, Joshua-Tor L, Hannon GJ (2004) Argonaute2 is the catalytic engine of mammalian RNAi. Science 305:1437–1441

Lu LF, Boldin MP, Chaudhry A, Lin LL, Taganov KD, Hanada T, Yoshimura A, Baltimore D, Rudensky AY (2010) Function of miR-146a in controlling Treg cell-mediated regulation of Th1 responses. Cell 142:914–929

Lu MC, Yu CL, Chen HC, Yu HC, Huang HB, Lai NS (2014) Increased miR-223 expression in T cells from patients with rheumatoid arthritis leads to decreased insulin-like growth factor-1-mediated interleukin-10 production. Clin Exp Immunol 177:641–651

Lu LF, Thai TH, Calodo DP, Chaudhry A, Kubo M, Tanaka K, Loeb GB., Lee H, Yoshimura A, Rajewsky K, Rudensky AY (2009) Foxp3-dependent microRNA155 confers competitive fitness to regulatory T cells by targeting SOCS1 protein. Immun 30:80–91

Lu LF, Gasteiger G, Yu IS, Chaudhry A, Hsin JP, Lu Y, Bos PD, Lin LL, Zawislak CL, Cho S, Sun JC, Leslie CS, Lin SW, Rudensky AY (2015) A single miRNA-mRNA interaction affects the immune response in a context- and cell-type-specific manner. Immun. 43:52–64

Lund E, Guttinger S, Calado A, Dahlberg JE, Kutay U (2004) Nuclear export of microRNA precursors. Science 303:95–98

Luo X, Yang W, Ye DQ, Cui H, Zhang Y, Hirankarn N, Qian X, Tang Y, Lau YL, de Vries N, Tak PP, Tsao BP, Shen N (2011) A functional variant in microRNA-146a promoter modulates its expression and confers disease risk for systemic lupus erythematosus. PLoS Genet 7: e1002128

Ma F, Xu S, Liu X, Zhang Q, Xu X, Liu M, Hua M, Li N, Yao H, Cao X (2011) The microRNA miR-29 controls innate and adaptive immune responses to intracellular bacterial infection by targeting interferon-gamma. Nat Immunol 12:861–869

Mansour MR, Sanda T, Lawton LN, Li X, Kreslavsky T, Novina CD, Brand M, Gutierrez A, Kelliher MA, Jamieson CH, von Boehmer H, Young RA, Look AT (2013) The TAL1 complex targets the FBXW7 tumor suppressor by activating miR-223 in human T cell acute lymphoblastic leukemia. J Exp Med 210:1545–1557

Mehta A, Baltimore D (2016) MicroRNAs as regulatory elements in immune system logic. Nat Rev Immunol 16:279–294

Meisgen F, Xu Landen N, Wang A, Rethi B, Bouez C, Zuccolo M, Gueniche A, Stahle M, Sonkoly E, Breton L, Pivarcsi A (2014) MiR-146a negatively regulates TLR2-induced inflammatory responses in keratinocytes. J Invest Dermatol 134:1931–1940

Metzler M, Wilda M, Busch K, Viehmann S, Borkhardt A (2004) High expression of precursor microRNA-155/BIC RNA in children with Burkitt lymphoma. Genes Chromosomes Cancer 39:167–169

Mok Y, Schwierzeck V, Thomas DC, Vigorito E, Rayner TF, Jarvis LB, Prosser HM, Bradley A, Withers DR, Martensson IL, Corcoran LM, Blenkiron C, Miska EA, Lyons PA, Smith KG (2013) MiR-210 is induced by Oct-2, regulates B cells, and inhibits autoantibody production. J Immunol 191:3037–3048

Muljo SA, Ansel KM, Kanellopoulou C, Livingston DM, Rao A, Rajewsky K (2005) Aberrant T cell differentiation in the absence of Dicer. J Exp Med 202:261–269

O'Connell RM, Taganov KD, Boldin MP, Cheng G, Baltimore D (2007) MicroRNA-155 is induced during the macrophage inflammatory response. Proc Natl Acad Sci USA 104:1604–1609

O'Connell RM, Chaudhuri AA, Rao DS, Baltimore D (2009) Inositol phosphatase SHIP1 is a primary target of miR-155. Proc Natl Acad Sci USA 106:7113–7118

Pan W, Zhu S, Yuan M, Cui H, Wang L, Luo X, Li J, Zhou H, Tang Y, Shen N (2010) MicroRNA-21 and microRNA-148a contribute to DNA hypomethylation in lupus CD4+ T cells by directly and indirectly targeting DNA methyltransferase 1. J Immunol 184:6773–6781

Ponting CP, Oliver PL, Reik W (2009) Evolution and functions of long noncoding RNAs. Cell 136:629–641

Pratama A, Srivastava M, Williams NJ, Papa I, Lee SK, Dinh XT, Hutloff A, Jordan MA, Zhao JL, Casellas R, Athanasopoulos V, Vinuesa CG (2015) MicroRNA-146a regulates ICOS-ICOSL signalling to limit accumulation of T follicular helper cells and germinal centres. Nat Commun 6:6436

Pua HH, Steiner DF, Patel S, Gonzalez JR, Ortiz-Carpena JF, Kageyama R, Chiou NT, Gallman A, de Kouchkovsky D, Jeker LT, McManus MT, Erle DJ, Ansel KM (2016) MicroRNAs 24 and 27 suppress allergic Inflammation and target a network of regulators of T Helper 2 cell-associated cytokine production. Immunity 44:821–832

Rodriguez A, Vigorito E, Clare S, Warren MV, Couttet P, Soond DR, van Dongen S, Grocock RJ, Das PP, Miska EA, Vetrie D, Okkenhaug K, Enright AJ, Dougan G, Turner M, Bradley A (2007) Requirement of bic/microRNA-155 for normal immune function. Science 316:608–611

Schneider C, Setty M, Holmes AB, Maute RL, Leslie CS, Mussolin L, Rosolen A, Dalla-Favera R, Basso K (2014) MicroRNA 28 controls cell proliferation and is down-regulated in B-cell lymphomas. Proc Natl Acad Sci USA 111:8185–8190

Sebastiani G, Grieco FA, Spagnuolo I, Galleri L, Cataldo D, Dotta F (2011) Increased expression of microRNA miR-326 in type 1 diabetic patients with ongoing islet autoimmunity. Diabetes Metab Res Rev 27:862–866

Severin ME, Lee PW, Liu Y, Selhorst AJ, Gormley MG, Pei W, Yang Y, Guerau-De-arellano M, Racke MK, Lovett-Racke AE (2016) MicroRNAs targeting TGFbeta signalling underlie the regulatory T cell defect in multiple sclerosis. Brain 139:1747–1761

Sharma S, Findlay GM, Bandukwala HS, Oberdoerffer S, Baust B, Li Z, Schmidt V, Hogan PG, Sacks DB, Rao A (2011) Dephosphorylation of the nuclear factor of activated T cells (NFAT) transcription factor is regulated by an RNA-protein scaffold complex. Proc Natl Acad Sci USA 108:11381–11386

Sheedy FJ, Palsson-Mcdermott E, Hennessy EJ, Martin C, O'Leary JJ, Ruan Q, Johnson DS, Chen Y, O'Neill LA (2010) Negative regulation of TLR4 via targeting of the proinflammatory tumor suppressor PDCD4 by the microRNA miR-21. Nat Immunol 11:141–147

Simpson LJ, Patel S, Bhakta NR, Choy DF, Brightbill HD, Ren X, Wang Y, Pua HH, Baumjohann D, Montoya MM, Panduro M, Remedios KA, Huang X, Fahy JV, Arron JR, Woodruff PG, Ansel KM (2014) A microRNA upregulated in asthma airway T cells promotes TH2 cytokine production. Nat Immunol 15:1162–1170

So AY, Sookram R, Chaudhuri AA, Minisandram A, Cheng D, Xie C, Lim EL, Flores YG, Jiang S, Kim JT, Keown C, Ramakrishnan P, Baltimore D (2014) Dual mechanisms by which miR-125b represses IRF4 to induce myeloid and B-cell leukemias. Blood 124:1502–1512

Song JJ, Smith SK, Hannon GJ, Joshua-Tor L (2004) Crystal structure of Argonaute and its implications for RISC slicer activity. Science 305:1434–1437

Sonkoly E, Wei T, Janson PC, Saaf A, Lundeberg L, Tengvall-Linder M, Norstedt G, Alenius H, Homey B, Scheynius A, Stahle M, Pivarcsi A (2007) MicroRNAs: novel regulators involved in the pathogenesis of psoriasis? PLoS ONE 2:e610

Stanczyk J, Pedrioli DM, Brentano F, Sanchez-Pernaute O, Kolling C, Gay RE, Detmar M, Gay S, Kyburz D (2008) Altered expression of MicroRNA in synovial fibroblasts and synovial tissue in rheumatoid arthritis. Arthritis Rheum 58:1001–1009

Steiner DF, Thomas MF, Hu JK, Yang Z, Babiarz JE, Allen CD, Matloubian M, Blelloch R, Ansel KM (2011) MicroRNA-29 regulates T-box transcription factors and interferon-gamma production in helper T cells. Immunity 35:169–181

Taganov KD, Boldin MP, Chang KJ, Baltimore D (2006) NF-kappaB-dependent induction of microRNA miR-146, an inhibitor targeted to signaling proteins of innate immune responses. Proc Natl Acad Sci USA 103:12481–12486

Tagawa H, Seto M (2005) A microRNA cluster as a target of genomic amplification in malignant lymphoma. Leukemia 19:2013–2016

Tang Y, Luo X, Cui H, Ni X, Yuan M, Guo Y, Huang X, Zhou H, de Vries N, Tak PP, Chen S, Shen N (2009) MicroRNA-146A contributes to abnormal activation of the type I interferon pathway in human lupus by targeting the key signaling proteins. Arthritis Rheum 60:1065–1075

Teng G, Hakimpour P, Landgraf P, Rice A, Tuschl T, Casellas R, Papavasiliou FN (2008) MicroRNA-155 is a negative regulator of activation-induced cytidine deaminase. Immunity 28:621–629

van den Berg A, Kroesen BJ, Kooistra K, de Jong D, Briggs J, Blokzijl T, Jacobs S, Kluiver J, Diepstra A, Maggio E, Poppema S (2003) High expression of B-cell receptor inducible gene BIC in all subtypes of Hodgkin lymphoma. Genes Chromosomes Cancer 37:20–28

Wang P, Gu Y, Zhang Q, Han Y, Hou J, Lin L, Wu C, Bao Y, Su X, Jiang M, Wang Q, Li N, Cao X (2012) Identification of resting and type I IFN-activated human NK cell miRNomes reveals microRNA-378 and microRNA-30e as negative regulators of NK cell cytotoxicity. J Immunol 189:211–221

Wang H, Flach H, Onizawa M, Wei L, McManus MT, Weiss A (2014) Negative regulation of Hif1a expression and TH17 differentiation by the hypoxia-regulated microRNA miR-210. Nat Immunol 15:393–401

Wang Y, Zhong H, Xie X, Chen CY, Huang D, Shen L, Zhang H, Chen ZW, Zeng G (2015) Long noncoding RNA derived from CD244 signaling epigenetically controls CD8+ T-cell immune responses in tuberculosis infection. Proc Natl Acad Sci USA 112:E3883–E3892

Wightman B, Ha I, Ruvkun G (1993) Posttranscriptional regulation of the heterochronic gene lin-14 by lin-4 mediates temporal pattern formation in C. elegans. Cell 75:855–862

Willingham AT, Orth AP, Batalov S, Peters EC, Wen BG, Aza-Blanc P, Hogenesch JB, Schultz PG (2005) A strategy for probing the function of noncoding RNAs finds a repressor of NFAT. Science 309:1570–1573

Xiao C, Calado DP, Galler G, Thai TH, Patterson HC, Wang J, Rajewsky N, Bender TP, Rajewsky K (2007) MiR-150 controls B cell differentiation by targeting the transcription factor c-Myb. Cell 131:146–159

Xu S, Guo K, Zeng Q, Huo J, Lam KP (2012) The RNase III enzyme Dicer is essential for germinal center B-cell formation. Blood 119:767–776

Yamanaka Y, Tagawa H, Takahashi N, Watanabe A, Guo YM, Iwamoto K, Yamashita J, Saitoh H, Kameoka Y, Shimizu N, Ichinohasama R, Sawada K (2009) Aberrant overexpression of microRNAs activate AKT signaling via down-regulation of tumor suppressors in natural killer-cell lymphoma/leukemia. Blood 114:3265–3275

Yi R, Qin Y, Macara IG, Cullen BR (2003) Exportin-5 mediates the nuclear export of pre-microRNAs and short hairpin RNAs. Genes Dev 17:3011–3016

Zhang B, Liu SQ, Li C, Lykken E, Jiang S, Wong E, Gong Z, Tao Z, Zhu B, Wan Y, Li QJ (2016) MicroRNA-23a curbs necrosis during early T cell activation by enforcing intracellular reactive oxygen species equilibrium. Immunity 44:568–581

Zhao X, Tang Y, Qu B, Cui H, Wang S, Wang L, Luo X, Huang X, Li J, Chen S, Shen N (2010) MicroRNA-125a contributes to elevated inflammatory chemokine RANTES levels via targeting KLF13 in systemic lupus erythematosus. Arthritis Rheum 62:3425–3435

Zhou X, Jeker LT, Fife BT, Zhu S, Anderson MS, McManus MT, Bluestone JA (2008) Selective miRNA disruption in T reg cells leads to uncontrolled autoimmunity. J Exp Med 205:1983–1991

Zhu E, Wang X, Zheng B, Wang Q, Hao J, Chen S, Zhao Q, Zhao L, Wu Z, Yin Z (2014) miR-20b suppresses Th17 differentiation and the pathogenesis of experimental autoimmune encephalomyelitis by targeting RORgammat and STAT3. J Immunol 192:5599–5609

MIX
Papier aus verantwortungsvollen Quellen
Paper from responsible sources
FSC® C105338

If you have any concerns about our products,
you can contact us on
ProductSafety@springernature.com

In case Publisher is established outside the EU,
the EU authorized representative is:
Springer Nature Customer Service Center GmbH
Europaplatz 3, 69115 Heidelberg, Germany

Printed by Libri Plureos GmbH
in Hamburg, Germany